Food Security
and
Global Environmental Change

Food Security
and
Global Environmental Change

Edited by

John Ingram, Polly Ericksen
and Diana Liverman

publishing for a sustainable future

London • Washington, DC

First published in 2010 by Earthscan

Earthscan Ltd, Dunstan House, 14a St Cross Street, London EC1N 8XA, UK
Earthscan LLC, 1616 P Street, NW, Washington, DC 20036, USA
Earthscan publishes in association with the International Institute for Environment and Development

For more information on Earthscan publications, see www.earthscan.co.uk or write to earthinfo@earthscan.co.uk

ISBN: 978-1-84971-127-2 hardback
ISBN: 978-1-84971-128-9 paperback

Typeset by 4word Ltd, Bristol
Cover design by Clifford Hayes
Photo courtesy of AA World Travel Library

A catalogue record for this book is available from the British Library

Library of Congress Cataloging-in-Publication Data

Food security and global environmental change / edited by John Ingram, Polly Erickson, and Diana Leverman.
    p. cm.
  Includes bibliographical references and index.
  ISBN 978-1-84971-127-2 (hardback) – ISBN 978-1-84971-128-9 (pbk.) 1. Food security–Environmental aspects. 2. Food supply–Environmental aspects. I. Ingram, J. S. I. II. Erickson, Polly. III. Liverman, Diana.
  HD9000.5.F5966 2010
  338.1'9–dc22

                                    2010027614

Printed and bound in the UK by CPI Antony Rowe.
The paper used is FSC certified.

Contents

List of Figures, Tables and Boxes

Figures

Tables

Boxes

Editorial Committee

List of Contributors

Molly Anderson, College of the Atlantic in Bar Harbor, Maine, USA
Jens Andersson, Wageningen University, the Netherlands
Gabriele Bammer, The Australian National University, Australia
David Barling, City University, UK
Hans-Georg Bohle, University of Bonn, Germany
Mike Brklacich, Carleton University, Canada
Molly Brown, NASA, USA
Bruce Campbell, CCAFS, Denmark
Andrew Challinor, University of Leeds, UK
Stephen Devereux, Institute of Development Studies, UK
Jane Dixon, The Australian National University, Australia
Scott Drimie, International Food Policy Research Institute, South Africa
Hallie Eakin, Arizona State University, USA
Polly Ericksen, University of Oxford, UK
Siri Eriksen, Norwegian University of Life Sciences, Norway
Ken Giller, Wageningen University, the Netherlands
Ronald Gordon, University of Florida, USA
Peter Gregory, Scottish Crop Research Institute, UK
Jim Hansen, International Research Institute for Climate and Society, USA
Thomas Henrichs, European Environment Agency, Denmark
Claudia Hiepe, UN Food and Agriculture Organization, Italy
John Holmes, University of Oxford, UK
John Ingram, University of Oxford, UK
Anne-Marie Izac, CGIAR, Italy
James W. Jones, University of Florida, USA
Kamal Kapadia, University of Oxford, UK
Anna Krithis, Arizona State University, USA
Gabriela Litre, IHDP, Germany
Diana Liverman, University of Oxford, UK
Philip Loring, University of Alaska Fairbanks, USA
Alison Misselhorn, University of KwaZulu-Natal, South Africa
Siwa Msangi, International Food Policy Research Institute, USA
Michael Obersteiner, IIASA, Austria
Laura Pereira, University of Oxford, UK
Veronique Plocq-Fichelet, SCOPE, France
Anette Reenberg, University of Copenhagen, Denmark

Martin Rice, ESSP, France
Thomas Rosswall, CCAFS, France
Winston Rudder, Cropper Foundation, Trinidad and Tobago
Rachel Sabates-Wheeler, Institute of Development Studies, UK
Miriam Saxl, University of Oxford, UK
Rüdiger Schaldach, University of Kassel, Germany
Rutger Schilpzand, Schuttelaar & Partners, the Netherlands
Elisabeth Simelton, University of Leeds, UK
Mark Stafford Smith, CSIRO, Australia
Beth Stewart, University of Oxford, UK
David Tecklin, University of Arizona, USA
Philip Thornton, International Livestock Research Institute, Kenya
Emma Tompkins, University of Leeds, UK
Petra Tschakert, Penn State University, USA
Keith Wiebe, UN Food and Agriculture Organization, Italy
Steve Wiggins, ODI, UK
Stanley Wood, International Food Policy Research Institute, USA
John Young, Overseas Development Institute, UK

Preface

Human activities, including those related to producing, processing, packaging, distributing, retailing and consuming food, are partly responsible for changing the world's climate through emissions of greenhouse gases and changes in land use. These activities are also contributing to other aspects of global environmental change (GEC), including changes in freshwater supplies, air quality, nutrient cycling, biodiversity, land cover and soils. Simultaneously, increases in population and wealth are leading to ever-growing demands by society for food, while increasing urbanization is leading to proportionally fewer people producing food; the next few decades are likely to see more of conditions contributing to the 'perfect storm' caused by the need to simultaneously provide 50 per cent more food, 50 per cent more energy and 30 per cent more fresh water – without further degrading the natural resource base upon which our food security largely depends. Furthermore, there is growing concern that GEC will threaten food security, particularly for those more vulnerable sections of society.

Attaining food security for all is clearly more complicated than just producing more food; the world produces more than enough food for everyone, yet – even today – over 1 billion people do not have access to sufficient food and go to bed hungry. The fundamental issue, therefore, concerns *access* to food rather than food production. This concept is well captured by the working definition of food security adopted by the 1996 World Food Summit held under the auspices of the Food and Agriculture Organization of the United Nations: food security is the state achieved such that 'all people, at all times, have physical and economic access to sufficient, safe, and nutritious food to meet their dietary needs and food preferences for an active and healthy life'. While many other definitions of food security exist, central tenets are the notion of access and sustainability.

Concerns about the additional challenges that GEC – and particularly climate change – will bring to meeting food security have risen sharply on political and policy agendas in recent years. Of course, the underlying twin issues of food security (vis-à-vis food production) and GEC have been on science agendas for much longer, and indeed research on the interactions of the two emerged well over a decade ago. Many national and international research organizations and groups have thus been addressing various aspects of the food security–GEC agenda. As the research interest has spread beyond the conventional GEC and agricultural communities, it has become increasingly clear that both socio-economic and biogeophysical factors determine food security. More recent research rightly includes, for example, studies on societal perceptions of GEC; vulnerability of food systems to GEC; seasonal weather forecasting and risk mitigation; and identifying the spatial and temporal levels of climate information needed by the range of stakeholders involved in the food security debate.

As the importance of interdisciplinarity emerges ever more strongly, research needs to build on the wealth of disciplinary studies that have characterized most GEC and food-related research to date. New interdisciplinary agendas need to be set to move research forward based on an integrated framing of the issues and challenges involved in the interactions between food security and GEC. These research agendas need to be determined by a range of stakeholders that includes the policy- and decision-makers who struggle daily with meeting both food security and environmental objectives.

Challenges for research include responding to new needs for developing adaptation agendas and facilitating communication amongst policy-makers, resource managers and researchers working at a range of levels on spatial, temporal and jurisdictional scales. Further, developing research agendas in support of food security policy formulation needs to recognize that setting such policy is complicated, needs systematic analysis that cuts across these scales and levels, and is only going to become more complicated under the pressure of global environmental change. Drawing on experiences of the Global Environmental Change and Food Systems (GECAFS) international research project of the Earth System Science Partnership, and a wide range of other research endeavours, this book aims to help set the stage for further research on this critical set of issues.

The book is structured in five parts. Part I introduces the background and context for research on food security and GEC and makes the case for taking a systems approach (the 'food system') for researching what appears to be an otherwise dauntingly complex set of issues. Drawing on and synthesizing a wide range of literature, the fundamental issues of food system vulnerability to GEC and the broad food system adaptation agenda are then covered in Part II. While Parts I and II set a broad conceptual and state-of-the-science foundation for thinking about *research content*, the next two parts review and discuss a range of more practical considerations about the *research process*. Part III reviews and synthesizes lessons learnt from a range of stakeholder engagement approaches, given that stakeholder interaction is crucial at all stages of the food security–GEC research endeavour. Part IV makes the case for, and discusses lessons learnt from, efforts to undertake research at the regional level. This is between the global and local levels which characterize most food security studies. Collectively, these first four parts synthesize the knowledge and experiences derived from a great many stakeholders including policy-makers, resource managers, business interests, donors, civil society and researchers. Finally, Part V looks ahead by reviewing a number of emerging 'hot topics' in the food security–GEC debate. These topics include many of the emerging food security challenges we will face, and help set a new agenda for the research community at large.

The central message of this synthesis is clear: an innovative research approach that integrates a wide range of concepts and methods is needed if science is to support policy formulation and resource management more effectively; as the requirement for more equitable access to food for all grows, so too grows the need for best science to underpin the development of more effective food systems. The food security–GEC community worldwide has clearly already made an impressive contribution, but many challenges remain. It is essential to address these in a structured and systematic manner.

John Ingram, Polly Ericksen and Diana Liverman
Oxford
May 2010

Acknowledgements

This book brings together work from a wide range of national and international food security research projects with that of the 'Global Environmental Change and Food Systems' (GECAFS) Joint Project of the Earth System Science Partnership (ESSP). We are indebted to the many individuals world-wide who have participated in these wide-ranging research efforts.

We are grateful for the sponsorship of the International Geosphere-Biosphere Programme (IGBP), the International Human Dimensions Programme on Global Environmental Change (IHDP) and the World Climate Research Programme (WCRP), all ESSP partners, for initiating GECAFS and for providing support both before and during the development of this book. We similarly acknowledge with thanks the UK Natural Environment Research Council (NERC) and the UK Economic and Social Research Council (ESRC) for providing core funds to GECAFS for developing this work. We thank the Environmental Change Institute, University of Oxford, for hosting the GECAFS International Project Office.

The book would not have been possible without the advice, help and encouragement of the Editorial Committee; and we are also particularly grateful to the chapter lead authors and co-authors for all their thoughtful work and effort. We also acknowledge with thanks the very valuable advice provided by external reviewers Richard Bissell, Louise Fresco, Mark Howden, Louis Lebel, Elizabeth Malone, Simon Maxwell, Jon Padgham, Will Steffen and Coleen Vogel. We also note with thanks editorial assistance from Miriam Saxl and Beth Stewart.

Finally, we would like to pay particular thanks to Anita Ghosh for all her hard work and diligence in readying the manuscript for publication.

John Ingram, Polly Ericksen and Diana Liverman

List of Acronyms and Abbreviations

AIACC	Assessments of Impacts and Adaptation to Climate Change
AMPRIP	Agricultural Marketing Promotion and Regional Integration Project
ASB	Alternatives to Slash and Burn Project
CAADP	Comprehensive Africa Agriculture Development Programme
CAP	Common Agricultural Policy
CARICOM	Caribbean Community
CAWMA	Comprehensive Assessment of Water Management in Agriculture
CCAFS	Climate Change, Agriculture and Food Security
CCCCC	Caribbean Community Climate Change Centre
CFS	Committee on World Food Security
CGIAR	Consultative Group on International Agricultural Research
CIMH	Caribbean Institute for Meteorology and Hydrology
COMESA	Common Market for Eastern and Southern Africa
CRN	Collaborative Research Network Program
CRSP	Collaborative Research Support Program
CSA	commodity systems analysis
CSR	corporate social responsibility
Danida	Danish International Development Agency
DFID	UK Department for International Development
DG SANCO	EU Directorate General for Health and Consumer Affairs
EFSA	European Food Safety Authority
ENSO	El Niño Southern Oscillation
ESCR-Net	International Network for Economic, Social and Cultural Rights
ESSP	Earth System Science Partnership
EU	European Union
FANR	Directorate for Food, Agriculture and Natural Resources
FANRPAN	Food, Agriculture and Natural Resources Policy Analysis Network
FAO	Food and Agriculture Organization
FEWS NET	Famine Early Warning System
FSC	Forest Stewardship Council

GEC	global environmental change
GECAFS	Global Environmental Change and Food Systems
GEO	Global Environmental Outlook
GFSA	Global Food Security Assessment
GHG	greenhouse gas
GHI	Global Hunger Index
GIEWS	Global Information and Early Warning System
GM	genetically modified
GMO	genetically modified organism
GPN	global production network
GSC	global supply chain
HDR	Human Development Report
IAASTD	International Assessment of Agricultural Knowledge, Science and Technology for Development
IAI	Inter-American Institute for Global Change Research
IAM	integrated assessment models
IFPRI	International Food Policy Research Institute
IGBP	International Geosphere-Biosphere Programme
IGP	Indo-Gangetic Plains
IHDP	International Human Dimensions Programme on Global Environmental Change
IIASA/GAEZ	International Institute for Applied Systems Analysis Global Agroecological Zones
IPCC	Intergovernmental Panel on Climate Change
IRGC	International Risk Governance Council
LCA	life-cycle analyses
LFS	local food systems
MA	Millennium Ecosystem Assessment
MDG	Millennium Development Goal
MRC	Mekong River Commission
MSC	Marine Stewardship Council
NAFTA	North American Free Trade Agreement
NGO	non-governmental organization
ODI	Overseas Development Institute
OECD	Organisation for Economic Co-operation and Development
PAGE	Pilot Assessment of Global Ecosystems
RAPID	Research and Policy in Development
REDD	Reduced Emissions from Deforestation and Degradation
RENEWAL	Regional Network on AIDS, Livelihoods and Food Security
RFID	radio-frequency identification
SAARC	South Asian Association for Regional Cooperation
SADC	Southern Africa Development Community
SAfMA	Southern African Sub-Global Assessment
SCOPE	Scientific Committee on Problems of the Environment
SES	social-ecological system
SOFA	State of Food and Agriculture
SOFI	State of Food Insecurity

SPS	Sanitary and Phytosanitary Standards
START	Global Change System for Analysis, Research and Training
TBT	technical barriers to trade
TNC	transnational corporations
UKFSA	UK Food Security Assessment
UN	United Nations
UNEP	United Nations Environment Program
UNFCCC	United Nations Framework Convention on Climate Change
USAID	US Agency for International Development
USDA	US Department of Agriculture
VAM	Vulnerability Assessment and Mapping
WB	World Bank
WCRP	World Climate Research Programme
WDR	World Development Report
WFP	World Food Programme
WHS	World Hunger Series
WRI	World Resources Institute
WRR	World Resources Report
WSSD	World Summit on Sustainable Development
WTO	World Trade Organization
WWF	World Wide Fund for Nature
WWI	Worldwatch Institute

PART I

FOOD SECURITY, FOOD SYSTEMS AND GLOBAL ENVIRONMENTAL CHANGE

1
Food Systems and the Global Environment: An Overview

Diana Liverman and Kamal Kapadia

Introduction

Food has always been linked to environmental conditions with production, storage and distribution, and markets all sensitive to weather extremes and climate fluctuations. Food production and quality are also sensitive to the quality of soils and water, the presence of pests and diseases, and other biophysical influences. Over millennia people have adjusted the production and consumption of food to the spatial and temporal variation in the natural environment, and the growing size and complexity of food systems have in turn transformed the landscapes that humans inhabit. The scope and scale of interaction are changing dramatically, particularly in relation to the risks of climate change, biodiversity loss and water scarcity; in terms of linkages to energy systems; and as food systems become more global in their networks of production, consumption and governance. This has started to raise concerns about food security, not only in governments, but also for private sector and non-governmental organizations (NGOs).

Global food security has been defined as 'when all people, at all times, have physical and economic access to sufficient, safe, and nutritious food to meet their dietary needs and food preferences for an active and healthy life' (FAO, 1996). Food security is underpinned by food systems that link the food chain activities of producing, processing, distributing and consuming food to a range of social and environmental contexts (see Chapter 2 and Figures 2.1a and 2.1b). The environmental context includes large-scale changes in land use, biogeochemical cycles, climate and biodiversity that collectively constitute global environmental change (GEC) now occurring at an unprecedented scale of human intervention in the earth system. These are discussed below.

What are some of the most important trends in food security and GEC and the interactions between them? Why are these interactions important and what steps are being taken to manage and govern them? This chapter provides background and initial insights into these questions – elaborated in subsequent chapters of the book – and illustrates the importance of these changes through case studies of the recent food price crisis and of the new proposals for integrating food systems into international climate governance. The recent warnings and debates about the interaction between GEC and food security are then summarized through a review of major reports, the media and shifts in the agendas of key development organizations.

Trends in food production, food consumption and food security

Food security is a function of many factors, of which the most important is access to food. This in turn depends on the balance between food prices, trade, stocks, employment and markets, and patterns of production and consumption of both food and non-food crops, livestock and fish.

Food production

While overall agricultural production has grown over the last century, trends since 1990 suggest that growth has slowed, especially in the developed countries, and that per capita production has levelled off in many regions. World food supply is very dependent on a few crops, especially cereals (e.g. wheat, rice, maize), oilseeds, sugar and soybeans. Just over 10 per cent of these are produced and traded from exporting regions including North America, Brazil, Argentina, Europe and Australia. Production of these commodities, as well as meat, rose over the last 50 years, but varied with weather conditions and with prices, which until recently were low, suppressing production. Meat and dairy production have grown faster than crops overall, driven by changing demand and price signals. In Africa, where smallholder agriculture still underpins food security, production is still low in many regions, especially in terms of yield per unit area.

Yields are only partly dependent on environmental conditions such as climate, soils and pests, and higher yields are usually dependent on the use of inputs such as improved germplasm, fertilizer, labour, pesticides, irrigation and machinery. Many inputs have become more expensive for farmers in recent years as a result of higher energy prices (which affect the cost of fuel and agri-chemicals) and reduced government subsidies, and because low commodity prices make inputs less affordable.

The gaps between actual and potential yields are considerable, especially in sub-Saharan Africa where maize yields of less than 2 tonnes per hectare are only 50 per cent of what could be achieved (World Bank, 2007). Closing the 'yield gap' requires not only a favourable climate (or good insurance and reliable and affordable irrigation) but also support for improved fertilizer use, technologies, access to markets and credit, and investment in physical and institutional infrastructure.

The overall amount of food produced is, however, also a function of the area under production which is limited by, for instance, land suitability and the availability of irrigation, and other demands on land-use including non-food crops such as biofuels and fibre. It is also very sensitive to price signals. For example, when prices rose rapidly in 2007, producers expanded the area under production and cereal production increased by 11 per cent in the developed countries between 2007 and 2008 (FAO, 2009a). However, production increased by less than 3 per cent in developing countries, due in part to poorer access to markets and inputs. The key point here is that production is dependent on both environmental and economic conditions (as well as other factors), and that although there is potential to increase production in many regions this may depend on higher price incentives to farmers that would reduce access to food by consumers (FAO, 2009a).

Food consumption

At the other end of the food chain, changes in consumption are having considerable impacts on food systems and food security. Overall consumption is likely to increase as global population rises, with estimates ranging from 8.5 to 11 billion by around 2050 – with recent UN projections suggesting a levelling off at around 9 billion. The UN Food and Agriculture Organization (FAO) projects that demand for cereals will increase by 70 per cent by 2050, and will double in developing countries. Depending on economic conditions per capita demand is also likely to grow, especially in the developing world, with income growth in Asia especially important. The structure of food demand has changed substantially in the last two decades, especially in developing countries where food consumption is shifting away from basic cereals to fruits, vegetables, meats and oils (see Figure 1.1). Further, evidence from Latin America and Asia indicates that the pace of change in the structure of diets is speeding up. This shift in dietary preferences has far-reaching ramifications for the entire food chain: it is transforming the structure of production systems, the ways in which consumers obtain their food, and the nature and scope of food-related health and environmental issues facing the world. The reasons for these dietary shifts include income growth, urbanization and the

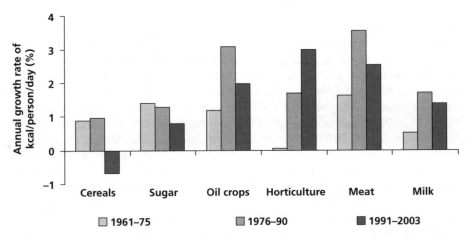

Source: World Bank, 2008. Original data source: FAOSTAT, Statistics Division, Food and Agriculture Organization of the United Nations

Figure 1.1 *Per capita food consumption in developing countries is shifting to fruits and vegetables, meat and oils*

spread of global processing and retail companies. Such dietary shifts are not, of course, uniform across all regions: total meat consumption in Asia has grown dramatically in the last decade, whereas it has increased only very slowly in North and South America (where it was already high), and increased very little or remained virtually unchanged in the rest of the world. On a per capita basis, meat consumption in most of the world, including Asia, is still less than per capita meat consumption in North America. Of all types of meat, the consumption, production and trade of poultry are growing the fastest, especially in developing and transition countries (Mack et al, 2005).

Demand for biofuels produced from agricultural feedstocks (whether food crops or cellulosic sources) is also likely to increase demand for agricultural commodities and resources, although the precise magnitude of this demand is highly uncertain, depending on energy prices as well as policy measures, especially in the developed countries.

Another trend in food preferences is the growth in processed food sales, which now accounts for about three-quarters of total world food sales. In developed countries, processed food makes up about half of total food expenditures; in developing countries it comprises only a third or less (Regmi and Gehlhar, 2005). Excessive consumption of meat, dairy and processed foods are contributing to the rapid spread of food-related health problems like obesity and diabetes, especially in the developed world. Nevertheless, such dietary transitions are not uniform across income groups. For the majority of the world's population and especially those living in poverty, cereals remain an extremely important source of calories: 90 per cent of the world's calorific requirement is provided by only 30 crops, with wheat, rice and maize alone providing about half the calories consumed globally (MA, 2005b). And for many of those living in poverty a modest increase in meat and dairy consumption can significantly improve nutritional status.

Food security

Overall growth in production and consumption has allowed some regions and people to become more food-secure but there are still millions of people who are undernourished and hungry and who do not have physical, social and economic access to sufficient, safe and nutritious food that meets their dietary needs and food preferences for an active and healthy life. A rapid rise in food prices in 2007–8 (discussed below) increased the number of hungry people to 923 million (FAO, 2009b), but even before the food price shock, progress on food security had been sluggish, and although the proportion of the population that was hungry dropped from almost 20 per cent of the population in the developing world in 1990–92 to just above 16 per cent in 2004–6 the total number of chronically hungry people stayed fairly constant through this period (see Figure 1.2). The number of hungry people increased again in 2008–9 (to more than 1 billion) because of the impacts of the financial crisis on access to food through loss of incomes. The 'food crisis' is discussed in more detail below. Further, increasing urbanization is changing the relationship between food demand and food supply, and the prevalence of food insecurity is increasing in urban areas (Frayne et al, 2010).

Number of undernourished in the world (millions)

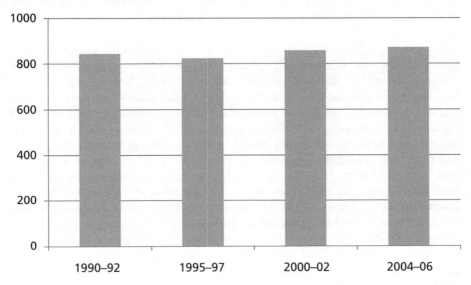

Number of undernourished (millions)

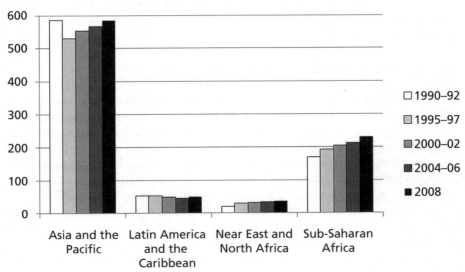

Source: FAO, 2009b. Original data source: FAOSTAT, Statistics Division, Food and Agriculture Organization of the United Nations

Figure 1.2 *Number of undernourished people*

These numbers and trends also hide large regional disparities; Africa has the largest number of hungry people as a proportion of the population but Asia has the majority in terms of absolute numbers. However, in general, the overall availability of food is rarely a cause of food insecurity: 78 per cent of all malnourished children under five in developing countries live in countries with food surpluses.

There are also many different ways in which particular groups become food insecure – for subsistence producers the loss of access to productive land or the vulnerability of their crops, fish and livestock to climate, land degradation and pests may be the main threat to the food security of their families. For those farmers and fishers producing for the market, their incomes and food security will vary with commodity and input prices as well as environmental conditions, and they may be vulnerable to shifts in government policies and to default on debts should their harvests fail. Those employed in agricultural and food systems – including farm workers, distributors, labourers in food processing and small-scale retailers – are also at risk if they become unemployed or incomes fall as a result of environmental or economic conditions. All poor households that are net buyers of food (i.e. they consume more than they produce) are vulnerable to increases in food prices. It is also important to remember that food security is not just about the amount of food but also depends on the nutritional quality, safety and cultural appropriateness of foods (see Chapter 2 and Ericksen, 2008).

Food emergencies add to food insecurity when conflict, natural disasters or failed policies create production failures, interrupt distribution, or exacerbate poverty and ill health. The hotspots for this type of insecurity include several countries in Africa (Zimbabwe, Ethiopia, Democratic Republic of Congo) and Haiti.

Trends in global environmental change

GEC includes changes in the physical and biogeochemical environment, either caused naturally or influenced by human activities such as deforestation, fossil fuel consumption, urbanization, land reclamation, agricultural intensification, freshwater extraction, fisheries over-exploitation and waste production. It includes changes in land cover and soils, biogeochemical cycles and atmospheric composition, biodiversity, climate and extreme weather events, sea level and ocean chemistry and currents, and freshwater quality and availability.

International assessments, while missing key trends in food systems (see Chapter 3), provide many useful insights into changes in the global environment.

The Millennium Ecosystem Assessment (MA) argues, for example, that humans have changed ecosystems more in the last 50 years than in any other comparable period of history, causing degradation of water, fisheries, air quality and land cover in ways that undermine food and water security. Key trends include rapid increases in cultivated land, loss of 20 per cent of coral reefs and 35 per cent of mangrove area, doubling of water withdrawals, doubling of nitrogen and phosphorus use, and declines in the size and range of many species (MA, 2005a). Figure 1.3 shows the conversion and loss of different ecosystems over history, since 1950, and projected into the future, while Box 1.1 illustrates how agricultural production has changed nutrient cycles and key ecosystem services.

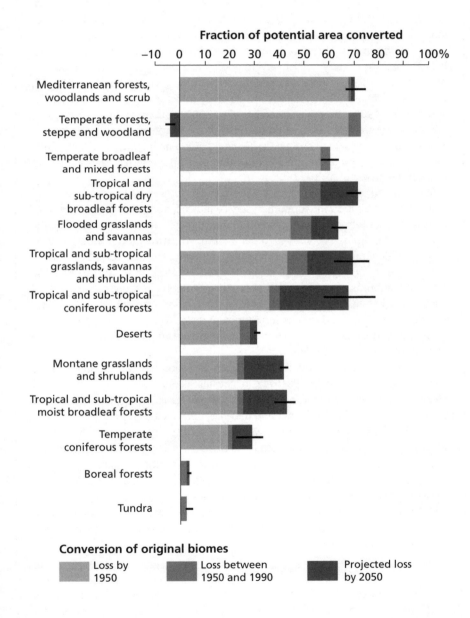

Source: Millennium Ecosystem Assessment (MA, 2005a)

Figure 1.3 *Conversion of biomes over time*

Box 1.1 *Intensive agriculture and nitrogen and phosphorus runoff*

Plant nutrients are essential for food production. Following the Second World War, agriculture underwent much change with the increased use of inputs, particularly of nutrients, leading to increased crop production and food security in many areas. Yet the increased production has come with a tradeoff, as the increased use of inputs has had detrimental effects on ecosystems both on- and off-farm.

Both nitrogen and phosphorus inputs have increased massively in the second half of the 20th century (whilst varying enormously with geography). Global use of nitrogen fertilizers increased roughly eight-fold between 1961 and 2002, whilst use of phosphorus fertilizers increased by a factor of three between 1961 and the late 1980s, before levelling off (IFA, 2004; data as reported in the MA, 2005a). It's estimated that only around half of the nitrogen added in agriculture is taken up by crops, with around 20 per cent ending up in aquatic systems (Smil, 1999). Similarly, only about 20–30 per cent of phosphorus in fertilizer is taken up by crops (Sharpley et al, 1993).

Such human inputs have had a great effect on nutrient cycles, the balance of which forms an important supporting service in ecosystems. For example, in pre-industrial times nitrogen cycling from the atmosphere into the land and aquatic systems was estimated at 90–140Tg per annum, which was roughly balanced by a reverse denitrification cycle. Nowadays, human activities contribute about 210Tg a year, only part of which is denitrified (Vitousek et al, 1997a). The increased amounts of nitrogen and phosphorus entering nutrient cycles threaten aquatic ecosystems with eutrophication – the over-enrichment of waters with nutrients which can lead to over-fertilization and algal blooms, robbing the water of oxygen and suffocating many of the organisms in those ecosystems (Bennett et al, 2001; McIsaac et al, 2001).

In extreme cases of eutrophication, oxygen depletion has caused huge marine dead zones of reduced productivity in coastal marine waters. In the Gulf of Mexico excessive nutrient runoff, primarily from fertilizer use, has led to the world's largest dead zone (Turner and Rabalais, 1991; Rabalais et al, 1999; UNEP, 2008). The Millennium Ecosystem Assessment reports that the number of such dead zones has doubled each decade as a result of increased use of nitrogen fertilizer in agriculture (2005a). The degradation in eutrophic lakes and marine areas can have a great impact on the ecosystem services that aquatic systems would normally serve; in particular, restricting an area's use for fisheries (Smith, 1998), drinking water (Carpenter et al, 1998), industry and recreation (MA, 2005a). This can lead to a degradation of an area's ecological, economic and recreational value (Bennett et al, 2001).

The latest report of the Intergovernmental Panel on Climate Change (IPC) (2007) concluded that the warming of the planet is unequivocal, with very high confidence that human activities such as fossil-fuel use and defor-estation are responsible for the warming that has occurred since 1750. Iconic graphics documenting these changes include those of rising concentrations of greenhouse gases and of increases in global temperature (Figures 1.4 and 1.5). Many of these trends are either driven by or have impacts on food systems with, for example, agricultural production associated with land-use change and greenhouse gas emissions, and climate change and water scarcity placing food security at risk. Non-production aspects of food systems also interact with GEC: 40 per cent of greenhouse gas emissions from the US food system come from food processing (10 per cent), distributing (7 per cent), consuming (15 per cent) and disposing of waste (8 per cent) (IATP, 2009); and physical infrastructure for food storage and distribution is also exposed to damage from weather extremes. Further, as urbanization grows in the developing world these percentages will likely increase as a higher proportion of food has to be transported into urban centres, with concomitant fuel usage and waste management challenges. Although climate change tends to dominate discussions about GEC there are many other reasons for concern. Rockström et al (2009) analyse nine 'planetary boundaries' which, if crossed, could generate unacceptable environmental change: for example, 350ppm in terms of greenhouse gas emissions and estimates of thresholds in rate of biodiversity loss (terrestrial and marine); interference with the nitrogen and phosphorus cycles; stratospheric ozone depletion; ocean acidification; global freshwater use; change in land use; chemical pollution; and atmospheric aerosol loading (Figure 1.6). They suggest that three of the boundaries have already been transgressed – climate change, rate of biodiversity loss and interference with the nitrogen cycle – and that boundaries for global freshwater use, change in

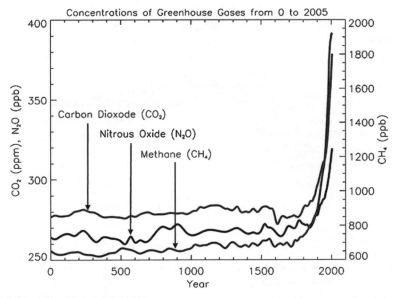

Source: Adapted from Figure 1, IPCC, 2007a

Figure 1.4 *Atmospheric concentrations of important long-lived greenhouse gases over the last 2000 years*

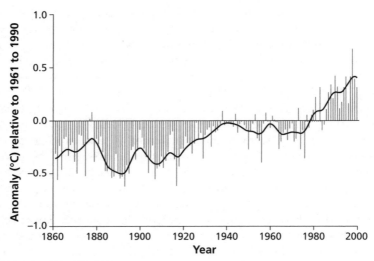

Source: Figure 2.1a within IPCC, 2001

Figure 1.5 *Annual anomalies of global average land-surface air temperature (°C), 1861–2000, relative to 1961–1990 values*

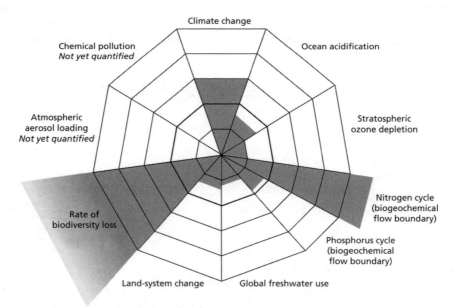

Source: Adapted from Rockström et al, 2009

Figure 1.6 *Nine 'planetary boundaries' which, if crossed, could generate unacceptable environmental change. The area inside the heavy line represents the proposed safe operating space for nine planetary systems. The shaded wedges represent an estimate of the current position for each variable. The boundaries in three systems (rate of biodiversity loss, climate change and human interference with the nitrogen cycle) have already been exceeded*

land use, ocean acidification and interference with the global phosphorus cycle are close to thresholds. Many of these planetary boundaries are closely linked to food system activities and pose risks to food security (Table 1.1; and see also Vitousek et al, 1997b; Steffen et al, 2004).

Growing attention to the food system–environment interactions

Recent high-profile reports are starting to pay more attention to the links between GEC and food security, especially between climate change and agricultural systems. The World Bank (e.g. World Development Report; World Bank, 2010) and the UN Development Programme (e.g. UN Human Development Report; UNDP, 2007) focus on climate and development, looking at the role of developing countries in both greenhouse gas emissions and as the most serious focus of climate change impacts and adaptation. FAO has focused several reports on environmental issues including a State of Food and Agriculture report on environmental services (FAO, 2007a, 2008a) and reports on carbon sequestration and mitigation (e.g. Müller-Lindenlauf, 2009), and impacts on fisheries and food safety.

International assessments such as those of the IPCC (2007b) include chapters on climate change and agriculture and include discussions of agricultural mitigation. The MA had extensive analysis of agriculture as a driver of ecosystem change and the International Assessment of Agricultural Knowledge, Science and Technology for Development (IAASTD) takes several different

Table 1.1 *Planetary boundaries (adapted from Rockström et al, 2009).*
MT = million tonnes

Process	Boundary	Examples of food system/security links
Biodiversity loss	Extinction rate no more than ten species per million per year	Agriculture one major cause of biodiversity loss Agricultural genetic base needs diversity Biodiversity enhances resilience of rural poor
Nitrogen cycle	No more than 35MT per year removed for human use	Agriculture, transport and manufacturing use/produce nitrogen (fertilizer) Fisheries degraded by nitrogen runoff
Phosphorus cycle	No more than 11MT to oceans each year	Agriculture major source of phosphorus pollution Fisheries damaged by phosphorus loading
Ocean acidification	No more than 2.75 global mean saturation of aragonite in sea water	Agricultural CO_2 emissions contribute to acidification Fisheries degraded by acidification
Global freshwater use	No more than 4000km³ per year consumption by human activity	Agriculture and food-processing consumer water Food production depends on adequate water and food processing and consumption requires good water quality
Change in land use	No more than 15 per cent of global land cover converted to cropland	Agriculture drives land-use conversion Forests provide environmental services that include food

perspectives on food security–environment interactions including analysis of biofuels, climate change and resource management (IAASTD, 2009).

The development community has also started to place GEC – especially climate change – higher on its agenda, with government international development departments (such as the UK Department for International Development (DFID), US Agency for International Development (USAID) and Danish International Development Agency (Danida)) expanding climate change expertise and programming, and major development NGOs such as Oxfam and Christian Aid focusing campaigns on climate change. One of the main reasons for this focus is the fear that climate impacts will undermine decades of investment and local efforts to improve livelihoods for those living in poverty and also the opportunities for new funding for low carbon development and adaptation. The expansion of interest in the GEC–food security nexus is rapidly expanding the number and range of concerned stakeholders and actors (see Chapters 11 and 18).

A brief analysis of coverage of food systems and GEC in major US and UK newspapers (*New York Times*, *Washington Post*, *LA Times*, *Guardian*, *Times*, *Telegraph* from Access World News database (NewsBank, undated)) suggests that media attention to the links between GEC and food systems was very low for the decade prior to 2005 with less than 100 articles a year in most newspapers and very little coverage of issues such as climate change or biofuels. From 2006 coverage increases to a peak of more than 200 articles in most papers including 600 articles across the group on biofuels and food prices and 1300 on food and climate change in 2008. An examination of coverage on food and climate in 16 leading US newspapers between September 2005 and January 2008 found a total of 4582 articles and a growing attention to the links between food and climate change over time but with only 103 focusing explicitly on the ways in which food systems contribute to climate change through greenhouse gas (GHG) emissions (Neff et al, 2009).

The food crises of 2007–9

A dramatic increase in food prices over just a few years from 2007 to 2008 provides a powerful example of how a 'perfect storm' of factors can rapidly threaten the food security of millions of people around the world. The FAO food price index doubled from 100 in 2002 to almost 200 in 2008 (Figure 1.7), and by mid-2008 real food prices were about 64 per cent above their 2002 levels, the highest in 30 years (FAO, 2008b).

The chief effect of the food price rise was on food security through reduced access to food. According to FAO estimates, an additional 40 million people fell into the category of chronically hungry by 2008 due to food price increases, bringing the total number of undernourished people to 915 million worldwide (FAO, 2008b). And a further 100 million were added to the number of hungry people worldwide due to the effects of the global financial and economic crisis – primarily through loss of employment, income and remittances – bringing the total to just over 1 billion people in 2009. The largest increases in numbers of undernourished people occurred in Asia and

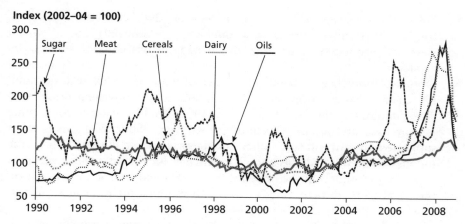

Source: FAO, 2009a. Original data source: FAOSTAT, Statistics Division, Food and Agriculture Organization of the United Nations

Figure 1.7 *Evolution of monthly FAO prices indices for basic food commodity groups 1990–2008*

sub-Saharan Africa, but hunger also increased in Latin America, where food security had improved considerably over the last decade. For many people who already spent a large proportion of their income on food, the price increases meant cutting back on food purchases or on other essentials.

The food price increase had far-reaching effects – many food-importing countries faced balance-of-payment problems, and virtually all developing countries implemented policies to reduce the effects of the price increase. An FAO survey found that nearly 40 countries reduced grain import tariffs and more than 20 countries implemented some form of export control (FAO, 2007b). As policies varied, so too did the effects of the food price rise, which varied between countries and even between different regions of a country. Chapter 7 includes further discussion on the impact of the food price spike on food systems.

Why did food prices start rising from 2006? Authoritative analysis from FAO (FAO, 2008b, 2008c, 2009a) and others (Msangi and Rosegrant, 2009; Piesse and Thirtle, 2009) have identified the following driving forces:

- Production declines in several key exporting regions – extreme weather events affecting major cereal exporters, including drought in Australia and floods in Canada, reduced cereal production by 10 per cent in 2005–6, and was accompanied by concern that this was a signal of impending climate changes. Markets responded with price increases.
- Falling stock levels – major cereal producers such as the USA, EU and China all chose to reduce holdings of food stocks, with world cereal reserves falling from 24 per cent of utilization in 2004 to 19 per cent in 2007. Because food reserves provide a buffer against production shortfalls the fall in stocks increased market volatility when uncertainty increased as to the reliability of supply.

- Oil prices – a steep rise in energy prices from 2003 to 2008 triggered a tripling of fertilizer prices and doubling of transport costs, which increased production costs and threatened the viability of small-scale producers.
- Biofuels – biofuels were blamed for much of the food price rise as demand for sugar, oilseeds, palm oil and maize increased, increasing prices and diverting land to the production of fuel that could substitute for costly, insecure and greenhouse-gas-emitting fossil fuels. Shifts to biofuels were supported by government policy in places such as the EU, USA and Indonesia and were a longstanding practice in Brazil. Estimates of the significance of biofuels in the price increases vary widely from 5 per cent of the trend to more than 50 per cent, with most estimates around 30 per cent.
- Increased consumption and dietary changes – economic growth, urbanization and higher incomes increased overall and per capita demand for food, especially in Asia, where there has also been a shift to diets with a higher proportion of cereal-fed meat and dairy products. FAO suggests, however, that these changes in consumption have been occurring over a longer period of time, have mostly been met from domestic production and may not have had a major impact on global commodity prices during 2006–8 (FAO, 2008b).
- Trade policies – in attempts to protect people living in poverty and respond to political pressure to control food prices some countries imposed restrictions on exports which exacerbated overall supply problems and dampened price incentives to producers.
- Financial speculation – several crises in overall financial markets in 2007–8 prompted investors to move into commodities and then to respond to rising prices with some speculation in futures and options, which may have caused prices to rise, persist and become more volatile.

By early 2009 food prices had fallen from the heights of 2008 as harvests improved, energy prices fell, speculation calmed and, most importantly, producers responded to higher prices. Nevertheless, food security declined further as a result of the global economic crisis, as loss of employment, income and remittances reduced access to the food by people living in poverty. FAO projected that prices would remain relatively high over the next decade because of growing demand, biofuels and input costs (FAO, 2009b), although the recession brings new uncertainties into this assessment.

The food crisis highlights the important link between food and energy as petroleum prices and biofuels interacted to create new tensions and volatility. Because of links between the use of fossil fuels and climate change, as well as concerns about energy security, countries are shifting to lower carbon-energy sources or domestic renewables, including biofuels. The role of climate variability in key exporting regions is another factor highlighted by the crisis, together with the dramatic links between poverty and food security as food prices forced millions into chronic hunger.

Climate change and food security in the aftermath of Copenhagen

Another powerful example of the intersections between GEC and food security has emerged in the efforts of the United Nations Framework Convention on Climate Change (UNFCCC) to establish an international agreement on climate change once the current version of the Kyoto protocol expires in 2012. Until recently, agriculture and food systems have been rather peripheral to the climate negotiations but are now of growing importance on both the mitigation and adaptation agendas. At the COP15 in Copenhagen (in December 2009) international networks organized the first 'Agricultural Day', modelled on the success of 'Forest Day' in raising the profile of forests in relation to climate change at previous conferences (CGIAR, 2009). The participants included major institutional players such as the World Bank, the Consultative Group on International Agricultural Research (CGIAR), FAO and the US Department of Agriculture (USDA), as well as scientists and farmers who called on the international community to recognize agriculture's vital role in climate change adaptation and mitigation. Increasing food production without increasing greenhouse gas emissions and ensuring food security in the face of climate change for the rural poor were identified as critical challenges.

The food system is a notable producer of greenhouse gases including carbon dioxide (25 per cent of all emissions if agriculture's role in causing deforestation is included), methane from rice and livestock (50 per cent of all emissions) and nitrogen dioxide from fertilizer (75 per cent of the global total). Non-agricultural activities are also important sources of greenhouse gases, as discussed above. These emissions have been growing, especially as diets switch to meat and dairy. Controlling greenhouse gas emissions from food systems has considerable food security and equity implications. Because industrialized countries are responsible for most of the historical emissions of greenhouse gases and still have high per capita emissions, it has been suggested that they should cut emissions earlier and more steeply than developing countries, where per capita emissions in a country such as Ghana are less than a tenth of those of a European. Cutting emissions in food systems can be partly achieved through switching to alternative energy sources such as renewables, at least for electricity used in food production, processing and consumption as for pumping irrigation water. Reducing emissions in food transport systems is more difficult, although in some countries delivery vehicles are switching to lower carbon energy. Biofuels, often proposed as a low carbon-energy source and supported by subsidies in regions such as the USA and EU, have the problem of competing with food production and may thus reduce food security while perhaps providing a solution to the climate change problem. However, some biofuels are energy- and land-intensive and may not provide a low carbon option. Soil management also has possibilities for sequestering carbon through practices such as biochar and agroforestry as discussed more in Chapter 19. Perhaps the greatest challenge is reducing methane emissions from livestock where changes in diet may produce slight reductions but significant reductions are unlikely without reducing the population of ruminants.

Climate change has the potential to transform the geography of food systems, especially the patterns and productivity of crop, livestock and fishery systems. The IPCC (2007b) suggested that without steep cuts in greenhouse gas emissions the world would warm by between 1.8 and 4°C by 2100, with drier conditions in regions such as southern Africa, Australia, the Mediterranean and western North America. Studies released prior to the Copenhagen conference suggested that emissions had continued to increase despite mitigation efforts, and that a slight weakening in the biosphere's ability to absorb carbon dioxide meant that emission trends were at the high end of IPCC projections. Models were showing an increased chance of higher temperatures sooner than previously anticipated with a possibility of global temperature increases as high as 4°C by 2070 if policies were not put in place to reduce emissions (New et al, 2009; Parry et al, 2009). The food system implications of a 4°C increase are very worrying, with research suggesting shifts in crop zones by hundreds of kilometres, the abandonment of cropping in parts of Africa and severe water shortages in many regions: the IPCC (2007b) projected that although warmer temperatures could lead to an initial increase in agricultural production at higher latitudes as growing seasons and land suitability expanded in countries such as Russia and Canada, the tropics would see a decrease in agricultural productivity with yields falling by up to 20 per cent and increased stress on livestock, and if warming continued all regions would see declines in yields. The IPCC also projected changes in the frequency and severity of extreme weather and climate events (such as increasing frequencies of heatwaves, droughts and floods) are expected to have adverse effects on food systems in all latitudes. Sea level rise, water supply reductions and changes in pests and diseases were also identified as posing risks to agriculture.

As noted in Chapter 3, the IPCC does not provide a sophisticated analysis of how these changes in agriculture translate into impacts on food systems and food security. Climate change poses risks to many aspects of food systems beyond production – including transport (e.g. through extreme weather) and food quality (e.g. spoilage at higher temperatures) – with food security consequences. If climate change alters crop mix across complete landscapes there are also risks to the livelihoods of producers, as well as to cultural traditions and preferences which are tied to regional varieties and diets. Food security also faces risks from changes in vulnerability that intersect with climate change such as poverty associated with the loss of employment in climate-sensitive sectors or loss of livelihoods in natural disasters (see Chapter 7). The impacts of climate change on producers are also more complicated than some studies suggest – because prices often emerge from a global trading system where a decline in yields can lead to higher prices for those who can still produce.

In terms of climate impacts on food security the overarching debate in Copenhagen focused on adaptation – what were the possibilities, needs and funds to adapt food systems to a warmer world? The most vulnerable countries – small island states, sub-Saharan Africa, supported by key NGOs – were demanding that temperature rise be capped at no more than 1.5°C and that

US$100 billion a year be made available to support adaptation efforts. Adaptation negotiations and discussions were fraught with tensions and uncertainties, including:

- uncertainty about how climate, especially soil moisture, would change at the local level and thus whether adaptation should be to drier or wetter conditions;
- the limited number of models that have been used to analyse climate change impacts on food systems;
- the lack of detailed studies on how sea-level rise would affect food systems, especially the interaction between sea-level rise, more frequent storms, the vulnerability of large cities (many located on coasts) and food security;
- inadequate research base for understanding how climate change would affect fisheries and the 200 million people who depend on them for their livelihoods;
- disagreements about who would be eligible for adaptation funds and how to measure vulnerability and impacts so as to demonstrate eligibility;
- lack of consensus about what sorts of adaptation should be taken in agriculture, especially the potential role of genetically modified (GM) crops and insurance and the balance of investments in breeding and inputs for temperate versus tropical agriculture;
- the potential conflicts and synergies between food security and proposals for an agreement on REDD (reduced emissions from deforestation and degradation), where using carbon finance to halt deforestation would limit new land available for agriculture and require intensification; and
- the need to maintain agricultural landscapes as cultural systems even when their productivity is at risk from climate change.

Agricultural and food system advocates argued for a specific negotiating track for agriculture and food security within the international climate regime, which would focus on agriculture as a solution to both the causes (emissions) and consequences (impacts) of climate change, with several possibilities for win–win outcomes that both reduce emissions and vulnerability to impacts such as conservation agriculture, drip irrigation and cover crops. While Copenhagen raised the profile of agriculture and food systems in the climate agenda, the negotiations ended in confusion and at the point of writing it is not clear how things will move forward. But the engagement of major food actors and representatives of the food and agricultural sector in key countries is likely to sustain links between the climate and food security agenda.

Changes in the governance of food systems and the earth system

A focus on the balance of supply and demand in food systems or on trends in environmental conditions such as land use and climate can overlook some of the most important ways in which global and local food and environmental systems are changing, particularly the profound changes that are occurring in the governance of food and environmental systems. Governance can be defined as the systems of rules, authority and institutions that coordinate, manage or steer society. Governance is more than the formal functions of government but also includes markets, traditions and networks, and non-state actors such as firms and civil society. Defining governance broadly to include the influence of both state and non-state actors at a variety of spatial levels reveals a number of important trends that are altering relationships within food systems and between food and environment (Biermann et al, 2009). The research literature and policy discussions about both food and environmental governance have grown rapidly in the last decade.

Because food is considered by many to be a basic need and human right there are traditions of creating institutions to protect and promote access to food by those in greatest poverty. These include the World Food Programme (which provides food assistance in emergencies), national food assistance programmes and humanitarian non-governmental organizations, as well as goals and standards such as the Millennium Development Goals and the *Codex Alimentarius*. Threats to food security underlie some of the debates about international environmental policy, including those about the magnitude and distribution of climate adaptation funds, the potential conflict between food and biofuels, or the relative priority of agricultural and forest lands in ecosystem protection. Health issues also dominate food governance in many countries, including concerns about food contamination and the use of biotechnology. One of the other major trends in food governance has been international restructuring of trade regimes including the role of the World Trade Organization (WTO), and trade agreements such as those of the North American Free Trade Agreement (NAFTA) and the EU, which seek to break down trade barriers.

One of the most significant factors in the governance of food is the importance of the private sector, and especially in recent decades of large transnational companies who produce, process and retail food (see Chapter 18). Consumer and environmental groups have also gained influence in food governance through campaigns that pressure for food safety, environmental protection, fair trade and more locally-focused food systems. For example, the organic, fair trade and slow food movements have grown from local activism to representation in commonly used food labelling and in the shelves of major supermarkets. The increasing numbers of actors in the chains that produce and market food (although there is also concentration in some large companies) have generally made the governance of food and the nature of its environmental and social impacts far more complex than in the past. Interactions between these non-state actors and governments have resulted in new public–private partnerships, voluntary certification schemes and supply chains that create new sets of norms, rules and international relations for food systems.

Environmental governance has also become more global with new forms of public–private partnerships and markets. Global institutions such as the UN have promoted international agreements to manage climate change, biodiversity, fisheries, forests and land degradation, and the environment has become increasingly important within the operations of the World Bank and a concern within trade negotiations. Market-based solutions to environmental problems have also become popular in sectors such as water, which were traditionally managed in the public domain, but which are now being privatized, priced and managed by the private sector. Cap and trade systems have been used to manage the environment by issuing permits to pollute or emit greenhouse gases which can be traded in new carbon and other environmental markets. NGOs have become important actors in environmental management at all scales, delegated to manage protected areas and working in partnership with the private sector to manage forests or watersheds.

In the chapters that follow these changes in governance contribute to explanations and contexts for changes in food systems and food security including those in vulnerability, regional networks, food conflicts and surprises and the growing importance of non-state actors.

The tendency to focus on climate change and on agriculture – and especially on agricultural production – misses many important links between food security and GEC. As noted earlier there are other GEC risks to food security including widespread pollution, ocean and land degradation, and biodiversity loss, as well as other dimensions of food security with which to be concerned, particularly the access and stability dimensions. The focus on producing food means that other major elements of food systems are overlooked including the interactions between GEC and processing, distributing, retailing and consuming food, as well as the livelihoods associated with these activities (see Chapter 2). The remainder of this book explores these linkages in much greater detail.

References

Bennett, E. M., S. R. Carpenter and N. F. Caraco (2001) 'Human impact on erodable phosphorus and eutrophication: A global perspective', *BioScience,* 51, 3, 227–34

Biermann, F., M. M. Betsill, J. Gupta, N. Kanie, L. Lebel, D. Liverman, H. Schroeder and B. Siebenhüner, with contributions from K. Conca, L. da Costa Ferreira, B. Desai, S. Tay and R. Zondervan (2009) 'Earth system governance: People, places and the planet', Science and Implementation Plan of the Earth System Governance Project, Bonn, IHDP

Carpenter, S. R., N. F. Caraco, D. L. Correll, R. W. Howarth, A. N. Sharpley and V. H. Smith (1998) 'Nonpoint pollution of surface waters with phosphorus and nitrogen', *Ecological Applications*, 8, 3, 559–68

CGIAR (2009) *Global Climate Change: Can Agriculture Cope?*, Washington, DC, CGIAR

Ericksen, P. J. (2008) 'What is the vulnerability of a food system to global environmental change?', *Ecology and Society*, 13, 2

FAO (1996) *Rome Declaration and World Food Summit Plan of Action*, Rome, FAO

FAO (2007a) 'The state of food and agriculture 2007 (SOFA): Paying farmers for environmental services', *FAO Agriculture Series*, Rome, FAO

FAO (2007b) 'Falling food prices: A window of opportunity to address the long-term challenges', *Food Outlook: Global Market Analysis*, Rome, FAO

FAO (2008a) 'Climate change and food security: A framework document', Rome, Interdepartmental Working Group on Climate Change of the FAO

FAO (2008b) *The State of Food Insecurity in the World 2008 (SOFI): High Food Prices and Food Security – Threats and Opportunities*, Rome, FAO

FAO (2008c) *The State of Food and Agriculture 2008 (SOFA): Biofuels: Prospects, Risks and Opportunities*, Rome, FAO

FAO (2009a) *The State of Agricultural Commodity Markets (SACM): High Food Prices and the Food Crisis – Experiences and Lessons Learned*, Rome, FAO

FAO (2009b) *The State of Food Insecurity in the World (SOFI): Economic Crises – Impacts and Lessons Learned*, Rome, FAO

Frayne, B., W. Pendleton, J. Crush, B. Acquah, J. Battersby-Lennard, E. Bras, A. Chiweza, T. Dlamini, R. Fincham, F. Kroll, C. Leduka, A. Mosha, C. Mulenga, P. Mvula, A. Pomuti, I. Raimundo, M. Rudolph, S. Ruysenaar, N. Simelane, D. Tevera, M. Tsoka, G. Tawodzera and L. Zanamwe (2010) The State of urban food insecurity in Southern Africa. In Crush, J. and B. Frayne (eds) *Urban Food Security Series No. 2*. Cape Town and Kingston, African Food Security Urban Network

IAASTD (2009) *Agriculture at a Crossroads*, Washington, DC, Island Press

IATP (2009) *Identifying our Climate 'Foodprint': Assessing and Reducing the Global Warming Impacts of Food and Agriculture in the US*, Minneapolis, Institute for Agricultural Trade and Policy

IFA, IFADATA statistics. www.fertilizer.org/ifa/statistics.asp (accessed: 2009)

IPCC (2001) IPCC, 2001: Climate Change 2001: The Scientific Basis. Contribution of Working Group I to the Third Assessment Report of the Intergovernmental Panel on Climate Change. In Houghton, J. T., Y. Ding, D. J. Griggs, M. Noguer, P. J. van der Linden, X. Dai, K. Maskell and C. A. Johnson (eds) Cambridge, United Kingdom and New York, NY, Cambridge University Press

IPCC (2007a) Climate Change 2007: The Physical Science Basis. Contribution of Working Group I to the Fourth Assessment Report of the Intergovernmental Panel on Climate Change. Frequently Asked Question 2.1. In Solomon, S., D. Qin, M. Manning, Z. Chen, M. Marquis, K. B. Averyt, M. Tignor and H. L. Miller (eds) Cambridge, United Kingdom and New York, NY, Cambridge University Press

IPCC (2007b) Fourth Assessment Report of the Intergovernmental Panel on Climate Change, Cambridge, United Kingdom and New York, NY, Cambridge University Press

MA (2005a) *Ecosystems and Human Well-being: Synthesis,* Washington, DC, Island Press

MA (2005b) Volume 1: Current state and trends. *Ecosystems and Human Well-being*, Washington DC, Island Press

Mack, S., D. Hoffmann and J. Otte (2005) 'The contribution of poultry to rural development', *World's Poultry Science Journal*, 61, 1, 7–14

McIsaac, G. F., M. B. David, G. Z. Gertner and D. A. Goolsby (2001) 'Eutrophication: Nitrate flux in the Mississippi River', *Nature*, 414, 6860, 166–67

Msangi, S. and M. Rosegrant (2009) World Agriculture in a dynamically-changing environment: IFPRI'S long-term outlook for food and agriculture under additional demand and constraints. *FAO Expert Meeting on 'How to feed the World in 2050'*, Rome, FAO

Müller-Lindenlauf, M. (2009) *Organic Agriculture and Carbon Sequestration: Possibilities and Constraints for the Consideration of Organic Agriculture within Carbon Accounting Systems*, Rome, Natural Resources Management and Environment Department, FAO

Neff, R. A., I. L. Chan and K. C. Smith (2009) 'Yesterday's dinner, tomorrow's weather, today's news? US newspaper coverage of food system contributions to climate change', *Public Health Nutrition,* 12, 7, 1006–14

New, M., D. Liverman and K. Anderson (2009) Mind the gap Periodical, Mind the gap Nature Publishing Group, www.nature.com/doifinder/10.1038/climate.2009. 126 (accessed: undated)

NewsBank, Access World News, www.newsbank.com/colleges/product.cfm?product=24 (accessed: undated)

Parry, M., J. Lowe and C. Hanson (2009) 'Overshoot, adapt and recover', *Nature,* 458, 7242, 1102–3

Piesse, J. and C. Thirtle (2009) 'Three bubbles and a panic: An explanatory review of recent food commodity price events', *Food Policy,* 34, 2, 119–29

Rabalais, N. N., R. E. Turner and W. J. Wiseman (1999) Hypoxia in the northern Gulf of Mexico: Linkages with the Mississippi River. In Kumpf, H., K. Steidinger and K. Sherman (eds) *The Gulf of Mexico Large Marine Ecosystem – Assessment, Sustainability and Management,* Oxford, Blackwell Science

Regmi, A. and M. Gehlhar (2005) *New Directions in Global Food Markets,* Washington, DC, Economic Research Service/USDA

Rockström, J., W. Steffen, K. Noone, Å. Persson, F. S. Chapin, III, E. Lambin, T. M. Lenton, M. Scheffer, C. Folke, H. Schellnhuber, B. Nykvist, C. A. de Wit, T. Hughes, S. van der Leeuw, H. Rodhe, S. Sörlin, P. K. Snyder, R. Costanza, U. Svedin, M. Falkenmark, L. Karlberg, R. W. Corell, V. J. Fabry, J. Hansen, B. Walker, D. Liverman, K. Richardson, P. Crutzen and J. Foley (2009) 'Planetary boundaries: exploring the safe operating space for humanity', *Ecology and Society,* 14, 2, art 32

Sharpley, A. N., T. C. Daniel and D. R. Edwards (1993) 'Phosphorus movement in the landscape', *Journal of Production Agriculture,* 6, 4, 492–500

Smil, V. (1999) 'Nitrogen in crop production: An account of global flows', *Global Biogeochemical Cycles,* 13, 2, 647–62

Smith, V. H. (1998) Cultural eutrophication of inland, estuarine and coastal waters. In Pace, M. L. and P. M. Groffman (eds) *Successes, Limitations and Frontiers of Ecosystem Science,* New York, Springer-Verlag

Steffen, W., A. Sanderson, J. Jäger, P. D. Tyson, B. Moore III, P. A. Matson, K. Richardson, F. Oldfield, H.-J. Schellnhuber, B. L. Turner II and R. J. Wassn (2004) *Global Change and the Earth System: A Planet under Pressure,* Heidelberg, Germany, Springer Verlag

Turner, R. E. and N. N. Rabalais (1991) 'Changes in Mississippi river water quality this century', *BioScience,* 41, 3, 140–47

UNDP (2007) Human Development Report 2007/2008: Fighting climate change: Human solidarity in a divided world. New York, United Nations Development Programme

UNEP (2008) In dead water – Merging of climate change with pollution, over-harvest, and infestations in the world's fishing grounds. In Nellemann, C., S. Hain and J. Alder (eds) GRID-Arendal, Norway, United Nations Environment Programme

Vitousek, P. M., J. D. Aber, R. W. Howarth, G. E. Likens, P. A. Matson, D. W. Schindler, W. H. Schlesinger and D. G. Tilman (1997a) 'Human alteration of the global nitrogen cycle: sources and consequences', *Ecological Applications,* 7, 3, 737–50

Vitousek, P. M., H. A. Mooney, J. Lubchenco and J. Melillo, M. (1997b) 'Human domination of earth's ecosystems', *Science,* 277, 5325, 494–99

World Bank (2007) World Development Report 2007: Development and the Next Generation. Washington, DC, World Bank

World Bank (2008) *World Development Report 2008: Agriculture for development.* Washington, DC, World Bank

World Bank (2010) *World Development Report 2010: Development and Climate Change.* Washington, DC, World Bank

2
The Value of a Food System Approach

Polly Ericksen, Beth Stewart, Jane Dixon, David Barling,
Philip Loring, Molly Anderson and John Ingram

Food systems, food security and global environmental change

The challenge is significant: enhancing food security without further compromising environmental and social welfare outcomes. The nature and direction of many food system trends suggest that meeting this challenge will be daunting (see Chapter 1; and Ericksen, 2008). Highly connected commodity markets affect the global environment and food security. Solving problems of food insecurity and loss of ecosystem services must therefore be based upon understanding complex interactions among multiple processes. Systems-based approaches are needed to help deliver this understanding.

Food security: A complex outcome

Food security is the outcome of multiple factors, operating at household up to international levels. It depends upon not only availability from production, but a suite of entitlements that enable (or protect) economic and social access to food. Thus historical famines have occurred where supply was not the issue, but rather poverty, conflicts or an inadequate social contract to protect people from hunger (Devereux, 2000; Maxwell, 2001). Poverty has long been associated with under- and malnutrition; the ability of individuals to obtain adequate nutritional value from food is also now embedded in definitions of food security. So as Box 2.1 exemplifies, evaluating whether or not food security exists for given households or communities requires analysis of multiple social, economic and political factors, as well as purely agronomic issues.

Complex food systems

Food systems include a range of activities from planting seeds through to disposing of household waste. As explained by Maxwell and Slater (2003) and others, the nature of food production and consumption transformed in the late 20th century, and more sophisticated analytical lenses are needed

---

Box 2.1 *Misreading the 2005 Niger food security crisis*

In 2005, a serious food security crisis occurred in the agricultural areas of Niger. The failure of the national government and the international relief community to prevent the crisis makes the case an important one for food security theory and practice. In 2004, Niger experienced a drought, followed by a reduction of its per capita staple grain (millet and sorghum) production of 12 per cent as compared to the ten-year average (Aker, 2008). Millet prices were 25 per cent higher than the ten-year average. In November 2004, the international community early warning system (USAID's Famine Early Warning Systems Network, FEWS NET, see Boxes 10.4 and 11.6; and FAO's Global Information and Early Warning System, GIEWS) began issuing messages of concern, and the national government issued a request for 78,100 tonnes of emergency food. Yet by June 2005, an estimated 2.4 million Nigerians were affected by severe food shortages, with more than 800,000 of these classified as critically food insecure (Aker, 2008). A number of factors are thought to have contributed to this. One is that the international early warning systems are more focused on production shortfalls from weather anomalies than on tracking market signals; in addition, these systems are plagued by delays and disagreements (Clay, 2005). Second, as Niger regularly reports global acute malnutrition levels of between 14 and 20 per cent, the development community considers this to be normal, and so paid little attention (Harrigan, 2006). But perhaps the most overlooked factor was the role that grain markets in Nigeria play in both affordability and availability of staple grains in Niger. The price of cereals was far too high by July 2005 (30,000 CFA, ca. US$60, for a 100kg sack) for the average household to afford to make up for production deficits. This is because the price of staple grains in Nigeria, upon which Niger depends for about 75 per cent of its millet and sorghum imports, were at record levels from June to August 2005, at the height of the hungry period in Niger (Aker, 2008). Compounding this, many local areas experienced production failures (over 25 per cent of departments had greater than 50 per cent failures; Aker, 2008). The chronic levels of poverty and malnutrition in Niger leave almost no buffer capacity for households when price shocks arise (Harrigan, 2006). Interestingly, 2009 was another year of poor agricultural production in Niger, yet FEWS NET and ECHO (European Commission Humanitarian Aid department) were already tracking high food prices, low livestock prices and low wages in early 2010, demonstrating a more comprehensive assessment of food security (Investor Relations Information Network, www.irin.com, accessed 28 January 2010).

---

to comprehend both how food makes its way from 'field to fork', and how to frame policy that corrects for the negative social and environmental outcomes of food system activities. Agriculture is no longer the primary income generating (or labour employing) activity in food supply chains globally and in many developed countries. Processing and packaging and distributing and retailing activities have grown. However, many developing countries still do depend upon agriculture for economic growth. Processes of economic globalization have connected commodity markets and food security outcomes across geographies and over time (von Braun and Diaz-Bonilla, 2008). Much more agricultural production is traded than 30 years ago. Food-price shocks in one country or region have ripple effects elsewhere. In order to assess sustained and equitable access to food security, appropriate research approaches need to be capable of capturing the interlinked relationships that comprise a food system. These include the biophysical resources which make food production possible, the resource-use demands of food processors and retailers, and consumer behaviour, including food preferences, preparation and intra-household distribution patterns. Such a comprehensive approach will also contribute to understanding the multiple ways in which food systems interact with global environmental change, and the consequences of these interactions for food security.

Global environmental change and feedbacks to ecosystem services

Food systems contribute to global environmental change (GEC) through several processes. The most basic is land-use change that occurs when land is cleared for agriculture, or converted from cropping to pasture, or replanted with trees, or when natural mangroves are replaced with aquaculture ponds. These land-use changes themselves drive changes in biodiversity, surface and subsurface hydrology, and nutrient cycles. The addition of fertilizers introduces further changes in nutrient cycles (see Box 1.1). As documented in a number of studies (e.g. Tilman et al, 2002; De Fries et al, 2004; Cassman et al, 2005), many of the change processes associated with producing food have increased food availability at the expense of key ecosystem services. As discussed in Chapter 1, other food system activities also contribute to changes in atmospheric composition. These feedbacks from food systems to GEC processes pose a real dilemma for future food security, especially given the increasing demand for food from a diminishing natural resource base (Godfray et al, 2010).

The GECAFS 'food system' approach

The Global Environmental Change and Food Systems (GECAFS) project set out to foster research on ways to enhance food security without further degrading ecosystem services. This required a broad framework to comprehensively describe all of the activities, processes and outcomes involved in modern food systems and all possible interactions with GEC. Ericksen (2008) built upon the original GECAFS Science Plan to elaborate one such framework, shown in Figure 2.1a.

At a minimum, a food system includes the set of activities involved in producing food, processing and packaging food, distributing and retailing food, and consuming food (i.e. linking commodity chains to consumers). To analyse the dynamic interactions among GEC processes, food systems and feedbacks from food system outcomes, the drivers of these activities and their social, environmental and food security outcomes are also included. The drivers comprise the interactions between and within biogeophysical and human

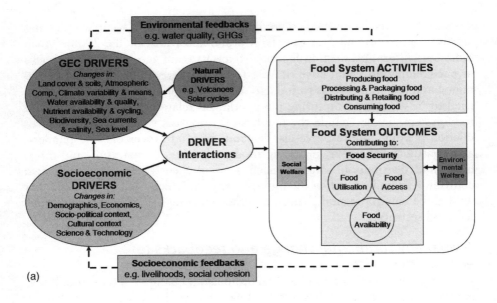

(a)

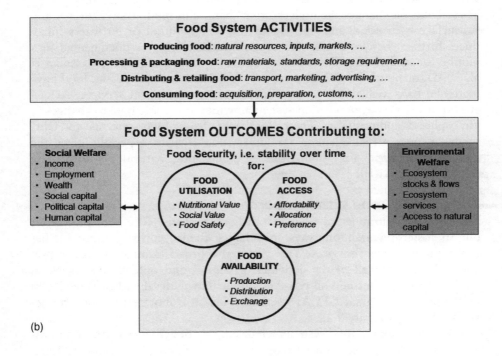

(b)

Source: GECAFS, 2009

Figure 2.1 *(a) Food systems, their drivers and feedback; (b) components of food systems*

environments which determine how food system activities are carried out (see Chapter 1 for a summary of recent trends). These activities lead to a number of outcomes, some contributing to food security and others contributing to environmental and social concerns (see Figure 2.1b). Some drivers also affect food system outcomes directly (e.g. household income levels or disease status). There are also interactions between the different categories of food system outcomes; for example, between regulating ecosystem services and availability from local production, or between income levels and access to food. Finally, food system activities and outcomes result in processes which feedback to environmental and socioeconomic drivers; food systems themselves are drivers of global change.

Food security is a principal outcome of any given food system, although in many cases food security is not achieved; instead, people are undernourished and face regular hungry seasons, or struggle with a host of non-communicable diet-related diseases, or spend the major portion of their income on poor or inadequate diets.

To help analyse the factors underpinning food security, the food security situation of a given unit of analysis can be explained in terms of three components, each of which has a number of elements. Food availability is the amount, type and quality of food a unit has at its disposal to consume; food can be available through local production; availability can rely on *distribution* channels to get food where it needs to be; and availability depends upon mechanisms to *exchange* money, labour or other items of value for food. Access to food is the ability to gain access to the type, quality and quantity of food required, and it can be analysed in terms of the *affordability* of food that is available, how well *allocation* mechanisms such as markets and government policies work, and whether consumers can meet their social and other food *preferences*. Finally, the utilization of food refers to ability to consume and benefit from food; it thus depends upon the *nutritional* and *social* values of food, and the *safety* of available and affordable food. Although the food system activities have a large influence on food security outcomes, these outcomes are also determined by socio-political and environmental drivers directly, as shown in Figure 2.1b. Stability is an important dimension of all these components of food security.

The environmental outcomes of food systems include both the stocks of available natural capital and ecosystem services – this recognizes the significant impact that food system activities have on ecosystems (see Chapter 3 for more evidence). The social welfare outcomes arise because many people rely on food systems as sources of livelihoods; thus these outcomes include income and wealth, as well as health status.

The rest of this chapter draws on other systems approaches to looking at the linkages among food security, food systems and global environmental change. This is still an emerging area of research, and examples of integrated analysis are few, both within the major agricultural development and food security agencies (e.g. the World Bank, the United Nation's Food and Agricultural Organization (FAO), and the Consultative Group on International Agricultural Research (CGIAR)), as well as the academic research community. GECAFS hosted an international conference in April 2008, and the publication emerging from that (a special issue of *Environmental Science and Policy* in 2009) features some examples of integrated research on food systems and global environmental change. Many of those authors have contributed to chapters in this book. Yet as Fresco (2009), Thompson and Scoones (2009) and Ericksen et al (2009) all point out, the research community still has a long way to go and the challenges are large.

Theoretical concepts for framing food systems and global environmental change

Accepting that food systems encompass social, economic and political issues as well as ecological, acknowledges contributions of different disciplines. However, in bridging disciplines we must recognize the importance of framing these systems when devising appropriate management interventions, development strategies and policies (Thompson et al, 2007). Different framings or narratives of how food systems function and what the key drivers are result in very different outcomes being valued and different solutions being posed. For example, economists will emphasize markets as key to food security, climate scientists worry about the greenhouse gas emissions from intensive agriculture, agronomists emphasize yields, and political scientists focus on governance arrangements as the solution to undesirable outcomes. Researchers must acknowledge that food systems serve different 'functions' for different actors, who also value their outcomes differently. This is at the heart of the tradeoffs inherent to the relationship between modern food systems, food security and ecosystem service outcomes (Rodríguez et al, 2006; Scoones et al, 2007); both the framings as well as the specific context influence how tradeoffs are evaluated and hence policy and other decisions made.

There is a rich and diverse body of literature discussing theoretical concepts which can enhance the 'food system' approach. Concepts elaborating on how to approach the complexity of food systems, from both social and ecological perspectives, are discussed initially, followed by discussion of the evolution of food system studies.

Interactions across scales and levels

Like all complex and dynamic systems, the processes and components within food systems are highly interconnected. Key to systems or complexity analysis is an emphasis on dynamics, interactions and feedbacks, many of which occur at multiple levels and scales (Ramalingam et al, 2008; Thompson and Scoones, 2009). Thus there are many feedbacks among activities, outcomes and drivers, as shown in Figure 2.1a. Feedbacks arise when social, economic and political actors respond to changes, as well as when ecosystems respond to a variety of drivers of change. Although feedback processes are inherent to coupled social–ecological systems (Carpenter et al, 2001; Holling, 2001), in food systems feedbacks cause concern because they often have negative and unintended consequences which are difficult to manage or govern, especially if they occur across different levels and scales (as they necessarily do in highly globalized food systems). Of primary concern for GEC research are the feedbacks from food system activities to ecosystem stocks and services, related to, for example, land-use and land-cover change, changes in water quality and quantity, and greenhouse gas emissions. Feedbacks can also be social, as, for example, people draw down their assets below critical thresholds (after a shock) and fall into poverty traps (Barrett and Swallow, 2006; Swallow et al, 2009); or when decisions UK consumers make about purchasing air-freighted vegetables affect the incomes of farmers in Kenya, or changes in international coffee markets benefit Vietnamese farmers but hurt Central American farmers (Eakin et al, 2009). In complex systems feedbacks are not predictable or regular, because unexpected and undesirable outcomes result (Gunderson, 2003). As most policy is not designed for surprise, unanticipated feedbacks create policy challenges (see Chapter 20).

A predominant feature of 21st-century food systems is that they are inherently cross-level and cross-scale. In a key publication, Cash et al (2006) define 'scale' as the spatial, temporal, quantitative or analytical dimensions used to measure and study any phenomenon, and 'levels' as the units of analysis that are located at different positions on a scale (Figure 2.2). GEC and food security issues span a number of different scales (e.g. spatial, temporal, jurisdictional, institutional, management) and a number of levels along each of these. Food systems are inherently multiscale and multilevel (Ericksen et al, 2009).

For example, household food security is influenced not only by factors operating at the local level; district, national and even international factors (e.g. grain prices) are also very influential. In terms of scales, food systems span institutional, informational, biophysical and cultural scales, to name but a few. These cross-level and cross-scale interactions add to the complexity of interactions and feedbacks, and also mean food systems are complex to

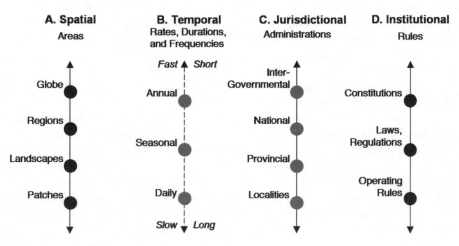

Source: Cash et al, 2006

Figure 2.2 *Different scales and levels critical in understanding and responding to food system interactions*

govern. For example, the commodity price increases of 2007–8 were compounded when some national governments imposed export restrictions to stabilize domestic prices, as such restrictions prolong international market failures and high prices (see Chapter 1; and Lustig, 2009). The multiple perspectives on food system activities and outcomes, along with differences in power across levels and scales, also means that it is very difficult to agree on solutions to food system problems, as the debates about biofuels recently illustrated. For research and policy analysis, then, it is critical to analyse specific contexts across the relevant scales and levels (and see Chapter 13). Globalization has altered many cross-level and cross-scale interactions, in many cases increasing them in new ways (Sundkvist et al, 2005; Young et al, 2006). Changes in system dynamics or structure may cause fundamental shifts in their function and outcomes, undermining food security and ecosystem services in the long term, although enhancing some types of food provisioning and income (or profit) for certain groups in the short term.

Modifying traditional food systems analysis to incorporate political and ecological dimensions

A range of researchers concerned with trends in the food industry, consumerism, globalization and political economy of food embraced the concept of a food system as an analytical tool in order to link the multiple activities and discuss the political and social dimensions and arrangements (Dixon, 1999; McMichael, 2000; Lang and Heasman, 2004). Some usefully concentrate on the behind-the-scenes processes which lead food to be produced and consumed, and therefore highlight how food security is influenced by these processes. Such approaches provide insights as to the key points of leverage, where political pressure might be applied to achieve better food security outcomes. Explicitly linking outcomes to the activities of producers, retailers and distributors and consumers is an important research consideration, as food security results from a complex set of interactions in multiple domains that are often not highlighted in conventional food chain analysis with their focus on food yields and flows. There are three conceptually rigorous systems approaches with capacity for illuminating the interests vested in transforming food systems to better advance the goal of food security for all, now and into the future. The more recent of these highlight ecological dimensions.

The commodity systems analysis (CSA), as proposed by Professor William Friedland (1984), used the relationships surrounding a single commodity as a departure point. Highlighting the multiple activities of food systems, the CSA focused on production practices, grower organization and organizations, labour, science production and application, and marketing and distribution networks to capture the power dynamics operating with the industrial agriculture sector. In 2001, Friedland (2001) modified his original CSA, proposing three additional foci to better encapsulate the dynamics of globalization and consumer and community cultures: scale; sectoral organization and the relationship of the state to the commodity; and commodity culture.

The global production network (GPN) approach builds on the CSA by acknowledging global interdependency (Henderson et al, 2002). The approach begins with several principles: first, flows of capital, labour, technology, etc. are altering the interrelationships between people and food as they cross national borders; second, vertical integration from production to consumption is not the only model by which food systems are organized and that the network model, involving horizontal relationships that are more fluid and multidimensional, is equally valid; and third, the emergence of multinational firms who have contributed to the 'uncoupling' of food from its place of production complicate the determination of appropriate regulatory responses (and see Chapter 18).

The GPN is underpinned by the following key conceptual categories: the creation, capture and enhancement of value; the sources of power operating within the networks that are engaged in a variety of value activities (corporate, institutional and collective); and embeddedness (the territorial and network nature and reach of the activities). The GPN highlights the importance of socio-technical systems to any endeavour, including the various auditing approaches by which social and economic values are enumerated. Systems including corporate responsibility audit tools, corporate taxation schemes, lifecycle assessment of commodity inputs–outputs, consumer research, community focus groups and nutrient profiling are deemed worthy of scrutiny in their own right because their control influences how food systems are regulated. In terms of identifying points of leverage, the GPN highlights temporal and spatial specificity and the difficulty of generalizing across commodity sectors. The GPN approach lends itself to mapping the flow of a basket of nutritious food commodities to particular sub-populations.

The food regimes approach, as proposed by Friedmann and McMichael (1989), shared many of the underlying influences of the CSA but it began with a world historical perspective and was premised on the interrelationship between food systems and institutional power systems. The key to the methodological approach is how 'value relations' (of the type described above) unfold historically. Thus, it differs from CSA and GPN in not offering a snapshot of current circumstances but focuses on explaining the undercurrents and turbulences in the present.

In recent work revisiting the utility of the approach, Friedmann (2009) and Campbell (2009) have independently argued for the collection of data regarding the transitions in the ecology of agriculture, thereby acknowledging the linkages between environmental change and food systems. They address arguments used by 'alternative' food movements – peasant, green and slow food – that there is a need to account for the environmental and social externalities or consequences created by industrial food systems. They problematize the audit technologies listed earlier, based on evidence showing that those who control the technologies are in a powerful position.

Their respective analyses reveal current 'food crises' to be largely the result of the ecological consequences of intensive agricultural production and distribution: the 'ecologies at a distance' regime. Foods produced in one part of the world and eaten thousands of kilometres away make it more difficult for ecological feedbacks to register with decision-makers. For Campbell, a sustainable food system has the best chance of surviving when 'social-ecological systems

... can adapt and change in response to critical signals, have the redundancy or resilience to withstand shock'. The type of research approach then is one that builds on GPN to incorporate assessments of global–local food system resilience and ecological feedbacks.

Incorporating ecology into these social and political frameworks remains a challenge. The key concepts include: the ecological inputs, outputs and dynamics incorporated in the food production–distribution–consumption cycle; the cultural politics surrounding both the measurement/audit tools used and the way that food is thought about (as nutrition, economic good, social good, a pathway to national health and wealth, etc.); an account of the strategies adopted by multiple actors as they struggle over who gets what to eat and under what conditions. For food system theorists, a historical account of each of these factors is vital for illuminating the well-springs of change and where food system transitions might lead. Developing combined social–ecological approaches is an important area of work.

Some practical examples of using a food systems approach

In this section, the goal is to illustrate how a more comprehensive food systems approach can lead to better practice and richer analysis. However, these are selected cases, as this type of research is still quite new.

Motivating consumer activism

Local food systems (LFSs) have become popular among consumer movements aiming for a closer connection to where their food is produced and seeking a lower environmental footprint. The term 'local' is in explicit contrast to 'conventional' food systems, which in many cases rely on long supply chains in which the locus of food production is distant from consumers' homes (see Chapter 19). Many aspects of a food systems approach are often applied to LFSs (even if not always overtly), both by those involved in the practice of creating LFS, and by those studying them. The nature of LFSs, often consumer-led, generally with a shorter distance between production and consumption, makes this fertile ground for looking past a production-focused analysis of food systems, typical of much agro-food literature. There have been strong calls to bring together foci on production and consumption in LFS thinking (Goodman and DuPuis, 2002), and research attempting to do this (Selfa and Qazi, 2005; Ilbery and Maye, 2006). A fruitful area of research in LFSs is an examination of the social relationships between different actors in the different food system activities, with a belief that these are different to those in conventional food systems (Jarosz, 2000). Likewise, much LFS scholarship thinks about a wider range of outcomes of the food system than just food availability. Much of the popular and academic literature sets LFSs in the context of a reaction to mainstream or conventional food systems which are viewed to have a detrimental effect on social welfare and the environment. LFSs are examined in terms of these outcomes. Although most literature has not examined LFSs in the context of food security, there is a body of critical literature that has examined issues of availability and access; for example,

Hinrichs (2000) highlighting the way in which LFSs are often firmly situated in certain socio-economic groups.

One thing LFS literature has been keen to highlight is the complexity of globalized food systems and their multiple outcomes. Advocating a more holistic view in research, some researchers have warned against a valorization of the local – the assumption that local necessarily stands in opposition to the global and thus means better (Holloway and Kneafsey, 2000; Winter, 2003; DuPuis and Goodman, 2005). Here authors have examined the complex outcomes of LFSs, highlighting instances where LFSs do not reduce negative social (Allen et al, 2003; Winter, 2003) and environmental (van Hauwermeiren et al, 2007; Coley et al, 2009) impacts. These studies, by examining case-specific scenarios, serve to remind those actively involved in creating LFSs of the importance of attending to the complexity in the outcomes of food systems.

Embodied carbon

Another area where a food systems approach is increasingly being applied in academia and the policy world is to highlight the resources embodied in commodities throughout the production process. Much of this work has surrounded greenhouse gases, specifically carbon, calculating the total amount of greenhouse gases produced through the whole supply chain, from farm to fork, for different commodities. This is often referred to as the carbon footprint of a product. More recently the concept has been extended to water, nitrogen and land-use footprints. For example, Galloway et al (2007) have developed the MEAT model to calculate the virtual trade in environmental degradation as meat and the animal feed required to produce it travel the globe. Such approaches take account of an array of food system activities, specifically examining the impact of food systems on the environment.

Policy-makers, as well as researchers, are increasingly embracing the idea of ecological footprints, using life-cycle analyses (LCAs) to calculate the impact of foodstuffs throughout food production on GEC processes, chiefly carbon and nitrogen emissions (Garnett, 2009) or uptake of water. Recently, some researchers have examined the potential impacts of labelling the 'embedded' carbon, water or nitrogen in commodities. This is an important area of research as LCAs are embraced more and more in policy-making arenas (both governmental and non-governmental), having very real effects. Such research echoes the focus of GPN on the impacts of socio-technical systems on food systems. Edwards-Jones et al (2009) examine the potential impact of carbon labelling schemes (reporting carbon footprints to consumers and businesses in order to inform choices) on the vulnerability of developing countries exporting to the UK, thereby examining the interaction between environmental and socio-economic outcomes. Similarly highlighting the performative nature of LCAs, Garnett (2009) argues that LCAs need to be considered within a broader framework that includes, amongst other things, consideration of second-order effects of production on land-use change and the resulting greenhouse gas emissions – in effect, incorporating the idea of feedbacks into LCAs. These are all conceptually appealing approaches to making the 'hidden' feedbacks more obvious, but implementing LCAs in practice will be complicated.

A food systems approach to integrated policy formulation

The institutional framework for a coherent approach to food policy is fragmented between different policy sectors. Each policy sector has its own institutional and regulatory arrangements, and constituency of attentive interests, often effectively client groups, and its own policy prerogatives and horizons. Any 'joined-up' food policy must traverse agriculture, trade, energy, industrial, labour, science and technology, health, environment, education policies and more (Barling et al, 2002; Lang et al, 2009). This list stands at the national level where the nation state is the primary locus of policy-making. But the state has been hollowed out to some extent over recent decades, with legal authority being ceded upwards to international and regional institutions and governance regimes, notably the EU, and downwards to more national–regional and local levels of government (Lang et al, 2001). Furthermore, there has been a shift from state-controlled governing to more soft forms of governance where the state effectively enrols private actors, such as corporations, sectoral groups such as farmers, and civil society organizations to undertake the administration of policy. In the case of the move of governance outwards, public forms of governing mingle and merge with more private forms of governance, providing a more complex policy picture. Food supply chains see the private setting of standards by industry, corporate retailers and manufacturers, and by civil society organizations (e.g. fair trade, animal welfare; Barling, 2008). These developments in private governance can lead the state in terms of providing private regulation over food supply. Indeed, one of the international or global bodies with authority to police food standards, the World Trade Organization (WTO), has become concerned that private standards governing international commerce in food and agriculture might effectively bypass the international legally-binding rules on food and agriculture standards and supports (Stanton and Wolff, 2009). These standards and permitted state supports are set out under the WTO's Sanitary and Phytosanitary Standards (SPS), Technical Barriers to Trade (TBT), and the Agreement on Agriculture.

The WTO standards provide a departure point for explaining the nature of multilevel governance of food policy. The development of the WTO agreements have framed subsequent agricultural support regime reforms in member states, notable examples being the reforms to the EU's Common Agricultural Policy and the EU member states' farming policies. However, such changes are contingent and based upon strategic evaluation of the scope for manoeuvre within these different international regimes, such as the Agreement on Agriculture. The rules are themselves contested as witnessed in the current halt to the revision of the Agriculture Agreement and the Doha Round of trade liberalization negotiations. International regime outcomes are the result of negotiations and tradeoffs between national states, and associated interests, as the outcomes of the Global Climate Change Protocols illustrate (e.g. Kyoto).

Integrating food policy to address effectively the twin challenges of GEC and food security will take place at these different levels of governing and across these different policy sectors. Achieving both horizontal and vertical or multilevel policy integration and compliance is a staggering political challenge. At the national level policy, priorities will need to shift and institutional means be set up to ensure such change is maintained and disseminated throughout the public policy process and the private governance of food supply (Barling et al, 2002). Conceptually, the notion of global governance suggests that the framing of policy change is achieved through the establishment of new hegemonic forms of discourse, such as the recent historical case of trade liberalization. Integration of policy can be driven by the terms of such a dominant discourse. The growing international consensus to act upon climate change and the interest in a low carbon economy is a sign of such a possible shift. The recent food price peak and renewed international attention to the security and sufficiency of the global food supply is a complementary political current. However, as observed above, the response to the 2007–8 price hike crisis was the setting up of national tariff barriers and reduction in exports of grain so exacerbating the sense of the crisis as national constituency demands came (understandably) before global coordination. Similarly, there is the previously cited example of the large-scale planting of biofuels to meet energy needs, in turn using land needed for food production. The development of a dominant policy discourse around meeting the demands of climate change, and food security, will need to continually shape the more particularistic demands of sectional economic and social interests, and their brokering by the national states who vote in the international regimes where the directions of global governance are decided.

Modelling food systems complexity

Many systems analysts stress approaching complexity by looking for patterns and typologies. Thus holistic frameworks are useful because they help to identify the full range of interactions, as well as provide an organizing framework to understand change (Reynolds et al, 2007). For example, proponents of 'syndrome' approaches to develop multiple typologies of human–environment interactions in different contexts (Petschel-Held et al, 1999); those who work on scenarios and models are keen to determine the 'key' drivers of food security and ecosystem service outcomes (Nelson et al, 2006), or to analyse the key processes controlling interactions and feedbacks among food system activities, food security and ecosystem outcomes (Plummer and Armitage, 2007). Integrated assessment models (IAMs) are the tool of choice for many analysts looking at the interactions between environmental processes and agricultural or food security outcomes (Bland, 1999; van Ittersum et al, 2008). As Schmidhuber and Tubiello (2007) note, however, these models only include some aspects of food systems, chiefly economic and land use, which misses key issues for food security and GEC feedbacks and impacts. Chapter 3 reviews recent global environmental and food security assessments, many of which have relied on IAMs.

Typologies of food systems can help to collapse their complexity into a manageable analytical framework, and draw attention to the types of food systems that show most promise for their capacity to provide particular goods or services (e.g. sustainable livelihoods, carbon sequestration), or to provide them in ways that minimize undesirable feedbacks. Typologies are also useful to show changes in food systems over time, as socio-economic or environmental drivers favour certain types of food systems and may lead to the expansion of some spaces in the typology and collapse of others.

The choice of dimensions for food system typologies is critical to their value. Two obvious choices are the attributes of food security (availability, access and utilization) or important trends in the development of food systems (e.g. globalization, increasing use of petroleum products, increasing attention to multifunctionality). Figure 2.3 shows one possible typology of the food security space by plotting components of food security against each other. Food security exists when utilization is high (the entire front face of Figure 2.3), but some food systems may enable high utilization despite low food access and availability (e.g. when a limited food supply is targeted to people with the lowest consumption rates). This is done when providing food aid in complex humanitarian emergencies.

Typologies using current trends as dimensions can assist knowledge discovery for analysts, investors or policy-makers. Dimensions of a food system typology might include household-level engagement in food production and consumption; resilience-enhancing properties such as access to and use of knowledge or technology (e.g. seeds, irrigation, nutrients, extension services); and the capacity to buffer supply and demand through storage or trade. By assigning value chains to their appropriate spaces in the typology, the consequences of different policy and investment choices can be clearly shown. While developing typologies requires careful attention to the dimensions and purpose of the exercise, their development is likely to be an essential step for analytical comparison of food systems and clear depiction of the consequences of different options.

Identifying the real food security issues of concern

Adopting a holistic food systems approach has been valuable in moving discussions about GEC impacts on food security beyond agricultural production. This is important to move beyond assumptions that may mask what is actually going on, especially cause and effect. Thus the approach can be used to

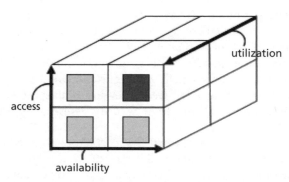

Figure 2.3 *Possible typology of the food security space*

understand how a given GEC driver transmits through a food system to affect food security; for example, a flood may have the most important impact on distribution channels (e.g. in Nepal) rather than on destroying yield. Food availability is not directly linked to production capacity in many places throughout, for example, the Caribbean, where much food is imported. Thus a hurricane which disrupts transport routes from the USA is a major concern, even if production on a given island is not affected.

In the high-latitude North, a food systems analysis can provide a useful window for examining the myriad impacts of a changing environment on both individual and community health and viability. Livelihoods in the North tend to be tightly connected to climate, weather and ecosystems, having relied for millennia upon the landscape for food, whether through hunting, herding, gathering, fishing, small-scale gardening, or a mix of all of the above. However, the impacts of climate change as currently understood threaten to undermine the viability of these essential ecosystem services (White et al, 2007; Hovelsrud et al, 2008; Loring and Gerlach, 2009). In Alaska, for instance, residents have observed changes in the landscape such as landslides and lakes drying, in some cases resulting in the complete or temporary loss of important subsistence harvest locations. Many also report that 'the world is not the way it used to be', referring to observed changes to weather and seasonality, and to the distribution, abundance and migration patterns of fish and game (Krupnik and Jolly, 2002; McNeeley and Huntington, 2007).

Access to these traditional 'country foods' is thus confounded in myriad ways, and alternatives tend to be quite limited, especially in remote, bush communities not typically connected to urban centres by roads or other infrastructure (Colt et al, 2003; Goldsmith, 2007; Gerlach et al, 2008; Martin et al, 2008). Finding that their food needs cannot be met with locally available wild food resources, many now fill their cupboards with foods of far-lesser quality and cultural relevance, purchased either from the meagre selections available at village stores (for those communities with a store) or from costly periodic provisioning trips to urban supply centres (Receveur et al, 1997; Kuhnlein et al, 2004; Ford, 2008; Loring and Gerlach, 2009).

This trajectory of change, away from traditional foods and towards industrially produced ones, has been described as a 'nutrition transition' by some (Kuhnlein et al, 2004; Popkin and Gordon-Larsen, 2004), and comes at great economic, physical and psychosocial expense. Near-epidemic increases are being observed and projected for Type II diabetes, obesity, coronary heart disease and cancer, as well as for depression, substance abuse, alcoholism and violence (McLaughlin et al, 2004; Graves, 2005; ADHS, 2006; Segal and Saylor, 2007; Wolsko et al, 2007). The extent and manner to which these health trends are directly and indirectly linked to changes in community food systems, climate-driven or otherwise, still need extensive research and quantification, however, as does the question of whether or not these trends are duplicated elsewhere in the circumpolar North. The goal is that future research will identify current patterns and distributions of risk and vulnerability, and strive to understand the many pathways through which fundamental changes to food systems can undermine individual as well as community social and cultural and ecological health outcomes, so that communities can themselves understand, plan for and effectively manage these changes.

Conclusion

A holistic approach to describing and analysing food systems allows the direct linking of ecosystem services to a critical part of human well-being (i.e. food security). Continuing to document the importance of ecosystem services for human well-being is still a research priority stemming from the Millennium Ecosystem Assessment and on the agenda of international conservation organizations (Carpenter et al, 2009). Chapter 3 explores this further, through an analysis of international global assessments and their take on food systems. This chapter has discussed the value of using a food systems approach for understanding not only the consequences of GEC for multiple aspects of food security, but also how GEC interacts with other drivers of food system activities and food security outcomes. A systems approach also helps to explain how food systems drive environmental change and the state of key ecosystem services. Such research is critical if we are to sustainably feed 9 billion people in 2050 without sacrificing other aspects of human well-being and ecosystems.

The most important policy implication of a food systems approach from an environmental change perspective is using the framework for analysis of tradeoffs between multiple aspects of food security and a range of ecosystem services (De Fries et al, 2004; Tomich et al, 2005). However, the empirical databases for such comprehensive analysis across ecosystem and food system types are insufficient, although the current debates around reducing the carbon footprint of food or about biofuels are bringing a renewed sense of urgency to these issues. Chapter 3 suggests ways to improve such databases through improved global assessments.

References

ADHS (2006) Health Risks in Alaska among adults. In Mandsager, R. (ed) *Alaska Behavioral Risk Factor Survey*, Anchorage, Alaska Department of Health and Social Services

Aker, J. C. (2008) *How can we Avoid Another Food Crisis in Niger?* Washington, DC, Center for Global Development

Allen, P., M. FitzSimmons, M. Goodman and K. Warner (2003) 'Shifting plates in the agrifood landscape: the tectonics of alternative agrifood initiatives in California', *Journal of Rural Studies*, 19, 1, 61–75

Barling, D. (2008) Governing and governance in the agri-food sector and traceability. In Coff, C., D. Barling, M. Korthals and T. Nielson (eds) *Ethical Traceability and Communicating Food*, Dordrecht, Springer

Barling, D., T. Lang and M. Caraher (2002) 'Joined up food policy? The trials of governance, public policy and food systems', *Social Policy and Administration*, 36, 6, 556–74

Barrett, C. B. and B. M. Swallow (2006) 'Fractal poverty traps', *World Development*, 34, 1, 1–15

Bland, W. (1999) 'Toward integrated assessment in agriculture', *Agricultural Systems*, 60, 157–67

Campbell, H. (2009) 'Breaking new group in food regime theory: Corporate environmentalism, ecological feedbacks and the "Food from Somewhere" Regime?', *Agriculture and Human Values*, 26, 4, 309–19

Carpenter, S., B. Walker, J. M. Anderies and N. Abel (2001) 'From metaphor to measurement: resilience of what to what?', *Ecosystems*, 4, 765–81

Carpenter, S. R., H. A. Mooney, J. Agard, D. Capistrano, R. S. De Fries, S. Díaz, T. Dietz, A. K. Duraiappah, A. Oteng-Yeboah, H. M. Pereira, C. Perrings, W. V. Reid, J. Sarukhan, R. J. Scholes and A. Whyte (2009) 'Science for managing ecosystem services: Beyond the Millennium Ecosystem Assessment', *Proceedings of the National Academy of Sciences of the United States of America*, 106, 5, 1305–12

Cash, D. W., W. Adger, F. Berkes, P. Garden, L. Lebel, P. Olsson, L. Pritchard and O. Young (2006) 'Scale and cross-scale dynamics: governance and information in a multilevel world', *Ecology and Society*, 11, 2

Cassman, K. G., S. Wood, P. S. Choo, H. D. Cooper, C. Devendra, J. Dixon, J. Gaskell, S. Khan, R. Lal, L. Lipper, J. Pretty, J. Primavera, N. Ramankutty, E. Viglizzo and K. Wiebe (2005) Chapter 26, Cultivated Systems. In *Ecosystems and Human Wellbeing: Conditions and Trends*, Washington, DC, Island Press

Clay, E. (2005) The Niger food crisis: how has this happened? What should be done to prevent a recurrence? *ODI Opinions*, London, Overseas Development Institute

Coley, D., M. Howard and M. Winter (2009) 'Local food, food miles and carbon emissions: A comparison of farm shop and mass distribution approaches', *Food Policy*, 24, 150–55

Colt, S., S. Goldsmith and A. Wiita (2003) *Sustainable Utilities in Rural Alaska: Effective Management, Maintenance and Operation of Electric, Water, Sewer, Bulk Fuel, Solid Waste*, Anchorage, AK, Institute of Social and Economic Research, University of Alaska Anchorage

De Fries, R. S., J. A. Foley and G. P. Asner (2004) 'Land-use choices: balancing human needs and ecosystem function', *Frontiers in Ecology and the Environment*, 2, 5, 249–57

Devereux, S. (2000) Famine in the Twentieth Century. *Working Papers*, Brighton, Institute of Development Studies, University of Sussex

Dixon, J. (1999) 'A cultural economy model for studying food systems', *Agriculture and Human Values*, 16, 151–60

DuPuis, E. M. and D. Goodman (2005) 'Should we go "home" to eat?: Toward a reflexive politics of localism', *Journal of Rural Studies*, 21, 359–71

Eakin, H., A. Winkels and J. Sendzimir (2009) 'Nested vulnerability: Exploring cross-scale linkages and vulnerability tele-connections in Mexican and Vietnamese coffee systems', *Environmental Science and Policy*, 12, 4, 398–412

Edwards-Jones, G., K. Plassmann, E. H. York, B. Hounsome, D. L. Jones, I. Mila and L. Canals (2009) 'Vulnerability of exporting nations to the development of a carbon label in the United Kingdom', *Environmental Science and Policy*, 12, 479–90

Ericksen, P. J. (2008) 'Conceptualizing food systems for global environmental change research', *Global Environmental Change*, 18, 234–45

Ericksen, P. J., J. S. I. Ingram and D. M. Liverman (2009) 'Food security and global environmental change: Emerging challenges', *Environmental Science and Policy*, 12, 4, 373–77

Ford, J. D. (2008) 'Vulnerability of Inuit food systems to food insecurity as a consequence of climate change: A case study from Igloolik, Nunavut', *Regional Environmental Change*, 9, 83–100

Fresco, L. O. (2009) 'Challenges for food system adaptation today and tomorrow', *Environmental Science and Policy*, 12, 378–85

Friedland, W. (1984) 'Commodity systems analysis: An approach to the sociology of agriculture', *Research in Rural Sociology and Development*, 1, 221–35

Friedland, W. (2001) 'Reprise on commodity systems methodology', *International Journal of Sociology and Agriculture*, 9, 1, 82–103

Friedmann, H. (2009) 'Discussion: Moving food regimes forward: reflections on symposium essays', *Agriculture and Human Values*, 26, 4, 335–44

Friedmann, H. and P. McMichael (1989) 'Agriculture and the State System: The rise and decline of national agricultures, 1870 to the present', *Sociologia Ruralis*, 29, 2, 93–117

Galloway, J. N., M. Burke, G. E. Bradford, R. Naylor, W. Falcon, A. K. Chapagain, J. C. Gaskell, E. McCullough, H. A. Mooney, K. L. L. Oleson, H. Steinfeld, T. Wassenaar and V. Smil (2007) 'International trade in meat: The tip of the pork chop', *AMBIO: A Journal of the Human Environment*, 36, 8, 622–29

Garnett, T. (2009) 'Livestock-related greenhouse gas emissions: impacts and options for policy makers', *Environmental Science and Policy*, 12, 491–503

GECAFS (2009) *GECAFS Food Systems Brochure*, Oxford, GECAFS

Gerlach, S. C., P. A. Loring and T. Paragi (2008) The future of Northern food systems. In Loring, A. J. K. (ed) *Arctic Forum 2008: Tipping Points – the Arctic and Global Change*, Washington, DC, The Arctic Research Consortium of the US (ARCUS)

Godfray, H. C. J., J. R. Beddington, I. R. Crute, L. Haddad, D. Lawrence, J. F. Muir, J. Pretty, S. Robinson, S. M. Thomas and C. Toulmin (2010) 'Food security: The challenge of feeding 9 billion people', *Science*, 327, 812–18

Goldsmith, S. (2007) *The Remote Rural Economy of Alaska*, Anchorage, University of Alaska Anchorage

Goodman, D. and E. M. DuPuis (2002) 'Knowing food and growing food: Beyond the production-consumption debate in the sociology of agriculture', *Sociologia Ruralis*, 42, 5–22

Graves, K. (2005) *Resilience and Adaptation Among Alaska Native Men*, Fairbanks, University of Alaska, Anchorage

Gunderson, L. H. (2003) Adaptive dancing: Interactions between social resilience and ecological crises. In Berkes, F., J. Colding and C. Folke (eds) *Navigating Social-Ecological Systems: Building Resilience for Complexity and Change*, Cambridge, Cambridge University Press

Harrigan, S. (2006) *The Cost of Being Poor: Markets, Mistrust and Malnutrition in Southern Niger 2005–2006*, London, UK, Save the Children UK

Henderson, J., P. Dicken, M. Hess, N. Coe and H. Yeung (2002) 'Global production networks and the analysis of economic development', *Review of International Political Economy*, 9, 436–64

Hinrichs, C. C. (2000) 'Embeddedness and local food systems: Notes on two types of direct agricultural market', *Journal of Rural Studies*, 16, 295–303

Holling, C. S. (2001) 'Understanding the complexity of economic, ecological and social systems', *Ecosystems*, 4, 390–405

Holloway, L. and M. Kneafsey (2000) 'Reading the space of the farmers' market: a case study from the United Kingdom', *Sociologia Ruralis*, 40, 285–99

Hovelsrud, G. K., M. McKenna and H. P. Huntington (2008) 'Marine mammal harvests and other interactions with humans', *Ecological Applications*, 18, S135–47

Ilbery, B. and D. Maye (2006) 'Retailing local food in the Scottish-English borders: A supply chain perspective', *Geoforum*, 37, 352–67

Jarosz, L. (2000) 'Understanding agri-food networks as social relations', *Agriculture and Human Values*, 17, 279–83

Krupnik, I. and D. Jolly (eds) (2002) *The Earth is Faster Now: Indigenous Observations of Arctic Environmental Change*, Fairbanks, Arctic Research Consortium of the United States

Kuhnlein, H. V., O. Receveur, R. Soueida and G. M. Egeland (2004) 'Arctic indigenous peoples experience the nutrition transition with changing dietary patterns and obesity', *Journal of Nutrition*, 134, 6, 1447–53

Lang, T. and M. Heasman (2004) *The Food Wars: The Global Battle for Mouths, Minds and Markets*, London, Earthscan

Lang, T., D. Barling and M. Caraher (2001) 'Food, social policy and the environment: Towards a new model', *Social Policy and Administration*, 35, 5, 538–58

Lang, T., D. Barling and M. Caraher (2009) *Food Policy: Integrating Health, Environment and Society*, Oxford, Oxford University Press

Loring, P. A. and S. C. Gerlach (2009) 'Food, Culture and Human Health in Alaska: An integrative health approach to food security', *Environmental Science and Policy*, 12, 466–78

Lustig, N. (2009) Coping with rising food prices: Policy dilemmas in the developing world. *Working Papers*, Washington, DC, Center for Global Development

Martin, S. M., M. Killorin and S. Colt (2008) *Fuel Costs, Migration and Community Viability*, Anchorage, Institute of Social and Economic Research (ISER), UAA

Maxwell, S. (2001) The evolution of thinking about food security. In Devereux, S. and S. Maxwell (eds) *Food Security in Sub-Saharan Africa*, London, ITDG

Maxwell, S. and R. Slater (2003) 'Food policy old and new', *Development Policy Review*, 21, 5–6, 531–53

McLaughlin, J. B., J. P. Middaugh, C. J. Utermohle, E. D. Asay, A. M. Fenaughty and J. E. Eberhardt-Phillips (2004) 'Changing patterns of risk factors and mortality for coronary heart disease among Alaska natives', *Journal of the American Medical Association*, 291, 2545–46

McMichael, P. (2000) 'The power of food', *Agriculture and Human Values*, 17, 21–33

McNeeley, S. and O. Huntington (2007) Postcards from the (not so) frozen North: Talking about climate change in Alaska. In Moser, S. C. and L. Dilling (eds) *Creating a Climate for Change*, Cambridge, Cambridge University Press

Nelson, G. C., E. Bennett, A. A. Berhe, K. Cassman, R. DeFries, T. Dietz, A. Dobermann, A. Dobson, A. Janetos, M. Levy, D. Marco, N. Nakicenovic, B. O'Neill, R. Norgaard, G. Petschel-Held, D. Ojima, P. Pingali, R. Watson and M. Zurek (2006) 'Anthropogenic drivers of ecosystem change: An overview', *Ecology and Society*, 11, 2, 29

Petschel-Held, G., A. Block, M. Cassel-Gintz, J. Kropp, M. K. B. Ludeke, O. Moldenhauer, F. Reusswig and H. J. Schellnhuber (1999) 'Syndromes of global change: A qualitative modelling approach to assist global environmental management', *Environmental Modelling and Assessment*, 4, 295–314

Plummer, R. and D. Armitage (2007) 'A resilience-based framework for evaluating adaptive co-management: linking ecology, economics and society in a complex world', *Ecological Economics*, 61, 62–74

Popkin, B. M. and P. Gordon-Larsen (2004) 'The nutrition transition: Worldwide obesity dynamics and their determinants', *International Journal of Obesity*, 28, S3, S2–S9

Ramalingam, B., H. Jones, with T. Reba and J. Young (2008) Exploring the Science of Complexity: Ideas and Implications for development and humanitarian efforts. *ODI Working Papers*, London, Overseas Development Institute

Receveur, O., M. Boulay and H. V. Kuhnlein (1997) 'Decreasing traditional food use affects diet quality for adult Dene/Métis in 16 communities of the Canadian Northwest Territories', *Journal of Nutrition*, 127, 2179–86

Reynolds, J. F., D. M. Stafford Smith, E. F. Lambin, B. L. Turner II, M. Mortimore, S. P. J. Batterbury, T. E. Downing, H. Dowlatabadi, R. J. Fernandez, J. E. Herrick, E. Huber-Sannwald, H. Jiang, R. Leemans, T. Lynam, F. T. Maestre, M. Ayarza and B. Walker (2007) 'Global desertification: Building a science for dryland development', *Science*, 316, 847–51

Rodríguez, J. P., T. D. Beard Jr, E. M. Bennet, G. S. Cumming, S. J. Cork, J. Agard, A. P. Dobson and G. D. Peterson (2006) 'Trade-offs across space, time and ecosystem services', *Ecology and Society*, 11, 1, art 28

Schmidhuber, J. and F. N. Tubiello (2007) 'Global food security under climate change', *Proceedings of the National Academy of Sciences of the United States of America*, 104, 19703–8

Scoones, I., M. Leach, A. Smith, S. Stagl, A. Stirling and J. Thompson (2007) Dynamic systems and the challenge of sustainability. *STEPS Working Papers*, Brighton, STEPS Centre

Segal, B. and B. Saylor (2007) 'Social transition in the North: Comparisons of drug-taking behavior among Alaska and Russian natives', *International Journal of Circumpolar Health*, 66, 1, 71–76

Selfa, T. and J. Qazi (2005) 'Place, taste or face-to-face? Understanding producer-consumer networks in "local" food systems in Washington State', *Agriculture and Human Values*, 22, 451–64

Stanton, G. H. and C. Wolff (2009) Private voluntary standards and the World Trade Organization Committee on Sanitary and Phytosanitary Measures. In Borot de Battisti, A., J. McGregor and A. Graffham (eds) *Standard Bearers: Horticultural Exports and Private Standards in Africa*, London, International Institute for Environment and Development

Sundkvist, A., R. Milestad and A. Jansson (2005) 'On the importance of tightening feedback loops for sustainable development of food systems', *Food Policy*, 30, 224–39

Swallow, B. M., J. K. Sang, M. Nyabenge, D. K. Bundotich, A. K. Duraiappah and T. B. Yatich (2009) 'Tradeoffs, synergies and traps among ecosystem services in the Lake Victoria basin of East Africa', *Environmental Science and Policy*, 12, 504–19

Thompson, J. and I. Scoones (2009) 'Addressing the dynamics of agri-food systems: An emerging agenda for social science research', *Environmental Science and Policy*, 12, 386–97

Thompson, J., E. Millstone, I. Scoones, A. Ely, F. Marshall, E. Shah and S. Stagl (2007) *Agrifood System Dynamics: Pathways to Sustainability in an Era of Uncertainty*, Brighton, STEPS Centre

Tilman, D., K. G. Cassman, P. A. Matson, R. Naylor and S. Polasky (2002) 'Agricultural sustainability and intensive production practices', *Nature*, 418, 6898, 671–77

Tomich, T. P., A. Cattaneo, S. Chater, H. Geist, J. Gockowski, D. Kaimowitz, E. Lambin, J. Lewis, O. Ndoye, C. Palm, F. Stolle, W. D. Sunderlin, J. Valentim, M. van Noordwijk and S. Vosti (2005) Balancing agricultural development and environmental objectives: assessing tradeoffs in the humid tropics. In Palm, C., S. Vosti, P. Sanchez and P. Ericksen (eds) *Slash and Burn Agriculture*, New York, Columbia University Press van Hauwermeiren, A., H. Coene, G. Engelen and E. Mathijs (2007) 'Energy lifecycle inputs in food systems: A comparison of local versus mainstream cases', *Journal of Environmental Policy & Planning*, 9, 1, 31–51

van Ittersum, M. K., F. Ewert, T. Heckelei, J. Wery, J. A. Olsson, E. Andersen, I. Bezlepkina, F. Brouwer, M. Donatelli, G. Flichman, L. Olsson, A. E. Rizzoli, T. van der Wal, J. E. Wien and J. Wolf (2008) 'Integrated assessment of agricultural systems – A component-based framework for the European Union (SEAMLESS)', *Agricultural Systems*, 96, 150–65

von Braun, J. and E. Diaz-Bonilla (2008) Globalization of agriculture and food: Causes, consequences and policy implications. In von Braun, J. and E. Diaz-Bonilla (eds) *Globalization of Food and Agriculture and the Poor*, Oxford, Oxford University Press

White, D. M., S. C. Gerlach, P. Loring, A. C. Tidwell and M. C. Chambers (2007) 'Food and water security in a changing arctic climate', *Environmental Research Letters*, 2, 4

Winter, M. (2003) 'Embeddedness, the new food economy and defensive localism', *Journal of Rural Studies*, 19, 1, 23–32

Wolsko, C., C. Lardon, G. Mohatt and E. Orr (2007) 'Stress, coping and well-being among the Yup 'ik of the Yukon-Kuskokwim Delta: the role of enculturation and acculturation', *International Journal of Circumpolar Health*, 66, 51–61

Young, O., F. Berkhout, G. C. Gallopin, M. A. Janssen, E. Ostrom and S. van der Leeuw (2006) 'The globalization of socio-ecological systems: An agenda for scientific research', *Global Environmental Change*, 16, 304–16

3
Lessons Learned from International Assessments

Stanley Wood, Polly Ericksen, Beth Stewart,
Philip Thornton and Molly Anderson

Introduction

This chapter reviews the goals and outputs of international assessments that have examined the linkages between environment and food. Focus in particular is on the treatment of food systems (as defined in Chapter 2), as well as on the extent to which the key implications of global environmental change (GEC) for global and local food security have been articulated and explored. The analysis suggests that such assessments have fallen short, sometimes significantly, of providing comprehensive and balanced evidence on the range and interdependence of environmental change phenomena and on the consequences of change on the many facets of food systems and security. As a consequence, it is concluded that most assessments have been limited in their ability to inform relevant science and development debates, shape food system strategies and policies and foster appropriate advocacy. The implications of these shortcomings and conclusions with lessons learned and proposals for improving the adequacy and relevance of future assessments are discussed. This includes aspects of the scope, governance, design, methods and outreach of such undertakings if they are to more satisfactorily assess the true scope and implications of GEC on food systems and food security.

Assessment landscape and typology

The recent past has seen a number of relevant assessments (Box 3.1) representing a variety of constituencies and goals that have addressed one or more of three driving concerns: How can accelerating GEC be contained? How can 9 billion people be adequately fed by 2050? How can the welfare of the world's poorest and most vulnerable people be improved? While these questions have been posed and assessments undertaken using a variety of lenses, most reviewed here have explicitly recognized the critical linkages that exist between GEC, food security and human welfare.

> **Box 3.1** *What is an assessment? (adapted from OMB, 2004)*
>
> An assessment is a judgement or decision based on an examination of available information about the amount, value, quality or importance of something. A 'scientific assessment' is an evaluation of a body of scientific or technical knowledge that typically synthesizes multiple factual inputs, data, models, assumptions and/or applies best professional judgement to bridge uncertainties in the available information. Where possible, each finding is presented with an associated level of certainty or confidence. These assessments include, but are not limited to, state-of-science reports; technology assessments; weight-of-evidence analyses; meta-analyses; health, safety, ecological or environmental risk assessments; integrated assessment models; hazard determinations; or exposure assessments.
>
> Scientific assessments are typically very influential in policy debates. It was the IPCC *First Assessment Report* in 1990 that led to the Framework Convention on Climate Change, and the IPCC *Second Assessment Report* in 1995 that led to the Kyoto Protocol.

A two-stage approach is taken in defining the scope of the review. An inventory of assessments of potential interest was first compiled, ranging from those driven by GEC concerns in which the consequential impacts traced through to food system activities and food security outcomes was sought, as well as food system- and security-related assessments in which multiple environmental change phenomena articulated as key drivers of change were expected to be seen. In surveying the landscape it became clear that assessments could also be distinguished by other criteria; whether instigated more by science or by development constituencies, whether designed to inform long-term strategic or short-term tactical needs and, highly conditioned by those factors, whether the assessments were periodic and systematic or, typically for larger international assessments, more idiosyncratic. The findings from organizing this expansive inventory of potentially relevant assessments are summarized in Table 3.1.

Table 3.1 identifies a number of assessment groupings. The first group includes the major international environmental change assessments, IPCC, GEO and MA (see Table 3.1 footnotes for complete assessment names) that have involved significant multistakeholder and in some cases intergovernmental forums. They are also typified by a heavy emphasis on scientific assessment of conditions and trends and the use of formal scenario-based evaluation of possible futures. These major initiatives are reviewed in greater detail below.

A second group of periodic assessments is geared towards alternative, long-term perspectives on food supply, and includes IAASTD, CAWMA, the FAO *World Agriculture Towards...* series, IFPRI's IMPACT model studies and IIASA's GAEZ framework. These assessment approaches are capable of accounting for a range of GEC drivers, but typically evaluate the impacts of any such change only on a narrow set of food production indicators and do not trace through impacts to other food system activities. Such food

Table 3.1 *Broad characterization of assessments of potential relevance to environmental change and food system linkages*

Assessment constituency/ driving perspective	Assessment frequency/time perspective		
	Idiosyncratic/ long term	Annual (or quasi-annual)/medium–short term	Sub-annual/ short term – crisis
Environmental change	**Climate:** IPCC (1990, 1995, 2001, 2007) **Environment:** GEO (1997, 1999, 2002, 2007) OECD/Env Outlook (2008) **Ecosystems and services:** MA (PAGE 2000, 2005)	**Environment:** WRI/WRR (2/6) WWI/State of the World (4/9)	
(Sustainable) food supply	**Agricultural R and D:** IAASTD (2009) **Water:** CAWMA (2007) **Food provision projections:** FAO/AT (1988, 1995, 2003, 2006) IFPRI/IMPACT (2001) Livestock (1999), Water (2002) Fisheries (2003) IIASA/GAEZ (2001) Climate Change (2002) Biofuels (2009)	**Food provision:** FAO/SOFA (2/8) OECD/FAO Agricultural Outlook (2/9)	
Food security/ human welfare		**Hunger:** FAO/SOFI (1/9) WFP/WHS (0/3) IFPRI/GHI (0/2) USDA/GFSA (3/10) **Human welfare:** UN/HDR (2/9) **Economic growth:** WB/WDR (2/10)	**Food security assessments:** USAID/FEWS FAO/GIEWS WFP/VAM

Notes: This inventory is not exhaustive and excludes other environmental assessments such as the Forest Resource Assessment, Global Biodiversity Outlook and the Global International Water Assessment with lesser focus on food system linkages or others such as the Land Degradation in Drylands (LADA) that have not yet fully reported. Full citations of assessments are provided in the references.

IPCC: Intergovernmental Panel on Climate Change; GEO: Global Environmental Outlook; OECD: Organization for Economic Cooperation and Development; MA: Millennium Ecosystem Assessment; PAGE: Pilot Assessment of Global Ecosystems; IAASTD: International Assessment of Agricultural Science and Technology Development; CAWMA: Comprehensive Assessment of Water Management in Agriculture; FAO/AT: Food and Agriculture Organization of the UN *Agriculture Toward* series; IFPRI/IMPACT: International Food Policy Research Institute International Model for Policy Analysis of Agricultural Commodities and Trade; IIASA/GAEZ: International Institute for Applied Systems Analysis Global Agroecological Zones; WRI/WRR: World Resources Institute World Resources Report; WWI: World Watch Institute; FAO/SOFA: State of Food and Agriculture; FAO/SOFI: State of Food Insecurity; WFP/WHS: World Food Programme World Hunger Series; IFPRI/GHI: Global Hunger Index; USDA/GFSA: United States Department of Agriculture Global Food Security Assessment; UN/HDR: United Nations Human Development Report; WB/WDR: World Bank World Development Report; USAID/FEWS: United States Agency for International Development Famine Early Warning System; FAO/GIEWS: Global Information and Early Warning System; WFP/VAM: World Food Program Vulnerability Assessment and Mapping.

Scoring food system/food security focus (X/N): In the case of the annual or quasi-annual assessments, a subjective scoring was made of how many reports (X) from all the reports examined (N) were considered to have significant content relating environmental change to food system or food security indicators. Many of these serial assessments have topical themes that occasionally focus on environmental change and food issues. The food system relevance of major international assessments is examined separately in the main chapter text.

perspective studies are typically the product of a single modelling group work-ing in relative isolation. The results of these model-based assessments are often persuasive in specific advocacy and policy arenas, but the lack of broad-based engagement can limit their wider acceptance (e.g. international assessments engage influential decision-makers and outreach specialists as part of their design and impact strategies and typically involve a blend of analytical philosophies and methods, whereas modelling studies typically engage only analysts from a single modelling group who apply and report the specific results from applying their proprietary model or approach).

By purpose and design most of these groups of longer-range assessments make extensive use of formal scenario formulation (Box 3.2) and evaluation approaches in describing and communicating alternative perspectives of the future. A notable development in this regard has been the increasing shift away from more traditional projection of likely future pathways change (e.g. 'business-as-usual', 'optimistic', 'pessimistic', as typified by Rosegrant et al (2001) and Fischer et al (2002)) to visions of idealized 'plausible futures', often polar world perspectives argued to enrich insights into major development

---

Box 3.2 *What is a scenario?*

Scenarios are carefully crafted stories that attempt to cut through uncertainty by articulating possible pathways into the future. They include an interpretation of the present, a vision of how current trends will unfold, a storyline of how critical uncertainties will play out and what new factors will come into play. Well-constructed scenarios provide coherent, comprehensive and internally consistent descriptions of plausible futures built on the imagined interaction of key trends. Scenarios are not forecasts of future events, nor are they predictions of what might or will happen in the future. Rather, they paint pictures of plausible futures, and explore the differing outcomes that might result if basic assumptions are changed (UNEP, 2002).

Scenarios serve many purposes. They can help map-out the range of potential outcomes associated with different pathways to the future, which in turn helps identify options and strategies to increase the chances of achieving preferred out-comes. Scenarios also help identify threats and challenge conventional, preconceived notions of the future, and are widely used in policy formulation and advocacy.

Scenarios are necessary since the future cannot be predicted with confidence; information on the current state of global systems or many of the drivers of change are far from complete. Even if precise information were available, predic-tion would be impossible as complex systems exhibit turbulent behaviour, extreme sensitivity to initial conditions and branching behaviours at critical thresh-olds. Furthermore, the future is unknowable because it is subject to human choices that have not yet been made. In the face of such indeterminacy, scenario analysis offers a means of exploring a variety of long-range alternatives, knowing that the uncertainty about the future increases with distance from the present (Raskin et al, 2002).

paradigm choices, rather than trying to second guess the actual pathway of future development (e.g. the Markets-, Policy-, Security- and Sustainability-First scenarios of GEO4, or the 'Global Orchestration', 'TechnoGarden', 'Order from Strength' and 'Adapting Mosaic' scenarios of the MA). To reap the benefits of both approaches, there is growing sentiment to include a 'business-as-usual' scenario with plausible future scenarios.

There are several clusters of relatively frequent (typically an annual or bi-annual cycle) assessments that serve to provide up-to-date interpretation and analysis around topical issues, provide near-term prognoses (often based on trend analyses), or report the tracking of key environmental change or food provision and related welfare indicators. One group of such assessments is borne primarily out of environmental and sustainability concerns (e.g. the *World Resources Report* (WRI, 2001, 2008) and the *State of the World* (Worldwatch Institute, 2004, 2005) assessments), another out of contemporary food production, trade and macro (national-level) food security concerns (e.g. FAO's status and trends-based *State of Food and Agriculture* and the FAO/OECD's annual short-term (ten-year) *Agricultural Outlook* assessments). The final two clusters are designed to respond more directly to human welfare and food security outcomes. The annual USDA *Global Food Security Assessment* provides annual tracking and prognoses for a number of national-level food production, trade, aid and consumption indicators, whereas the FAO's *State of Food Insecurity in the World* series and IFPRI's annual *Global Hunger Index* report and interpret assessments of the (national-, regional- and global-level) nutrition and health outcome status of individuals. FAO's indica-tor of 'share of population food energy deficient' has been adopted as a Millennium Development Goal (MDG) hunger metric. The WFP *World Hunger Series* (2009) provides topical reviews of major hunger drivers and determinants, and uniquely tracks an explicit set of national indicators of food availability and access.

Neither the UNDP's annual *Human Development Reports* nor the World Bank's *World Development Reports* (2008, 2009) have systematic linkages to concerns of GEC or food systems, but both address topical issues of relevance to human welfare and economic growth more broadly. Given the importance of the environment and food and agriculture to both those issues, however, especially to poorer nations, these flagship publications periodically might be expected to address them.

The third group comprises the operational assessments of food production and food security to inform market, logistical and relief efforts, particularly within and between growing seasons. Such assessments are necessarily time-sensitive and are based on rapid appraisal techniques linking past experiences with current trends to track and provide short-term projections of key indica-tors of productivity, food availability and food accessibility.

Table 3.2 provides a coded summary of salient features of each assess-ment of relevance to this review and a subjective scoring of the linking of environmental change and food systems activities or food security outcomes in the 'annual' assessments (see Table 3.2 footnotes for detailed description of both coding and scoring). The coded summary captures stakeholder and institutional scope, the importance placed on reviewing conditions and trends, the use of formal scenario-based approaches and projection models, the time line for projections, the importance of food-related variables or indicators in the assessment, and the emphasis given to tracking relevant and consistent food system or food security indicators over time.

Comparing and contrasting assessment types

Using the assessment typology and characterization of Table 3.1 reveals some structural patterns of interest. Broadly speaking there are four types of assessment, as follows:

- International scientific assessments: large-scale, multi-year, multi-sponsor, multi-stakeholder international scientific assessments, typically with long-term perspectives, that appear to be driven primarily by GEC concerns and constituencies (the upper part of the first column of Table 3.1, from IPCC to CAWMA). While the IAASTD focused on innovation systems for agriculture and food security, it drew much of its initial approaches, participants and momentum from the IPCC and MA processes. The CAWMA is the only assessment that is entirely concerned with links between a (threatened) global environmental resource (water) and food systems (although the ongoing LADA initiative will fall in this category when it reports).
- Serial institution-based assessments: quasi-annual or repeated periodic assessments by individual institutions that serve to review topical issues, make near-term projections or outlooks – typically based on trend analysis – and provide tracking updates on key environmental, food or welfare-related monitoring indicators (such assessments make up the central column of Table 3.1). By their nature and cycle times these assessments tend to be more limited and selective in scope, rely less on scientific assessment approaches (and on external peer review, the WDR and SOFA being notable exceptions), and are often targeted to support topical policy and advocacy needs rather than validate or consolidate knowledge.
- Operational food security assessments: rapid-cycle, time-sensitive assessments of the juxtaposition of local climatic, market and production situations and trends, food availability, accessibility and demand, as well as food supply needs – including food aid – to make up shortfalls (see the bottom right-hand column of Table 3.1). Such assessments are not truly global, but rather are generated by a quasi-global network of organizations undertaking national and regional assessments in poorer or crisis-hit parts of the world. Within (if not across) these networks harmonized procedures and indicators of food system activities and of food security are generated and tracked.
- Food supply perspective assessments: these are a distinct group of assessments undertaken by independent analytical/modelling groups to serve internal institutional or commissioned study needs. The groups and their

Table 3.2 Coded summary of salient features of assessments

| | Assessments | Institutional context | Status and trends | Future perspectives | | | Indicators | |
				Scenarios	Projections	Assessment periods	Food system indicators	Indicator tracking
Environmental change	**Climate:**							
	IPCC	E	R	S	P	50	f	T
	Environment:							
	GEO	E	r	S	P	50	F	t
	OECD Env. Outlook	I	r	S	P	25	F	t
	WRI/WRR	I	R	–	–	–	f	T
	WWI/State of the World	I	R	–	–	–	f	–
	Ecosystems and services:							
	MA	E	R	S	P	50	F	t
(Sustainable) food supply	**Agricultural R and D:**							
	IAASTD	E	R	s	p	50	F	t
	Water:							
	CAWMA	E	R	S	P	50	F	t
	Food provision projections:							
	FAO/AT	I	R	s	p	15/50	F	–
	IFPRI/IMPACT	I	r	S	P	20/50	F	–
	IIASA/GAEZ	I	r	s	P	50/80	F	–
	Food provision:							
	FAO/SOFA	I	R	–	–	–	F	T
	OECD/FAO Agricultural Outlook	I	R	S	P	10	F	t

Food security/human welfare	E/I	R/r/–	S/s/–	P/p/–	YY/mm	F/f	T/t/–
Hunger:							
FAO/SOFI	I	r	–	–		F	T
WFP/WHS	I	r	–	–		F	T
IFPRI/GHI	I	r	–	–		F	T
USDA/GFSA	I	r	–	–		F	T
Human welfare:							
UN/HDR	I	r	–	–		f	T
Economic growth:							
WB/WDR	I	R	–	–		f	T
Food security assessments:							
USAID/FEWS	E	r	–	P	3/6m	F	T
FAO/GIEWS	E	r	–	P	3/6m	F	T
WFP/VAM	E	r	–	P	3/6m	F	T

Assessment coding:

Institutional Context: E/I (E)ngaged with multi-stakeholder community or process, (I)nstitution or project based assessments.

Status and Trends: R/r/– (R)eview of current status and trends (some with stated confidence levels) is a major element of the assessment, (r)eview is cursory or specifically focused only on baseline definition, (–) no significant treatment of current status and trends.

Scenarios: S/s/– (S)cenario formulation, description and analysis is a major element of the assessment (typically with multiple scenarios defined), (s)cenarios are less formally defined or emphasized, (–) no or no significant use of scenarios.

Projections: P/p/– (P)rojections of future conditions are central to the assessment (most likely using formal predictive models), (p)rojections are made by qualitative or subjective means or are not given prominence, (–) no emphasis on predictions.

Assessment Periods: YY/mm (Y)ears over which projections were or are typically made (intermediate time periods are also usually reported). A range of years YY/YY is given when multiple assessments have been made using the same process or approach. For most food security assessments, projections are for coming weeks and months of the growing season and, less often, for the following growing season.

Food System Indicators: F/f (F)ood system related indicators are central to the assessment or are treated in depth, (f)ood system related indicators are limited or given less weight.

Indicator Tracking: T/t/– (T)racking of key performance indicators is central to the assessment for monitoring, evaluation or advocacy purposes, (t)racking of indicators is of lesser priority or assessment is attempting to establish new indicators to be tracked, (–) tracking indicators across assessments is not a priority or not relevant.

analyses are not independent of the other assessments. For example, the IFPRI IMPACT model provided analytical inputs to MA, GEO4 and IAASTD, and the IIASA GAEZ results were used in IPCC and CAWMA. But individual modelling groups also undertake assessments responding to their own research priorities (e.g. exploring future food perspectives focusing on water, livestock, fisheries, climate change and biofuel aspects).

Even a cursory examination of the material compiled to develop the typology and content of Table 3.1 is striking in its overall balance with respect to the goal of this review. First, where the links between GEC and food systems are given explicit treatment – most extensively in the international scientific assessments – the primary weight of evidence highlights the contribution of agricultural production to global environmental change (e.g. greenhouse gas emissions in IPCC, MA, GEO, IAASTD; freshwater overexploitation in MA, GEO, IAASTD, CAWMA; accelerated nutrient cycling in MA, GEO, IAASTD; land-use conversion (habitat and biodiversity loss, and accelerated greenhouse gas emission) in IPCC, GEO, MA and IAASTD). Evidence presented on how GEC impacts food systems and food security is much more limited, with two main exceptions. First is consideration of the constraints imposed by the growing scarcity of fresh water on future food production options (CAWMA, MA and IFPRI's IMPACT-WATER model, which is presented and applied by Rosegrant et al, 2002). The second, increasingly provided in more recent assessments, is consideration of the long-term impacts of climate change on crop productivity (IPCC, MA, IAASTD, CAWMA).

Other fairly consistent patterns are that scenario-based approaches (Box 3.2) dominate in the longer-term perspective assessments, but are largely absent elsewhere (the FAO/OECD annual Agricultural Outlook assessment and recent SOFAs being exceptions). Projections are an integral part of scenario-based assessments, but also of the short-term food security assessments. GEC-focused assessments tend to have longer projection horizons, 2050 being the time horizon of choice for assessments undertaken over the past decade. Food production-focused assessments tend to focus on shorter time lines (e.g. ten years for annually updated food projection models) and 2025/30 appear to represent the current time horizons of choice for this group of assessments.

Another striking feature is that the more assessments focus on food productivity and production through to welfare and growth, the less that GEC issues appear as major topical issues. In the serialized, quasi-annual set of assessments, thematic topics focus little on environmental change issues (as reflected in the low scores for significant treatment of environmental change linkages to food systems), and more on a broad range of topics of (presumably) higher perceived priority: trade barriers and market access, land tenure, smallholder financial services, women's control of resources, health (HIV/AIDS, malaria, zoonotic diseases), migration, genetically-modified organisms, market consolidation and speculation, etc. One perspective on this implicit prioritization of development issues is that as assessment time horizons become shorter, GEC factors are treated as increasingly exogenous to policy and investment choices.

Synoptic food security assessments, with their high reliance on short-term predictions, offer interesting insights. They present a paradox in that while they represent frontline assessments of some of the most direct effects of environmental change, such as more frequent and intense extreme events (e.g. droughts, floods, landslides, typhoons and tidal surges), they appear highly disconnected from the assessment work addressing longer time horizons. Furthermore, and of particular relevance, is that, of all the frameworks, it is the early warning and vulnerability assessments that most explicitly align with the GECAFS extended definition of food systems. The language and indicators of short-term food security assessments include notions of availability, access, stability and utilization (although the range and precise definitions of assessment methodologies and indicators vary across the VAM, GIEWS and FEWS approaches).

What was assessed?

Having surveyed the variety of assessments whose scope, by design or purpose, may be expected to span the linkages between environmental change, food systems and food security, their specific content are now explored, focusing on the MA, IPCC and GEO-4. To do this key volumes or chapters of these three major international assessments of environmental change and human well-being are systematically reviewed, in order to gauge the current state of knowledge about the interactions among global environmental change processes and impacts, food systems and food security. The assessment materials examined were the Millennium Ecosystem Assessment (Volume 1 – 'Current State and Trends', 2005a; and Volume 2 – 'Scenarios', 2005b, chapters 5 and 8), the contribution to the Fourth Assessment Report from Working Group Two of the Intergovernmental Panel on Climate Change (IPCC, 2007), and the UNEP's Fourth Global Environment Outlook report (UNEP, 2007). Each of these assessments involved the participation of several hundred scientists worldwide and each was intended to review all available scientific evidence. Since their specific goals and foci differed, each assessment is briefly described below.

The Millennium Ecosystem Assessment (MA)

The main task of the MA was to explore the linkages (tradeoffs or synergies) among maintenance of ecosystem services and dimensions of human well-being. The MA set out to document both the trends in and tradeoffs amongst the provision of ecosystem services in the context of their role in contributing to human well-being, including an extensive review of both published and grey literature. Ecosystem services are defined as the benefits that people obtain from ecosystems, and are classified as provisioning, regulating, cultural or supporting. Food is a provisioning service. Human well-being is defined as the opposite of poverty, and comprises security, the basic material for a good life, health, good social relations and freedom of choice and action. Two chapters in the 'Current State and Trends' volume (MA, 2005a) dealt specifically with food and agriculture, one dealt with vulnerability, and several MA sub-global assessments also included food or agriculture. The 'Scenarios' volume (MA, 2005b) has no specific chapters on food, although some outcomes pertaining to food security were evaluated across all four scenarios.

Intergovernmental Panel on Climate Change (IPCC): Fourth Assessment

The Working Group Two report of the IPCC Fourth Assessment Report, 'Impacts, Adaptation and Vulnerability' (IPCC, 2007; Parry et al, 2007), was designed to assess 'current understanding of the impacts of climate change on natural, managed and human systems, the capacity of these systems to adapt, and their vulnerability' (Parry et al, 2007, p8), building on previous IPCC reports. The report summarized an extensive volume of peer-reviewed published literature, and some grey literature, relating to the observed and modelled impacts of climate change, including changes in global mean temperature, precipitation patterns, extreme events, sea-level rise and increases in greenhouse gas emissions. Although the report emphasized empirical evidence, case study examples were also used to illustrate the broader range of issues involved with vulnerability to climate change, such as impacts on smallholder livelihoods and nutrition for which significant quantities of empirical evidence is lacking. The chapters most relevant to food security were: Chapter 5: 'Food, fibre and forest products'; Chapter 8: 'Human health'; and Chapter 3: 'Fresh water resources and their management'; as well as regional chapters for Africa, Latin America and Asia.

Global Environment Outlook 4 (GEO-4)

GEO-4 (UNEP, 2007) is centred on the premise of the interdependence between economic development, human well-being and the health of the environment. Building upon earlier GEO assessments, GEO-4 has a stronger focus than the MA on human well-being outcomes that are directly tied to the state of the environment, rather than just the ecosystem service outcomes, as its commissioning agency is the UN Environment Program (UNEP). The major chapters that summarized the current environmental 'conditions and trends' are described in Chapter 3: 'Land'; Chapter 4: 'Water'; and Chapter 5: 'Biodiversity'. Chapter 9: 'The future today – Towards 2015 and beyond' described and evaluated alternative scenarios of future outcomes relating to indicators of biodiversity, land and water use, as well as nutrition and other aspects of human well-being.

As part of the review of these materials, relevant factors were extracted that had been treated in one or more assessments, and which pertained to environmental conditions, environment-related stresses (that have relevance for food system functioning), food system measures of performance and food security outcomes. These factors are grouped and displayed in the four vertical columns of Figure 3.1. These are not meant to be completely exhaustive sets of all possible factors, rather of those that appeared in one or more of the assessments (though in some cases treatment was cursory). The second conceptual element of Figure 3.1 is an attempt to describe graphically both the

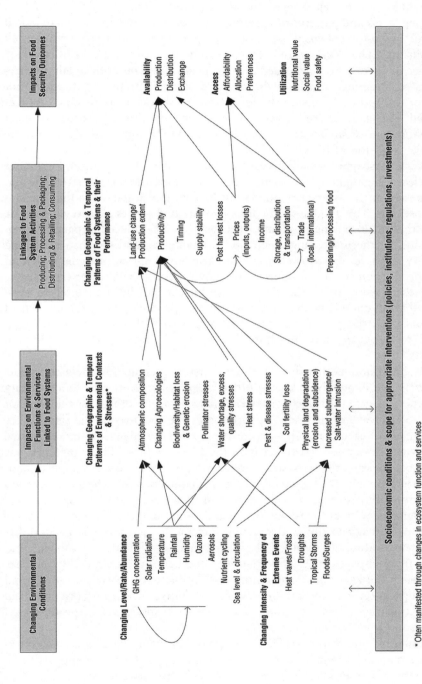

Figure 3.1 *Environmental change, food system and food security outcome components and dynamics: highlighting concentration of issues and pathways addressed by assessments*

pathways of relationships (linkages) flowing from left (environment condition) to right (food security outcomes) that received attention in the assessments. These linkages are represented by arrows connecting specific factors (e.g. changing levels and patterns of radiation, temperature and rainfall were linked to changing agro-ecological capacity to support agricultural production in all assessments). The thickness of each line is indicative of the degree of coverage the specific linkage was accorded (a judgement based on both depth and frequency of treatment). Although they are significant, the figure does not attempt to capture the backward links from food systems onto GEC. As recognized elsewhere those linkages have been the primary focus of most international assessments (e.g. MA, GEO-4, IIASTD and CAWMA), with the notable exception of IPCC.

While inescapably qualitative and subjective in its formulation, Figure 3.1 does clearly highlight a number of issues concerning the relative strengths and weaknesses of the assessments undertaken. First and foremost is that the single most dominant food system component treated across all assessments, by far, is that of producing food and, specifically, GEC-induced impacts on productivity. Furthermore, since the empirical evidence (and scenario modelling capacity) focused almost exclusively on crop-based agriculture, the metric of food production impact was most commonly cited as change in crop yields. While there were some separate evidence on changes in area suited to agricultural production (e.g. changing patterns of agro-ecology and the impacts of sea-level rise), these were typically not integrated with evidence of changing yields to provide more integrated perspectives of aggregate production impacts.

Of equal note is that there are many factors identified in the assessments that were not explicitly linked to other factors of relevance to food security outcomes. In general there are fewer linkages as one moves to the right of Figure 3.1. However, it is also apparent that the socio-economic significance of each factor is not equal, nor are the factors independent. Indeed, these issues were recognized, particularly in the MA in which initial emphasis was placed on assessing the ecosystem service valuation literature to support quantitative assessments (MA, 2005a), but it was concluded that the gaps were too many to attempt systematic valuation even of the core set of ecosystem services. Dependencies amongst factors in individual columns of Figure 3.1 are also apparent. For example, the state of greenhouse gas levels impacts radiation, temperature, rainfall and other surface climatic measures, and productivity impacts prices and quantities and hence trade. In this latter case, the explicit linkages documented amongst productivity, prices and trade, and their impacts on food availability and affordability, stem from both empirical observation as well as scenario-linked simulation modelling of productivity, market and trade systems.

A more subtle feature of Figure 3.1, particularly apparent from the intermediate materials compiled during its preparation, is that there appears to be systematic biases in knowledge and analytical capacity that are unrelated to the perceived importance of specific factors. For example, pests and diseases and post-harvest losses are anecdotally very significant factors influencing food availability (and sometimes food safety), but they receive relatively little treatment, especially in a quantitative manner. Similarly, questions of timing/seasonality and stability that feature prominently in day-to-day experience of food systems, particularly in poorer countries with less developed infrastructure and institutions, also receive very little attention. These shortcomings may well reflect persistent weaknesses in data gathering systems that translate, in turn, into more limited availability of empirical evidence and analytical capacity in such areas, perhaps exacerbated by insufficient representation of appropriate scientific expertise in the assessment process.

Data, knowledge and expertise gaps are even more apparent with regard to processes influencing the non-supply-related food security outcomes (witness the lack of impact pathway arrows reaching outcomes other than production, distribution, exchange and affordability). This likely reflects a less than adequate prior understanding or articulation of the potential pathways of impact between environmental change and access and utilization outcomes. Some of these missing linkages are considered in more depth elsewhere in this book, but it is also recognized that – by their nature – many of these factors are driven to a much greater extent by local socio-economic and intrahousehold conditions and customs.

Lessons learned

Perhaps the clearest message emerging from this review of environmental change and food systems linkages in assessment practice is that significant challenges remain in raising awareness, acceptance and adoption of the type of integrated (GEC and) food system conceptual frameworks articulated by the GECAFS initiative. These challenges include:

- A universally reductionist approach in representing food systems in their entirety by the activity of producing food coupled with aggregated treatment of commodity markets (that track or project international prices and trade).
- A primary focus on evaluating evidence of the impacts of agriculture as a primary driver of GEC, a long-standing perspective stemming from the emergence of sustainable development and environmental conservation as significant public policy issues in the 1970s and 1980s.
- A relatively narrow focus with regard to assessing GEC impacts on agriculture; primarily recognizing constraints imposed by changing agro-ecological conditions, increased water resource scarcity and, to a lesser extent and more recently, land degradation (e.g. long-term depletion of soil fertility). There is, however, rapidly expanding treatment of climate-change-induced impacts on crop productivity (but to a much lesser extent on livestock productivity and pest and disease stresses).
- A preoccupation with the 'long-term' impacts of climate change on future food provision but much less attention paid to the increasing variability of agricultural productivity in many parts of the world (driven by greater climate variability and more extreme events). This lack of attention likely

reflects the shortcomings of appropriate monitoring systems and analysis/modelling capacities rather than the high socio-economic costs engendered by increased variability and risk in food systems.

- A distinct gap in data, knowledge and tools between the conceptual models of, in particular, the international assessment initiatives (e.g. MA, GEO, IAASTD) and the ability to properly implement such comprehensive analytical assessments. This includes both gaps in scientific knowledge of the linkages between and interactions amongst environmental drivers of change and food system activities (e.g. the effects of changing temperatures, rainfall patterns and soil fertility on the incidence and severity of pests and diseases), and gaps in knowledge of the nature and value of such changes on food system-related livelihoods and on food security outcomes (e.g. how will sea-level rise displace coastal fishing livelihoods, or changing storm/surge intensities disrupt food distribution systems and, hence, food availability and affordability).

Recommendations for future assessments

Based on the review of a broad range of GEC and food-system related assessments a typology of such assessments developed (Table 3.1), and an overview of the relevant processes and impact pathways that they have addressed (Figure 3.1), have both been developed. Drawing on the findings of these contributions a number of potential actions are proposed worthy of consideration in the design or implementation of future assessments to provide a more balanced treatment of food systems and food security outcomes. In the absence of better balanced assessments there is a danger that the potentially most significant and cost-effective interventions for improving food security outcomes (e.g. design of production systems more resilient to water shortages and pest and disease stresses, generation of technologies to reduce post-harvest storage and processing losses, commercialization of food processing and preparation methods that waste less and reduce the loss of key nutrients) receive inadequate attention, and that a relative over-investment is made in boosting production potential per se, since that is the factor most readily observed and modelled.

Specific recommendations are as follows:

- Participation: there appears to be a need for broader disciplinary and sectoral engagement in assessments from the outset, so as to adequately represent the full range of food system and food security actors and issues. Evaluating food systems implies a broad disciplinary span; that is, breeders, agronomists, post-harvest specialists, food processors and commodity traders, retailers, economists, sociologists, nutritionists, finance and extension, health and education and food security/relief agencies.
- Comprehensive food system and food security frameworks: this review of the components and linkages examined by the major international assessments suggests that serious consideration be given by future assessments to basing their treatment of GEC on food systems and security

outcomes on a more rigorous and comprehensive framework. The framework presented and described at length in this book is offered as one such candidate.

- Resources to facilitate framework adoption and effectiveness: beyond the availability of appropriate conceptual and methodological approaches, more is needed to ensure that their proper implementation is feasible and cost-effective. This could be achieved by providing relevant resources (e.g. web-based materials) to support adoption and effectiveness. These might include:
  - a documented typology of food systems;
  - an inventory of pathways of impact from GEC phenomena to food system components and food security outcomes and vice versa (e.g. further elaboration of Figure 3.1);
  - an inventory of proposed or accepted indicators (and relevant data sources) linked to each process and impact pathway for different scales of assessment (time, space, theme); and
  - an inventory of case studies of best practice of GEC-food system integrated assessments.
- Variability/extremes versus long-term focus: greater attention is warranted on assessing evidence regarding impacts of environmental change – particularly climate-mediated change – on variability and extreme events. As part of this effort, there appears to be scope in exploring the linkage of long-term and short-term data and analytical perspectives, especially since the short-cycle food security assessments appear richer in their treatment of food security. Furthermore, early warning systems are investing in needed research (e.g. FEWS NET commissioned work on Indian Ocean climatology linked to changing growing season patterns in the Horn of Africa; Funk and Brown, 2009). It would be valuable for future assessments to shed light on a key development question: the appropriate balance of interventions (policies, investments, regulations) between addressing short-term food security concerns related to more variable and extreme environmental impacts on food security versus addressing long-term trends (where there would appear to be difficult adaptation requirements).

References

Fischer, G., M. Shah and H. Van Velthuizen (2002) Climate change and agricultural vulnerability. Special Report as a contribution to the World Summit on Sustainable Development, Laxenburg, IISA

Funk, C. C. and E. M. Brown (2009) 'Declining global per capita agricultural production and warming oceans threaten food security', *Food Security*, 1, 3, 271–89

IPCC (2007) Climate Change 2007: Impacts, Adaptation and Vulnerability. Contribution of Working Group II to the Fourth Assessment Report of the Intergovernmental Panel on Climate Change. In Parry, M., O. Canziani, J. Palutikof, P. van der Linden and C. Hanson (eds) Cambridge, Cambridge University Press

MA (2005a) Volume 1: Current state and trends. *Ecosystems and Human Well-being*, Washington, DC, Island Press

MA (2005b) Volume 2: Scenarios. *Ecosystems and Human Well-being*, Washington, DC, Island Press

OMB (2004) *Final Information Quality Bulletin for Peer Review*, Washington, DC, Office of Management and Budget

Parry, M., O. Canziani, J. Palutikof, P. van der Linden and C. Hanson (eds) (2007) *Climate Change 2007: Impacts, Adaptation and Vulnerability. Contribution of Working Group II to the Fourth Assessment Report of the Intergovernmental Panel on Climate Change*, Cambridge, Cambridge University Press

Raskin, P., T. Banuri, G. Gallopin, P. Gutman, A. Hammond, R. Kates and R. Swart (2002) *Great Transition: The Promise and Lure of the Times Ahead*, Boston, Stockholm Environment Institute

Rosegrant, M. W., X. Cai and S. A. Cline (2002) *World Water and Food to 2025: Dealing with Scarcity*, Washington, DC, International Food Policy Research Institute

Rosegrant, M. W., M. S. Paisner, S. Meijer and J. Witcover (2001) *Global Food Projections to 2020: Emerging Trends and Alternative Futures*, Washington, DC, International Food Policy Research Institute

UNEP (2002) *Global Environment Outlook 3: Past, Present and Future Perspectives*, London, Earthscan

UNEP (2007) *Global Environment Outlook 4: Environment for Development*, Nairobi, UNEP

WFP (2009) *World Hunger Series 2009: Hunger and Markets*, Rome, World Food Programme

World Bank (2008) *World Development Report 2008: Agriculture for Development*, Washington, DC, World Bank

World Bank (2009) *World Development Report 2009: Reshaping Economic Geography*, Washington, DC, World Bank

Worldwatch Institute (2004) State of the world 2004: Special Focus: The consumer society. *Worldwatch Institute Report*, Washington, DC, Worldwatch Institute

Worldwatch Institute (2005) State of the world 2005: Redefining global security. *Worldwatch Institute Report*, Washington, DC, Worldwatch Institute

WRI (2001) *World Resources 2000–2001: People and Ecosystems – The Fraying Web of Life*, Washington, DC, World Resources Institute

WRI (2008) *World Resources 2008: Roots of Resilience – Growing the Wealth of the Poor*, Washington, DC, World Resources Institute

4
Part I: Main Messages

1 The challenges related to global environmental change and food security are growing, and are ever-more-closely linked. This is particularly true in relation to the risks of climate change, biodiversity loss and water scarcity; in terms of linkages to energy systems; and as food systems become more global in their networks of production, consumption and governance.

2 Systems approaches can help improve our understanding of the interactions between global environmental change and food security, and thus of the range of policy options available to address them. Systems approaches connect the activities of food producers, processors, distributors, retailers and consumers involved in food systems to the food security and environmental outcomes, framing these activities as dynamic and interacting processes embedded in social, political, economic, historical and environmental contexts.

3 Explicitly linking food system activities to their food security and environmental outcomes is an important research consideration, as food security results from a complex set of interactions in multiple domains that are often not highlighted in conventional food chain analysis with their focus on food yields and flows. Feedbacks from food system activities are of particular concern because they may have unintended, and often negative, social as well as environmental consequences.

4 Food systems operate across multiple scales and on a range of levels within these scales. The food system activities, actors and interactions must be mapped at multiple levels and scales, so as to understand the diverse tradeoffs among actors' objectives, and food security and environmental outcomes. International environmental assessments conducted to date tend to focus narrowly on the impacts of changes in temperature and precipitation on agricultural production. These and other conventional analyses overlook key issues and linkages, including other dimensions of global environmental change and other dimensions of food security. As such, they miss a number of critically important social, economic and global environmental change interactions with food security.

PART II

VULNERABILITY, RESILIENCE AND ADAPTATION IN FOOD SYSTEMS

5

Vulnerability and Resilience of Food Systems

Polly Ericksen, Hans-Georg Bohle and Beth Stewart

Introduction

The unprecedented and long-term nature of environmental change in the 21st century has given rise to a new discourse on vulnerability to such change, where vulnerability implies harm from which it is difficult to recover. Understanding why people, ecosystems or food systems are vulnerable to shocks, stresses or long-term change such as the impacts of climate change is the key to developing adaptation options in the face of new threats, or to taking advantage of new opportunities from change. Although adaptation has long been the human reaction to change, the pace and scale of global environmental change (GEC) require proactive and planned adaptation, particularly to avoid further negative feedbacks to either ecosystems or food security. As shown in Part I of this book, food system activities have contributed to many of the environmental changes that now threaten food security. At the same time, food insecurity arises from deep-rooted structural problems such as chronic poverty, missing markets and social factors restricting access. Adaptation strategies to improve food security or ameliorate loss of ecosystem services are not guaranteed to work. Proponents of the 'sustainable adaptation' concept (Eriksen and O'Brien, 2007) argue that adaptation to climate change must also address poverty and development goals. Similarly, proponents of fostering or managing resilience in ecological systems are concerned with systems' abilities to cope with external change in a way that does not undermine the systems themselves (Cumming et al, 2006; Folke, 2006). Part II of this book will explore these approaches in relation to food systems and food security.

Social vulnerability

Most assessments of vulnerability to GEC have tended to favour either social or ecological outcomes (Ericksen, 2008). Social traditions of vulnerability focus on human welfare and livelihood strategies, emphasizing factors such as access to assets, inequity, and social and political determinants (Adger and

Kelly, 1999). These traditions have origins distinct from the discourse around GEC. From the very beginning, the introduction of the vulnerability concept in social sciences was closely linked to issues of poverty, famines and food security. Swift (1989), for example, when asking why poor people are particularly vulnerable to famine, used the new concept of social vulnerability to disaggregate the notion of poverty and translate it into a relational and dynamic concept. Chambers (1989) drew attention to what Bohle (2001) later called the 'double structure of vulnerability', with exposure to risks of hunger and famine on the one hand (the external side), and defencelessness, or the lack of coping with food insecurity, on the other hand (the internal side of vulnerability). Social vulnerability is a function of lack of choices, as well as societal inequity. As these concepts of social vulnerability have moved into the climate and GEC discourse, they have transformed the traditional impacts analysis by stressing that vulnerability is embedded in social, political and economic processes at multiple levels (Turner et al, 2003; O'Brien et al, 2004).

Adger (2006) has distinguished between two broad categories of vulnerability research, which he calls 'antecedent' traditions and 'successor' research frontiers. The former interpretations are basically people-centred approaches to vulnerability, in which the unit of analysis is a person or household, or occasionally a community. According to Adger, they comprise several distinctive traditions. The human ecology tradition, or what Hewitt and Griggs (2004) have termed 'human ecology of endangerment', stresses that vulnerable households tend to live in riskier areas; for example, in drought- or flood-prone regions that are particularly exposed to famine and food crises. GEC will certainly aggravate this tendency. However, entitlement-based interpretations of vulnerability can explain situations when particular populations have been vulnerable to hunger and famine, even where there are no absolute shortages of food or obvious environmental drivers at work (Devereux, 2000). The political economy stream notes that vulnerability to GEC does not exist in isolation from the wider political economy of resource use; thus food insecurity is a function of historical development patterns linked to colonialism and imperialism, embedded in modern political and economic inequities such as unequal trade barriers (McMichael, 2000). In the third Intergovernmental Panel on Climate Change (IPCC) report (IPCC, 2001), vulnerability to climate change is described as a function of exposure to change, sensitivity to change and adaptive capacity. Adaptive capacity is stressed as a quality that allows some people to be less vulnerable, in spite of exposure to stress or shocks (Yohe and Tol, 2002; Smit and Wandel, 2006). Adaptive capacity is a function of access to assets and capital (Adger, 1999); yet high adaptive capacity is insufficient to ensure successful management of change or reduction of vulnerability. Thus successful long-term adaptation requires broader level enabling institutions (Adger et al, 2007; Eakin et al, 2007).

In empirical economic explanations of poverty and food insecurity (both forms of vulnerability), research has stressed the importance of household assets and ability to use off-farm diversification as a strategy to accumulate wealth. Thus some families in a given community are less affected by a food production shock because they can afford to purchase food in markets, as they earn income either through selling their own farm agricultural products or they work off-farm in other wage-earning activities (Becker, 2000); these sources of income are particularly important in areas where grain production is regularly insufficient to last all year. As chronic poverty and chronic food insecurity are intertwined, food security analysts are increasingly recognizing the need for social protection programmes to insure against the impacts of repeated hungry seasons (Cromwell and Slater, 2004). A number of recent studies have examined the role that livelihood diversification plays for assisting rural families to move out of poverty; access to capital (whether assets or income) is critically important if such diversification is to contribute to wealth accumulation. The notion of 'poverty traps' is meant to explain why some groups or households are unable to move out of chronic poverty, often through a combination of ill health, malnutrition, lack of assets and also land degradation, which feed into one another (Barrett, 2007; Swallow et al, 2009).

Contrary to the people-centred 'antecedent' interpretations, 'successor' traditions and current research frontiers on social vulnerability in the context of GEC follow a more systems-oriented approach. Current research strands include social vulnerability to climate change (Brklacich and Bohle, 2005) as a product of multiple stressors or what Leichenko and O'Brien (2008) have called a 'double exposure' to both environmental and social change. More broadly, vulnerability to GEC is portrayed as a property of coupled human–environment systems, placing vulnerability analysis within the context of the newly emerging science of social ecology (Becker and Jahn, 2006) and resilience approaches to managing social–ecological systems (SESs), to be more adaptive and less vulnerable in the face of shocks and long-term change processes.

Ecosystem vulnerability

Studies of ecosystem vulnerability focus on ecological function and ecosystem services, and have traditionally taken a systems view. Thus land-use and land-cover change resulting from converting natural ecosystems into agricultural production systems are viewed as negative with respect to biodiversity, habitat integrity, soil structure and nutrient cycles. The structure and function of the ecosystems have changed to a less desirable state (Steffen et al, 2004, Chapter 4), which not only harms the viability of the previous ecosystem (e.g. a lake or a pasture), but has feedbacks to other ecosystem services as well as to human well-being (see Chapter 1). There is growing concern that medium- to long-term consequences of GEC will undermine human well-being. In the case of critical ecosystem services such as climate regulation and nutrient cycling, we may be close to 'planetary boundaries' (Rockström et al, 2009), or thresholds, that once crossed will threaten human life. At lower levels this is also occurring, as land degradation constrains agricultural production or rehabilitation of previously natural areas; water scarcity threatens not only ecosystems but also power generation; and cutting down trees threatens animal life and sources of fuel wood for cooking (De Fries et al, 2004; Cassman et al, 2005).

The notion of 'tipping points' or 'catastrophic regime shifts' highlights the non-linear nature of change in ecosystems (Lenton et al, 2008); slow changes can build to a point at which the system flips into a different state which is extremely difficult to change. Lenton et al (2008) and Smith et al (2009) discuss how in the case of climate change, tipping points could cause the Greenland ice sheets to melt, or the Indian Monsoon to shift. For ecological systems, eutrophication over time can cause clear lakes to flip to a cloudy state (Carpenter and Gunderson, 2001), or continued degradation can flip a grassland system into a shrub-dominated system (Walker and Abel, 2002).

The theory of ecological resilience highlights the failure of many ecosystem management paradigms as a major reason for changes in ecosystem function and loss of services (Carpenter and Gunderson, 2001; Gunderson, 2003). This emphasizes the role of human activities and agency in both causing and preventing environmental change. Resilience is defined as the capacity of a system to absorb change without shifting to an altered state with different properties. Thus resilient ecosystems can maintain their current functions without flipping into a new state. Introducing the role of human managers has fostered the concept of a coupled SES, in which ecological and social processes are co-evolving and interdependent. This is conceptually appealing for food systems, as neither social nor ecosystem factors are assumed to be paramount, and instead management focuses on the interactions of processes.

Vulnerability of food systems

Combining social approaches, which are more actor-oriented, with ecological approaches and their systems focus, remains a challenge to evaluating the vulnerability of food systems. A challenge stems from the many potential tradeoffs between food security enhancement and ecosystem service degradation, given the current design of food systems in many places. With the multiple and often tight connections between food system components, there is a danger that interventions to lessen one type of vulnerability could create a different vulnerability (Sundkvist et al, 2005; Young et al, 2006). However conceptually appealing it may be to approach food systems as coupled, an unanswered question is whether we really have a vulnerability concept for coupled SESs, especially since food systems have relied upon economic and technological processes to compensate for ecosystem decline, and in fact to use resources more intensively. Increasingly social processes drive environmental change and food security, yet we do not fully understand how far the ecosystem limits can be pushed before we undermine critical processes. Although the Millennium Ecosystem Assessment proposes that human well-being depends critically on ecosystem services, there are many unresolved questions involved in demonstrating this relationship empirically and describing the mechanics of this dependence (Carpenter et al, 2009).

Adger (2006), when speaking on systems-oriented approaches to vulnerability, criticizes the IPCC headline policy statements on vulnerability of regions and systems on the grounds that they do not reflect the richness and diversity of findings on causes and consequences of vulnerability to climate change and climate risks. In fact, the systems-oriented interpretations of vulnerability have remained more or less descriptive, mechanistic, deterministic and positivist, thus sacrificing the explanatory, differentiated, contextual and normative power that has always been the major strength of entitlement and political economy explanations of social vulnerability. The growing diversity and apparent lack of convergence in vulnerability research also calls for a more comprehensive theory of vulnerability to GEC that understands food system vulnerability in the context of GEC 'in a holistic manner in natural and social systems' (Adger, 2006).

A more sophisticated understanding of food system vulnerabilities based on multiple stressors has begun to emerge (Adger, 2006; Leichenko and O'Brien, 2008). The focus is shifting from GEC triggers of food insecurity to a more thorough assessment of other chronic or underlying factors such as institutional failures (Devereux, 2001; Adger et al, 2005), in combination with environmental stresses (Misselhorn, 2005). Increasingly such analyses highlight institutional and policy failures as limiting adaptive capacity. Many studies in the food security community sought to emphasize the importance of politics and economics rather than just environmental factors (Devereux, 2000; Moorehead and Wolmer, 2001). In the climate change community there has been a reaction against views of vulnerability solely determined by 'impacts' (i.e. just the exposure to environmental stress), rather than the structural and deeply-rooted issues (O'Brien et al, 2004; Eriksen and Silva, 2009). This all builds off the key idea that external shocks such as a climatic stress or event reveals underlying internal factors which actually contribute to vulnerability (Bohle et al, 1994). This has been expanded recently to include understandings about how coping strategies or adaptive responses affect future vulnerability by changing the context, outcomes or dynamic processes shaping vulnerability to multiple stressors (Leichenko and O'Brien, 2008). These processes operate at multiple levels, from the household up to international, and interact in both synergistic as well as conflicting ways. Recognition of these dynamics suggests a different approach to developing adaptation strategies for food systems that ensures food security (broadly defined) without furthering negative feedbacks to ecosystems and GEC processes.

GECAFS approach

Building upon all of these issues, Ingram and Brklacich (2002) and Ericksen (2008) describe the GECAFS vulnerability assessment approach. The key features originally were recognizing multiple stressors that interact across a variety of scales and levels (Chapter 2), and understanding adaptive capacity very much as a function of social and political issues; vulnerability is thus not determined by exposure and sensitivity (to GEC) alone (Figure 5.1). This approach was piloted in the GECAFS research in the Indo-Gangetic Plains, defining vulnerability as a function of exposure, sensitivity and adaptive capacity, and focusing on the vulnerability of food security outcomes to GEC. Focusing on the vulnerability of specific food security elements (e.g. nutritional diversity or affordability; Figure 2.1b) enabled a specific understanding of how changes in these outcomes arose.

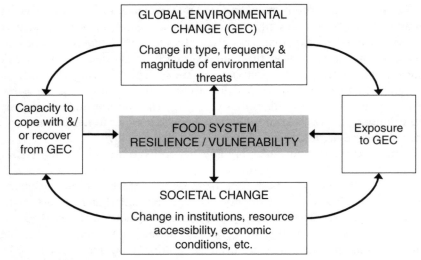

Source: Ingram and Brklacich, 2002

Figure 5.1 *Food system resilience/vulnerability*

However, larger questions remain before this approach can contribute to solutions to decrease food insecurity and maintain critical ecosystem services. Thus Chapter 6 elaborates on the issue of 'vulnerability *of* what' and the tensions between people-centred and system-centred approaches. Chapter 7 elaborates on the question of 'vulnerability *to* what', highlighting that any given environmental change is transmitted through a food system in combination with multiple stressors, and is dependent upon the food system structure and context.

GECAFS was one approach to combine social and ecological factors, and two other approaches are briefly discussed below. However, much more empirical and theoretical work needs to be done, in particular paying more attention to governance (as highlighted in Chapter 1) of the processes and actors affecting food system vulnerability.

Risk-governance approaches

In the field of hazards analysis, comprehensive risk and vulnerability assessment tools have been developed to identify vulnerable people, properties and resources that are at risk from hazardous incidents, and the respective political and policy perspectives have been addressed (Sarewitz et al, 2003). For food security analysis under the impact of multiple social–ecological stresses, the risk governance framework as developed by the International Risk Governance Council (IRGC) (2005) could be utilized to address the multiple causes that put food security at risk and to address the mechanisms that govern these risks, including the risk perceptions of the vulnerable, risk evaluation, risk management and, most importantly, risk communication (Renn, 2008). Conceptions of risk and risk governance could also help identify critical regions where food security is at stake under the impact of GEC (Kasperson et al, 1995).

The IRGC framework highlights both real and perceived risks. Its approach stresses the importance of communication by and among multiple food system actors in the good governance of risk. It highlights a need for an integrated approach not just from an interdisciplinary stance, but also one which engages stakeholders (in particular, policy-makers and the private sector). In light of future uncertainty (as with food systems in the context of GEC), this approach highlights the need for portraying uncertainty, complexity and ambiguity (known knowns, known unknowns, etc.) to stakeholders. The approach also highlights the role for non-technical knowledge ('lay' knowledge) in risk governance.

Finally, the IRGC approach sits governance in the social, institutional, political and economic contexts which also affect decision-making, highlighting that governing vulnerability is about more than a simple process of assessment, appraisal, evaluation, management and communication. Critical factors to consider are organizational capacity, actor networks, political and regulatory culture and social climate and the affects these can have on vulnerability in food systems.

Resilience-building for food security

The second need is to make vulnerability assessment more dynamic and proactive. Managing food systems for resilience offers a different approach to adapting food systems, focusing on more than just identifying vulnerability for a given place and time, but rather on dynamic processes of change which produce feedbacks. By stressing how to make food systems more resilient in some cases, transform them in others, and above all make them more adaptive in the face of shocks and stresses, managing for resilience possibly offers an approach to enhance food security and ecosystem services (and see Chapter 20). Folke (2006) has characterized resilience research as 'the emergence of a perspective for social-ecological systems analyses'. Food systems resilience, in this context, is not just the amount of GEC that a food system can undergo and still retain function and structure. Resilient food systems are also characterized by the degree to which the system is capable of self-organization, and by the ability to build the capacity for learning and adaptation (Berkes et al, 2003).

A resilient food system should therefore have the potential to create opportunities for doing new things, for innovation and development. In vulnerable food systems, even small disturbances through GEC may cause dramatic consequences (Folke, 2006). These propositions have induced Bohle et al (2009) to reframe food systems resilience from a systems-oriented to a more people-centred perspective, and to highlight the importance of agency-based and actor-oriented perspectives on food systems resilience. Taking the food system of Dhaka, Bangladesh, as a case, the authors framed resilience as the ability to support the buffering capacities of vulnerable livelihoods systems, to strengthen the adaptive capacities of people and their institutions to cope with the impacts of GEC, and to generate innovation and learning that allow for resilient transformations of food systems in the context of GEC.

Analysing resilience from a food systems perspective poses a number of research challenges. These include clarifying the feedbacks of interlinked SESs, the ones that cause vulnerability and those that build resilience, how they interplay, how they match and mismatch across scales, and how the role of adaptive capacity in this context can be explored (Folke, 2006). Conclusions from resilience studies are often pessimistic with regard to water and food, and poverty (Allison and Hobbs, 2004; Enfors and Gordon, 2008; Swallow et al, 2009). Such findings suggest that the underlying structure of modern food systems is a problem for both social and ecosystem service outcomes, as they are based in inequity across time and space, and feedbacks are ignored or masked (Sundkvist et al, 2005). As Lebel et al (2006) explain, resilience is a process which must be managed by actors with different goals, power and priorities. In food systems, given the multiple actors involved and the highly normative goals and perspectives each brings, this presents a considerable challenge, discussed more in Chapter 8. Ultimately, society has to struggle with the competing objectives of food systems and making difficult choices regarding which livelihood options in agriculture and type of agriculture to promote (Thompson and Scoones, 2009). The growing influence of private sector actors in all food system activities, but especially processing, distribution and retail, critically influences this struggle, as discussed in Chapters 1 and 18.

In general, the scientific community lacks empirical work on modern food systems to understand the complex interactions between food system activities, GEC and food security outcomes, and the vulnerability of food systems to future GEC (Carpenter et al (2009) echo this more broadly). This is why many analysts feel the 2008 food price crisis caught so many by surprise. These and other issues are explored in the rest of Part II. Chapters 6 and 7 explore the challenges of assessing the vulnerability of food systems in more detail. Chapter 8 turns to the new challenge of developing adaptation strategies to reduce vulnerability of food systems to GEC, and raises additional issues for applying resilience concepts to food systems.

References

Adger, W. N. (1999) 'Social vulnerability to climate change and extremes in coastal Vietnam', *World Development*, 27, 2, 249–69

Adger, W. N. (2006) 'Vulnerability', *Global Environmental Change*, 16, 268–81

Adger, W. N. and P. M. Kelly (1999) 'Social vulnerability to climate change and the architecture of entitlements', *Mitigation and Adaptation Strategies for Global Change*, 4, 253–66

Adger, W. N., N. W. Arnell and E. L. Tompkins (2005) 'Successful adaptation to climate change across scales', *Global Environmental Change*, 15, 77–86

Adger, W. N., S. Agrawala, M. M. Q. Mirza, C. Conde, K. O'Brien, J. Pulhin, R. Pulwarty, B. Smit and K. Takahashi (2007) Assessment of adaptation practices, options, constraints and capacity. In Parry, M. L., O. F. Canziani, J. P. Palutikof, P. J. van der Linden and C. E. Hanson (eds) *Climate Change 2007: Impacts, Adaptation and Vulnerability. Contribution of Working Group II to the Fourth Assessment Report of the IPCC*, Cambridge, UK, Cambridge University Press

Allison, H. F. and R. J. Hobbs (2004) 'Resilience, adaptive capacity and the "Lock-in Trap" of the Western Australian Agricultural Region', *Ecology and Society*, 9, 1, art 3.

Barrett, C. B. (2007) Food systems and the escape from poverty and hunger traps in sub-Saharan Africa. *The African Food System and its Interactions with Health and Nutrition*, New York, USA, United Nations University and Cornell University Symposium

Becker, J. and T. Jahn (eds) (2006) *Soziale kologie. Grundzuge einer Wissenschaft von den Gesellschaftlichen Naturverhaltnissen*, Frankfurt, Camput

Becker, L. C. (2000) 'Garden money buys grain: food procurement patterns in a Malian village', *Human Ecology*, 28, 2, 219–49

Berkes, F., J. Colding and C. Folke (eds) (2003) *Navigating Social-Ecological Systems: Building Resilience for Complexity and Change*, Cambridge, UK, Cambridge University Press

Bohle, H. G. (2001) Vulnerability and criticality: Perspectives from social geography. *IHDP Update*, 2

Bohle, H. G., T. E. Downing and M. J. Watts (1994) 'Climate change and social vulnerability: Toward a sociology and geography of food insecurity', *Global Environmental Change*, 4, 37–48

Bohle, H. G., B. Etzold and M. Keck (2009) Resilience as Agency. *IHDP Update*, 2, 8–13

Brklacich, M. and H. G. Bohle (2005) Assessing human vulnerability to global climatic change. In Ehlers, E. and T. Krafft (eds) *Earth System Science in the Anthropocene. Emerging Issues and Problems*, Heidelberg, Springer-Verlag

Carpenter, S. and L. H. Gunderson (2001) 'Coping with collapse: Ecological and social dynamics in ecosystem management', *BioScience*, 51, 451–57

Carpenter, S. R., H. A. Mooney, J. Agard, D. Capistrano, R. S. De Fries, S. Díaz, T. Dietz, A. K. Duraiappah, A. Oteng-Yeboah, H. M. Pereira, C. Perrings, W. V. Reid, J. Sarukhan, R. J. Scholes and A. Whyte (2009) 'Science for managing ecosystem services: Beyond the Millennium Ecosystem Assessment', *Proceedings of the National Academy of Sciences of the United States of America*, 106, 5, 1305–12

Cassman, K. G., S. Wood, P. S. Choo, H. D. Cooper, C. Devendra, J. Dixon, J. Gaskell, S. Khan, R. Lal, L. Lipper, J. Pretty, J. Primavera, N. Ramankutty, E. Viglizzo and K. Wiebe (2005) Chapter 26: Cultivated systems. In *Ecosystems and Human Well-being: Conditions and Trends*, Washington, DC, Island Press

Chambers, R. (1989) 'Vulnerability, coping and policy', *IDS Bulletin*, 20, 2, 1–7

Cromwell, E. and R. Slater (2004) *Food Security and Social Protection*, London, ODI/DfID

Cumming, G. S., D. H. M. Cumming and C. L. Redman (2006) 'Scale mismatches in social-ecological systems: causes, consequences and solutions', *Ecology and Society*, 11, 1, art 14

De Fries, R. S., J. A. Foley and G. P. Asner (2004) 'Land-use choices: balancing human needs and ecosystem function', *Frontiers in Ecology and the Environment*, 2, 5, 249–57

Devereux, S. (2000) Famine in the Twentieth Century. *Working Papers*, Brighton, Institute of Development Studies, University of Sussex

Devereux, S. (2001) 'Livelihood insecurity and social protection: a re-emerging issue in rural development', *Development Policy Review*, 19, 507–19

Eakin, H., M. Webhe, C. Avila, G. Sanchez Torres and L. Bojórquez-Tapia (2007) Social vulnerability of grain farmers in Mexico and Argentina. In Leary, N., C. Conde, J. Kulkarni, A. Nyong and J. Pulnin (eds) *Climate Change and Vulnerability*, London, Earthscan

Enfors, E. I. and L. J. Gordon (2008) 'Dealing with drought: The challenge of using water system technologies to break dryland poverty traps', *Global Environmental Change,* 18, 607–16

Ericksen, P. J. (2008) 'What is the vulnerability of a food system to global environmental change?', *Ecology and Society,* 13, 2

Eriksen, S. and K. O'Brien (2007) 'Vulnerability, poverty and the need for sustainable adaptation measures', *Climate Policy,* 7, 337–52

Eriksen, S. and J. A. Silva (2009) 'The vulnerability context of a savanna area in Mozambique: household drought coping strategies and responses to economic change', *Environmental Science and Policy,* 12, 33–52

Folke, C. (2006) 'Resilience: the emergence of a perspective for social-ecological systems analyses', *Global Environmental Change,* 16, 3, 253–67

Gunderson, L. H. (2003) Adaptive dancing: interactions between social resilience and ecological crises. In Berkes, F., J. Colding and C. Folke (eds) *Navigating Social-Ecological Systems: Building Resilience for Complexity and Change,* Cambridge, Cambridge University Press

Hewitt, C. D. and D. J. Griggs (2004) 'Ensembles-based predictions of climate changes and their impacts', *Eos,* 85, 566

Ingram, J. and M. Brklacich (2002) 'Global environmental change and food systems (GECAFS). A new, interdisciplinary research project', *Die Erde,* 113, 427–35

IPCC (2001) *Climate Change 2001: Impacts, Adaptation and Vulnerability,* Cambridge, Cambridge University Press

IRGC (2005) White Paper, 'Risk Governance – Towards an Integrative Framework', Geneva, International Risk Governance Council

Kasperson, J. X., R. E. Kasperson and B. L. Turner (eds) (1995) *Regions at Risk: Comparisons of Threatened Environments,* Tokyo, United Nations University Press

Lebel, L., J. M. Anderies, B. Campbell, C. Folke, S. Hatfield-Dodds, T. P. Hughes and J. Wilson (2006) 'Governance and the capacity to manage resilience in regional social-ecological systems', *Ecology and Society,* 11

Leichenko, R. M. and K. L. O'Brien (2008) *Environmental Change and Globalization: Double Exposures,* Oxford, Oxford University Press

Lenton, T. M., H. Held, E. Kriegler, J. W. Hall, W. Lucht, S. Rahmstorf and H. J. Schellnhuber (2008) 'Tipping elements in the Earth's climate system', *Proceedings of the National Academy of Sciences of the United States of America,* 105, 1786–93

McMichael, P. (2000) 'The power of food', *Agriculture and Human Values,* 17, 21–33

Misselhorn, A. A. (2005) 'What drives food insecurity in southern Africa? A meta-analysis of household economy studies', *Global Environmental Change,* 15, 33–43

Moorehead, S. and W. Wolmer (2001) Food security and the environment. In Devereux, S. and S. Maxwell (eds) *Food Security in Sub-Saharan Africa,* London, ITDG Publishing

O'Brien, K. and R. Leichenko (2008) *Environmental Change and Globalization: Double Exposures,* New York, Oxford University Press

O'Brien, K., S. Eriksen, A. Schjolden and L. Nygaard (2004) What's in a word? Conflicting interpretations of vulnerability in climate change research. *CICERO Working Papers,* Oslo, Norway, Center for International Climate and Environmental Research

Renn, O. (2008) *Risk Governance: Coping with Uncertainty in a Complex World,* London, Earthscan

Rockström, J., W. Steffen, K. Noone, A. F. Persson, I. Stuart Chapin, E. F. Lambin, T. M. Lenton, M. Scheffer, C. Folke, H. J. Schellnhuber, B. Nykvist, C. A. de Wit, T. Hughes, S. van der Leeuw, H. Rodhe, S. Sorlin, P. K. Snyder, R. Constanza,

U. Svendin, M. Falkenmark, L. Karlberg, R. W. Corell, V. J. Fabry, J. Hansen, B. Walker, D. Liverman, K. Richardson, P. Crutzen and J. A. Foley (2009) 'A safe operating space for humanity', *Nature*, 461, 472–75

Sarewitz, D., R. Piekle and M. Keykhah (2003) 'Vulnerability and risk: some thoughts from a political and policy perspective', *Risk Analysis*, 23, 4, 805–10

Smit, B. and J. Wandel (2006) 'Adaptation, adaptive capacity and vulnerability', *Global Environmental Change*, 16, 282–92

Smith, J. B., S. H. Schneider, M. Oppenheimer, G. W. Yohe, W. Hare, M. D. Mastrandrea, A. Patwardan, I. Burton, J. Corfee-Morlot, C. H. D. Magadza, H. M. Fussel, A. B. Pittock, A. Rahman, A. Suarez and J.-P. van Ypersele (2009) 'Assessing dangerous climate change through an update of the Intergovernmental Panel on Climate Change (IPCC) "reasons for concern"', *Proceedings of the National Academy of Sciences of the United States of America*, 106, 4133–37

Steffen, W., A. Sanderson, J. Jäger, P. D. Tyson, B. Moore III, P. A. Matson, K. Richardson, F. Oldfield, H.-J. Schellnhuber, B. L. Turner II and R. J. Wasson (2004) *Global Change and the Earth System: A Planet under Pressure*, Heidelberg, Germany, Springer Verlag

Sundkvist, A., R. Milestad and A. Jansson (2005) 'On the importance of tightening feedback loops for sustainable development of food systems', *Food Policy*, 30, 224–39

Swallow, B. M., J. K. Sang, M. Nyabenge, D. K. Bundotich, A. K. Duraiappah and T. B. Yatich (2009) 'Tradeoffs, synergies and traps among ecosystem services in the Lake Victoria basin of East Africa', *Environmental Science and Policy*, 12, 504–19

Swift, J. (1989) 'Why are rural people vulnerable to famine?', *IDS Bulletin*, 20, 2, 8–16

Thompson, J. and I. Scoones (2009) 'Addressing the dynamics of agri-food systems: an emerging agenda for social science research', *Environmental Science and Policy*, 12, 386–97

Turner, B. L., R. E. Kasperson, P. A. Matson, J. J. McCarthy, R. W. Corell, L. Christensen, N. Eckley, J. X. Kasperson, A. Luers, M. L. Martello, C. Polsky, A. Pulsipher and A. Schiller (2003) 'A framework for vulnerability analysis in sustainability science', *Proceedings of the National Academy of Sciences of the United States of America*, 100, 14, 8074–79

Walker, B. and N. Abel (2002) Resilient Rangelands – Adaptation in complex systems. In Gunderson, L. and C. S. Holling (eds) *Panarchy: Understanding Transformations in Human and Natural Systems*, Washington, DC, Island Press

Yohe, G. and R. S. J. Tol (2002) 'Indicators for social and economic coping capacity – moving toward a working definition of adaptive capacity', *Global Environmental Change*, 12, 25–40

Young, O., F. Berkhout, G. C. Gallopin, M. A. Janssen, E. Ostrom and S. van der Leeuw (2006) 'The globalization of socio-ecological systems: an agenda for scientific research', *Global Environmental Change*, 16, 304–16

6
What is Vulnerable?

Hallie Eakin

A shift in focus from vulnerable agriculture to food system vulnerability

Food system vulnerability to global environmental change (GEC) typically invokes vulnerability of agriculture: the sensitivity of food production, the resources on which production depends, and food producers to diverse aspects of GEC. The majority of research pertaining to foods and agricultural vulnerability generally adheres to this conceptualization. In this relatively large body of work, indicators of vulnerability tend to be associated with production and yields (e.g. Jones and Thornton, 2003; Thomson et al, 2005), farm income (e.g. Antle et al, 2004) and, more broadly, in terms of food availability, consumption and rural livelihood security (Rosegrant and Cline, 2003; Morton, 2007; FAO, 2008a). The human 'units of concern' of vulnerability analysis in this literature are typically producers (rural households, farm enterprises, agribusiness) or, alternatively, consumers (rural subsistence households or urban consumers). Vulnerability in this conceptualization also tends to be place-based, with the implicit assumption that outcomes such as a collapse in production or evidence of famine can be understood through an analysis of causal agents and impacts occurring in a contained and contiguous geographic space (see discussion in Eakin et al, 2009).

However, the need for a shift to a more systemic focus in vulnerability analysis has long been apparent. The analysis of vulnerability of an agricultural region, commodity group or farm population fails to adequately capture the complex spatial and temporal interdependencies and feedbacks that characterize contemporary modes of achieving food security. Far more attention is required on the vulnerability of food *access* and *utilization* in addition to the traditional focus on food *availability*. Food system vulnerability entails a move towards an integrated and simultaneous analysis of the sensitivity, exposure and capacities of the diverse activities, actors and inputs that characterize the activities and outputs of a food system (see Chapter 2, and Ericksen, 2008a). This shift in focus introduces numerous challenges in vulnerability analysis, not the least of which is the identification of 'what is vulnerable'.

The evaluation of food system vulnerability requires not only the assessment of the vulnerability of individual system elements (e.g. the activities, resources and actors involved in the production and processing of food) and the movement of food across space and time (e.g. trade, distribution and storage) and among food consumers, but also an understanding of how vulnerabilities are produced, exacerbated or mitigated through the synergistic or antagonistic interaction of the different food system elements and actors across spatial and temporal scales. For example, relatively little attention has been devoted to the vulnerability of food storage, processing, distribution and retailing to GEC, despite the importance of these elements in linking production to consumption across space and time. As critical links in food system functions, these elements can exacerbate or mute GEC signals, producing broader-scale unanticipated vulnerabilities. For example, research based on historical data from the UK also documents a significant relationship between the incidence of food-borne illness and temperature, suggesting that under certain climate change scenarios the incidence of food-borne illnesses will increase because food processing and storage facilities will be inadequate (Bentham and Langford, 1995). The Food and Agriculture Organization (FAO) also recently highlighted the potential problems associated with changes in microbial populations and mycotoxins in response to climatic change, leading to rising concerns over public health in relation to food storage, hygiene and safety (FAO, 2008b). Their report documents several cases of increased incidence of aflatoxin contamination associated with rising temperatures and heatwaves, with consequent implications not only for crop yields, but also the trade of contaminated products and the health of consumers (FAO, 2008b). The vulnerability of food transport and storage to GEC is particularly relevant for fisheries where changing temperatures in marine systems can affect the quantity and persistence of microbial agents, which in turn increase the probability of human exposure to food contamination through seafood consumption (Rose et al, 2001).

Similarly, the movement of food across space is subject to disruption by environmental extremes. In the USA, the railway system and sea ports are thought to be vulnerable to a variety of climatic disturbances, including flooding, increased severe hurricane activity and sea-level rise (NRC, 2008). As a primary player in global commodity markets, such disruptions will have implications not only for food prices in the USA, but also for food access globally (see discussion on maize below). The synergistic effects of vulnerability of food distribution and production subsystems can lead to unexpected food security outcomes. Woodward et al (1998), for example, document how the centralized food distribution system of the Democratic People's Republic of Korea was incapable of coping effectively with two years of anomalous flood events, crippling the timely delivery of food supplies in 1995 and 1996 and leading to unexpectedly high morbidity from the floods.

The dynamics of vulnerability within a food system also feed back to affect the ecosystem services on which food system activities depend. Multiple years of drought may, for example, encourage an expansion of irrigation to diminish the sensitivity of production to climate extremes in the immediate term, while over the long term cause an irreversible degree of salination and land degradation, effectively *increasing* sensitivity of farm livelihoods and production to extremes in the future (Anderies et al, 2006; Gordon et al, 2008). Such irreversible ecosystem changes may well affect food security locally, particularly in cases where production has a strong subsistence component. But quite often these complex interactions between ecosystem resilience, farm productivity, food supply, distribution and consumption characterize regions of commercial production, such that the vulnerabilities manifest in particular places have repercussions in broader regions and, in some cases, globally, through tightly connected markets and extended supply chains (Adger et al, 2009). For example, Vietnam is a recent and now major player in the global coffee market. The ecological sustainability of coffee production in Vietnam has implications for global coffee prices and, through market volatility, for the livelihood and food security of farmers in other coffee-producing regions (Eakin et al, 2009). In other words, while the vulnerability of production may be the most readily identifiable component of a vulnerable food system, it is not the only or necessarily the most important unit of concern. By taking a food system perspective, other components – such as the distribution, processing or availability of food to urban consumers – become potential issues of concern in vulnerability analysis.

Identifying vulnerable food systems

One of the initial steps in any vulnerability analysis is to identify the 'units of concern': what precisely is valued and subject to harm (Eakin and Luers, 2006). There are several distinct, and potentially complementary, approaches to this undertaking. Following trends in the broader literature, vulnerability may be identified either through the manifestation of harm (e.g. negative *outcomes*) or by innate characteristics of the system itself, whether or not the harm has yet to occur.

Identifying vulnerable food system components

The Global Environmental Change and Food Systems (GECAFS) framework categorizes system components into interactions between biophysical and human environments; food system activities (producing, processing, distributing and consuming food) and food system outcomes (food, environmental and social security) (Ericksen, 2008b). The distinct vulnerabilities of each component of a defined system can be qualitatively aggregated to inform an assessment of system vulnerability, or (preferably) the linkages between individual components of the food system can be 'mapped' conceptually to provide a qualitative picture of systemic vulnerability. For example, corn yields in the USA have been shown to be vulnerable to a diverse range of GEC stressors, but notably soil degradation (Pimentel and Kounang, 1998) and rising temperatures, drought and flooding (Hatfield et al, 2008). The vulnerability of US corn producers is also sensitive to spikes in global oil prices that translate into high costs of inputs and machinery. On a per-hectare basis, corn is the most fuel-intensive crop produced in the USA (Schnepf, 2004). In addition, the USA is the world's foremost producer of corn in terms of volume. The global distribution of corn through trade channels is highly dependent on the smooth functioning of US ports, particularly that of New Orleans, through which approximately two-thirds of US corn exports must pass (Schnepf and Chite, 2005). Hurricane Katrina, for example, caused considerable damage to the Mississippi River infrastructure (grain barges, silos and grain elevators) and to the port itself. The result was a sharp increase in transportation costs, which were already affected by the loss of oil refineries and production facilities in the Gulf of Mexico. Producer income declined as a result (Schnepf and Chite, 2005). The summation of the sensitivities and exposure of these different aspects of the US corn system suggests that the entire system is (potentially) vulnerable to diverse stressors.

Nevertheless, this approach is based on an assumption that the vulnerability of one or more components of the system translates directly and additively into the vulnerability of the food system more generally. As Ericksen (2008b) notes, in focusing only on specific components of the broader system there is a risk that important feedbacks and synergistic relations between components will be overlooked. For example, US corn production is not only sensitive to rising temperatures and associated climate changes, it also contributes to such changes through the high volume of synthetic nitrogen applied to US corn fields (Schnepf, 2004). Producing corn in this manner thus also indirectly leads to the vulnerability of yields as well as other components in the corn food system. For this reason it may well be that food system vulnerability is not adequately described by the summation of vulnerabilities in discrete components of a food system (e.g. the activities of production, processing, distribution and consumption). Rather, the system's vulnerability is manifest in the functions and mechanisms that connect activities and deliver outcomes.

Vulnerable system functions and structures

An alternative, more integrated approach is to begin with the aggregated conceptualization of the whole food system as a vulnerable entity, and either identify its vulnerability through an exploration of outcomes to be avoided or inherent characteristics of the system that predispose it to vulnerability. Here 'what' is vulnerable is the capacity of a collection of actors, activities and processes (not necessarily geographically contiguous) to deliver critical ecological and social services, conditioned on distinct drivers of GEC. Undesirable outcomes would indicate that somewhere in the system a critical capacity is failing and that the functions of the system are vulnerable.

One of the more critical indicator outcomes of food system vulnerability is the failure to provide food security, evident in malnutrition, hunger and famine (Ericksen, 2008a, 2008b). Other indicative outcomes might be the collapse or failure of facets of food system sustainability: irreversible soil degradation, widespread crop failure or highly volatile agricultural markets, disrupted storage and transport infrastructure. Structural characteristics of the system may predispose it to such negative outcomes. For example, Fraser et al (2005) argue that the characteristics such as biological wealth (e.g. high net primary productivity), connectivity and lack of diversity can make an urban food system particularly vulnerable to GEC. Tightly linked and concentrated global markets and resources enable the rapid communication of systemic shocks, whether the shock is a result of a climatic disturbance, food contamination or hike in oil prices. Conversely, participating in competitive markets with multiple suppliers can also serve to maintain a flow of goods and services, smoothing the impacts of any local extreme event or shock.

Again using corn for illustration, the 'tortilla crisis' of 2007 is indicative of underlying systemic vulnerability. An unprecedented spike in global corn prices that year, driven by high fuel prices, an expansion of corn-based ethanol production and adverse climatic events in corn-producing regions, translated into a spike in corn tortilla prices in Mexico. Mexicans consume over 10 million tonnes of maize grain annually, much of it in the form of 13 million tonnes of tortillas (Galarza Mercado, no date). Maize is culturally, politically and nutritionally the most important crop in Mexico. Over the last several decades, maize production has become increasingly geographically concentrated such that Mexican consumers are increasingly dependent on the formal maize flour industry for tortilla consumption, supplying approximately half of the tortilla industry (de Ita, 2008). Over a third of household cereal expenditure is on tortilla purchases, such that any change in tortilla prices has a direct effect on household budgets (Galarza Mercado, no date).

Although the majority of corn used for tortilla production is procured in Mexico, the Mexican corn market is tightly coupled with the USA where Mexico obtains the vast majority of the corn that it imports for livestock feed and other industrial uses (de Ita, 2008). Supply shocks in the USA affect not only global maize commodity prices (positively and negatively), but also the amount of corn exported to Mexico. Similarly, economic impacts of production failure in Mexico can be smoothed by the availability of US corn surpluses, as has occurred during some severe droughts in the 1990s.

The tight coupling of US and Mexican maize markets illustrates these complex vulnerability dynamics. In the recent 'tortilla crisis', low-income tortilla consumers were immediately identified as actors particularly vulnerable to food insecurity; nevertheless, media reports suggest other components of the system may also be vulnerable. In 2007 the Mexican National Confederation of Maize Producers demanded public support for expanding the area under maize to buffer Mexican consumers from commodity price volatility (Michel, 2007).

As other cases also illustrate, using outcomes (e.g. food security) to identify the vulnerability of a food system facilitates an understanding of food system vulnerability in terms of system functions and interconnectivities, rather than in terms of boundaries of political geographic or population that may or may not reflect the full scope of the system in question. For example, the outbreak of bovine spongiform encephalopathy (BSE) in 2001 in Europe resulted in the curtailment of the consumption of animal-protein-fed beef. According to research by Nepstad et al (2006), this policy indirectly contributed to the growth of the soybean and the beef industry in Brazil through increased demand for grass-fed beef, contributing to multiple existing pressures to expand the agricultural frontier. This expansion in turn has led to an increase in deforestation and consequently an increase in greenhouse gas emissions, water resource degradation, regional climatic disturbances and a loss in biodiversity (Nepstad et al, 2006, p1599). In this example, behind the undesired outcomes of food insecurity in Europe and environmental degradation in Brazil is a complex web of vulnerabilities involving consumers in both countries, tight connectivity through trade networks and interdependent grain and livestock markets and the natural resources on which producers depend (see Table 6.1).

Distinct issues surface in the southern Africa food system (GECAFS, 2006). Behind the manifestation of chronic hunger and malnutrition is a complex system involving degraded natural resources, a domestic maize sector highly sensitive to climate variability and water stress, and a high dependence on imports of wheat and rice. The capacity for storing grain is constrained and the limited transport infrastructure in the region often leads to problems in grain distribution. Livestock production and domestic maize farming depends on maintaining adequate water supplies for irrigation and sustaining the biodiversity and soils of the region's rangelands. The vulnerability of this region to GEC is multifaceted. Water availability is particularly threatened by climate change. Maize production and the livelihoods of smallholder farmers are vulnerable to increases in drought frequencies and soil degradation. The cultural preference for maize is thus threatened; alternative grains that are better adapted to drier conditions are not as highly valued in local diets. The high dependence on imports for rice and wheat makes the region susceptible to supply shocks and climate events affecting trading partners. As summarized in Table 6.1, in this case the vulnerability of the food system is significantly bounded by the physical and social geography of the region and some of the more vulnerable actors in the system are those who also produce food: poor rural subsistence farmers.

Conclusion

A shift in focus in vulnerability analysis from the discrete components of food systems to the system itself requires greater attention to the vulnerability of the structure and processes that drive the functioning of food systems. To date, the bulk of vulnerability analysis associated with food systems has focused on production/producers and consumption/consumers, and the natural resources on which production depends, providing important insights on the vulnerability of food availability to GEC. Equally important, however, are vulnerabilities associated with food access and utilization. These elements of food security are closely tied to the critical linkages in the function of food systems, associated with the mechanisms that distribute food globally and locally, and the ecological, cultural and social elements associated with food quality and content. These linkages can create synergies, mitigating effects and countervailing influences on the vulnerability of particular actors and places in the food system. Identifying and addressing the critical vulnerabilities of specific food systems suggests a need for greater attention to food system governance

Table 6.1 *What is vulnerable in a food system?*

System in question	Resources *Biophysical, economic, social/cultural*	Activities *Production, distribution, etc.*	Actors *Producers, governments, consumers*	Values *Cultural, environmental, political*	Undesired outcomes
US-Mexico maize system (de Ita, 2008)	Forests (exposed to agricultural expansion) Soil and water resources (degradation from agricultural intensification) Port and transport infrastructure	Mexican maize consumption US-Mexico corn imports	Poor urban consumers Small-scale tortilla processors and grain traders	Mexican food sovereignty	Urban food insecurity Political unrest Deforestation–environmental degradation
Southern Africa food system (GECAFS, 2006)	Water supplies Soil quality Rangeland biodiversity	Maize grain production Livestock ranching Grain storage capacity Grain imports and trade Regional grain distribution Consumption in rural areas	Rural poor (particularly subsistence households) Socially marginalized populations (female headed households, HIV affected households) Government agencies with limited capacity to manage food crises	Preference for white maize	Malnutrition Social conflict Unemployment and poverty Loss of biodiversity
Soy-beef food system (Nepstad et al, 2006)	Amazon forest, soils and water Climate system (through increase in GHG)	Beef production and processing in Europe Europe-Brazil trade in beef and soy Beef consumption in Europe	European beef consumers Amazonian displaced smallholders European beef producers affected by BSE	Concern for the Amazon as global environmental/ biodiversity heritage	Increase in GHG emissions Tropical deforestation Social conflict and poverty

and the importance of coordination of vulnerability assessments and vulnerability mitigation activities across scales and political boundaries.

References

Adger, W. N., H. Eakin and A. Winkels (2009) 'Nested and teleconnected vulnerabilities to environmental change', *Frontiers in Ecology and the Environment*, 7, 3, 150–57

Anderies, J. M., P. Ryan and B. Walker (2006) 'Loss of resilience, crisis and institutional change: Lessons from an intensive agricultural system in southeastern Australia', *Ecosystems*, 9, 6, 865–78

Antle, J., S. M. Capalbo, E. T. Elliot and K. H. Paustian (2004) 'Adaptation, spatial heterogeneity and the vulnerability of agricultural systems to climate change and CO_2 fertilization: An integrated assessment approach', *Climatic Change*, 64, 3, 289–315

Bentham, G. and I. H. Langford (1995) 'Climate change and the incidence of food poisoning in England and Wales', *International Journal of Biometerology*, 39, 81–86

de Ita, A. (2008) Americas Program Special Report: Fourteen Years of NAFTA and the Tortilla Crisis. Washington, DC, CIP Americas Program

Eakin, H. and A. L. Luers (2006) 'Assessing the vulnerability of social-environmental systems', *Annual Review of Environment and Resources*, 31, 365–94

Eakin, H., A. Winkels and J. Sendzimir (2009) 'Nested vulnerability: Exploring cross-scale linkages and vulnerability tele-connections in Mexican and Vietnamese coffee systems', *Environmental Science and Policy*, 12, 4, 398–412

Ericksen, P. J. (2008a) 'Conceptualizing food systems for global environmental change research', *Global Environmental Change*, 18, 234–45

Ericksen, P. J. (2008b) 'What is the vulnerability of a food system to global environmental change?', *Ecology and Society*, 13, 2

FAO (2008a) Climate change and food security: a framework document. Rome, Interdepartmental Working Group on Climate Change of the FAO

FAO (2008b) Climate change: implications for food safety. Rome, FAO

Fraser, E. D. G., W. Mabee and F. Figge (2005) 'A framework for assessing the vulnerability of food systems to future shocks', *Futures*, 37, 6, 465–79

Galarza Mercado, J. M. (no date) Situación actual y perspectivas del maíz en México 1996–2012. Mexico, SAGARPA (Secretaría de Agricultura, Ganadería, Desarrollo Rural, Pesca y Alimentación)

GECAFS (2006) GECAFS Southern Africa Science Plan and Implementation Strategy. *GECAFS Report No. 3*. Oxford, GEFCAS

Gordon, L. J., G. D. Peterson and E. M. Bennett (2008) 'Agricultural modifications of hydrological flows create ecological surprises', *Trends in Ecology and Evolution*, 23, 4, 211–19

Hatfield, J. L., K. Boote, P. A. Fay, G. L. Hahn, R. C. Izaurralde, B. A. Kimball, T. L. Mader, J. A. Morgan, D. R. Ort, H. W. Polley, A. M. Thomson and D. W. Wolf (2008) Agriculture. In a report by the US Climate Change Science Program and the Subcommittee on Global Change Research, *The Effects of Climate Change on Agriculture, Land Resources, Water Resources and Biodiversity in the United States*, Washington, DC, 362pp

Jones, P. G. and P. K. Thornton (2003) 'The potential impacts of climate change on maize production in Africa and Latin American in 2055', *Global Environmental Change*, 13, 1, 51–59

Michel, V. H. (2007) Piden abrir selvas a cultivo de maíz Periodical, Piden abrir selvas a cultivo de maíz, http://biodiv-mesoam.blogspot.com/2007/01/piden-abrir-selvas-cultivo-de-maz.htm (accessed: 18 January 2009)

Morton, J. F. (2007) 'The impact of climate change on smallholder and subsistence agriculture', *Proceedings of the National Academy of Sciences of the United States of America*, 104, 50, 19680–85.

Nepstad, D. C., C. M. Stickler and O. T. Almeida (2006) 'Globalization of the Amazon soy and beef industries: Opportunities for conservation', *Conservation Biology*, 20, 6, 1595–603

NRC (2008) *Potential Impacts of Climate Change on US Transportation*, Washington DC, Committee on climate change and US transportation, National Research Council

Pimentel, D. and N. Kounang (1998) 'Ecology of soil erosion in ecosystems', *Ecosystems*, 1, 5, 416–26

Rose, J. B., P. R. Epstein, E. K. Lipp, B. H. Sherman, S. M. Bernard and J. A. Patz (2001) 'Climate variability and change in the United States: Potential impacts on water- and foodborne diseases caused by microbiologic agents', *Environmental Health Perspectives*, 109, 2, 211–20

Rosegrant, M. W. and S. A. Cline (2003) 'Global food security: Challenges and policies', *Science*, 302, 5652, 1917–19

Schnepf, R. D. (2004) *Energy Use in Agriculture: Background and Issues*, Washington DC, Congressional Research Service

Schnepf, R. D. and R. M. Chite (2005) *US Agriculture After Hurricanes Katrina and Rita: Status and Issues*, Washington DC, Congressional Research Service

Thomson, A. M., R. A. Brown, N. J. Rosenberg, R. C. Izaurralde and B. Benson (2005) 'Climate change impacts for the conterminous USA: An integrated assessment – Part 3. Dryland production of grain and forage crops', *Climatic Change*, 69, 1, 43–65

Woodward, A., S. Hales and P. Weinstein (1998) 'Climate change and human health in the Asia Pacific Region: Who will be most vulnerable?', *Climate Research*, 11, 31–38

7
Vulnerability to What?

Alison Misselhorn, Hallie Eakin, Stephen Devereux,
Scott Drimie, Siwa Msangi, Elisabeth Simelton and
Mark Stafford Smith

Introduction

Many aspects of global environmental change (GEC) contribute to the vulnerability of food systems. These may stem from factors embedded in society, in the natural environment, or from social–environmental interactions and feedbacks. Environmental stresses and changes can also alter the 'internal' nature of food systems to undermine their capacity to respond or adapt, which can make them vulnerable to changes they could formerly cope with, as illustrated in Chapter 6. Such feedbacks give rise to additional causes of vulnerability.

Stresses and changes do not function in isolation from one another, but co-occur and interact in numerous ways. Climate change and economic globalization, for example, are understood to synergistically drive a particular kind of vulnerability in regions or among systems 'doubly exposed' to both; this can lead to compound negative outcomes, such as the simultaneous loss of agricultural productivity and social services (O'Brien and Leichenko, 2000; Leichenko and O'Brien, 2008).

In the context of food systems, the question of 'vulnerability to what' therefore requires not only the identification of 'what stresses and changes threaten food systems', but also some exploration of their interaction on a variety of scales and levels (see Chapter 2), and the mechanisms by which they generate vulnerability. Drivers of change may be thought of as having inherent characteristics and dynamics that in turn shape the kinds of risk and vulnerability they might herald. The potential for adaptation to change is one such characteristic, and issues of adaptation are largely covered in Chapter 8. Four further – not necessarily mutually exclusive – categories of characteristics are identified below, and are used to frame the discussions that address the title question of this chapter:

- *Temporal-scale dynamics* – the varying pace at which drivers act and varying vulnerabilities in food systems that can occur over different time frames.

- *Extremes and predictability* – severe perturbations that are infrequent or rare, and/or are situated some distance from the mean. Varying *levels of predictability* are also related to extremes in driver functioning; the likelihood of or confidence in their occurrence or timing, and the extent to which they are understood or recognized by different knowledge systems.[1] (Many characteristics of predictability are addressed in Chapter 20.)
- *Change and reversibility* – drivers may reach thresholds and tipping points in food systems, or exert a continuous but reversible impact. GEC alters food systems in a number of ways, including reorganization of production systems, supply and demand, and changes in ecosystem services.
- *Spatial-scale dynamics* – their magnitude and spatial extent of impact, as well as linkages between spatial levels and localities (e.g. local to global).

Temporal-scale dynamics

The time frames over which drivers work

A distinction is drawn here between drivers that act quickly, which may also (but not always) cause change over a short period of time, and those that act more steadily over a longer period of time that may at times (but not always) cause long-term changes in food systems. Many aspects of climate change, such as gradual changes in mean temperature and shifts in precipitation patterns, are slow-moving and will play a key role in determining the long-term evolution of food systems and the performance of the underlying ecosystems that support them.

Climate variability and extreme incidents of weather that are presently evident in many regions act over a much shorter time-scale, and in some areas may themselves be symptomatic of the underlying changing climate. Often, fast-acting drivers also have short-term effects on food systems; for example, weather shocks that lead to sudden drops in local or regional food production can push up market prices rapidly.

The dynamics of temporal scale and how they are perceived reach beyond the functioning of drivers to the very heart of the functioning of food systems – including policies and programming. Some of these complexities may be explored through the lens of food insecurity as a major symptom of food system failures.

Short-term versus long-term interactions

Literature on temporal dimensions of food security often draws a distinction between chronic and transitory food insecurity, both conceptually and in policy responses. Transitory food insecurity implies a shock to a food system – a disruption of food production, availability or access that will typically be followed by a return to pre-perturbation 'normality'. It refers to 'a sudden (and often precipitous) drop in the ability to purchase or grow enough food to meet physiological requirements for good health and activity' (Barrett and Sahn, 2001), and elicits responses such as emergency assistance: 'The major sources of transitory food insecurity are year-to-year variations in international food prices, foreign exchange earnings, domestic food production and household incomes' (World Bank, 1986). Chronic food insecurity can be defined as a situation 'when people are unable to meet their minimum food requirements over a sustained period of time' (DFID, 2004). It is a sustained outcome of a food system experiencing persistent structural failures (FIVIMS, 2002).

The temporal and severity dimensions of food insecurity are often conflated, with 'transitory' being equated with 'acute', and 'chronic' assumed to be 'mild' or 'moderate' food insecurity. The consequence of this misdiagnosis can be inappropriate policy responses (Darcy and Hofmann, 2003). During the Bosnian conflict of the early 1990s, a relatively minor (transitory) decline in food insecurity indicators, among a population that was previously food secure and had strong coping capacity, triggered larger per capita volumes of donor assistance than were delivered to highly (chronically) vulnerable African populations who had been presenting worse nutritional indicators for many years (Watson, 2007). Similarly, the southern African drought of 2001–2 caused malnutrition rates in one country to rise from 2.5 per cent to 5 per cent, precipitating an international appeal and a massive food aid intervention. At the same time, routine nutritional surveillance in southern Somalia recorded a malnutrition rate of 13 per cent, but no intervention was recommended because this figure fell within the 'normal' range for Somalia (Prendiville, 2003). This tendency to assign lower policy priority to 'severe chronic' than to 'moderate transitory' food insecurity has been labelled the 'normalization of crisis' (Bradbury, 1998) – aid responses are triggered by rapid *relative* changes in status indicators, whereas a persistently high *absolute* level of chronic food insecurity often becomes politically acceptable over time, and is ignored.

The sharp definitional differentiation that is drawn between chronic and transitory food insecurity obscures the dynamic interactions or negative synergy that characterize the two concepts in reality. Such interactions can occur both ways; from transitory into chronic, and vice versa.

In the first instance, chronic food insecurity is often the product of one or a series of 'transitory' shocks, such as a sequence of failed harvests in farming communities. Households most vulnerable to food insecurity lack the capacity to cope with repeated shocks – even moderate and predictable shocks such as the annual 'hungry season'. Their resource base erodes with each successive setback (e.g. if they are forced into 'distress sales' of assets for food) until they are left chronically food insecure or destitute. This is a form of 'poverty ratchet' or 'food insecurity ratchet'. This has happened to farming families affected by drought in Ethiopia and hurricanes in Honduras, and many were unable to recover their productive assets and remained chronically food insecure for several years (Carter et al, 2005). Figure 7.1 visually illustrates a vulnerable versus resilient household in terms of 'poverty ratchets'. It depicts a series of shocks or acute stressors a household (or households) experiences over time (X axis), and the impacts these have on hypothetical household resources (whether measured in terms of a group of capitals or in terms of wealth, or by some other means) as an indicator of resilience to food insecurity (Y axis).

In the Somali region of Ethiopia, a situation of moderate chronic food insecurity was transformed by a transitory livelihood shock into a 'chronic emergency' from which the region's human and livestock populations have yet to recover. In the six to eight years that preceded the devastating drought of 1999–2000, during which as many as 98,000 people may have died (Salama et al, 2001), less than 10 per cent of the region's population (<400,000) were judged to be 'at risk' in each annual emergency food needs assessment. After the drought-induced food crisis the population assessed as 'at risk' exceeded 25 per cent (>1 million) every year for several years (Devereux, 2006b).

A similar analysis could be applied to the effects of population growth on annual harvests. Over time, population pressure can act in the same way as a sequence of cumulative shocks, reducing a region from food secure to chronically food insecure. This has happened in highland Ethiopia (Sharp and

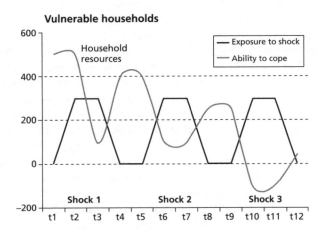

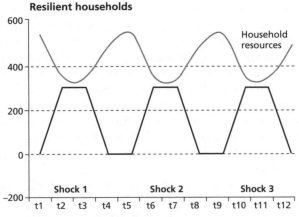

Source: Adapted from Haddad and Frankenberger, 2003

Figure 7.1 *Food insecurity ratchets (shocks and acute stressors are indicated over time on the X axis, while the Y axis measures household resources as an indicator of resilience to food insecurity)*

Devereux, 2004) and in southern Malawi where per capita food production fell steadily from 270kg in 1970 to 210kg by 1990, and in 2001–2 Malawi suffered its first famine (Devereux and Tiba, 2007).

In the second instance, chronic food insecurity is a robust predictor of vulnerability to more severe but transitory food insecurity crisis. People who are poor and food insecure are already living close to the margins of survival with few productive assets and no savings or asset buffers against livelihood shocks. Smallholder farmers and pastoralists in Africa are highly susceptible to food crises, despite producing much of their own food needs, because they subsist in marginal environments, in fragile agro-ecologies with low and unreliable rainfall, and in communities that are politically marginalized, often conflict-prone, and weakly integrated into domestic and global markets (Bankoff et al, 2004; Misselhorn, 2005). Systemic structural and political problems are exemplified in the case of Zimbabwe where the food crises are symptomatic of the underlying chronic food insecurity in Zimbabwe widely argued to have been at least exacerbated by a series of politically motivated actions among powerful groups at a number of levels. These included the refusal of genetically modified food aid by government (Coghlan, 2002) and the diversion of food aid from rightful beneficiaries by community leaders or other powerful groups (Taylor and Seaman, 2004). The roots of food system vulnerability can often be found in the very types of political corruption that withhold or impede needed infrastructure development, technical assistance and market developments in certain regions. Seeking, therefore, to address food insecurity through these same political structures may be inherently challenging.

Figure 7.2 introduces the concept of 'composite food insecurity' to illustrate this dynamic interaction between chronic and transitory food insecurity. Individuals who are subsisting below the food security threshold (say <2100kcal/day) are at greater risk of falling further below this threshold following a shock than are 'food secure' individuals who consume adequate diets (>2,100kcal/day), and have more asset buffers and more resilient livelihoods. Composite food insecurity is arguably more common than transitory food insecurity, as conventionally understood: 'Transitory food insecurity affects households that are able to meet their minimum food needs at normal times, but are unable to do so after a shock' (WFP, 2004). In reality, climatic and other shocks in marginal agro-ecologies typically subject poor rural people to a further deterioration of their food insecurity status, from 'moderate chronic' to 'severe chronic'. In such circumstances, chronic and transitory food insecurities are inextricably connected. Moreover, slow-onset processes such as rising population density or rising HIV prevalence can deepen chronic food insecurity and intensify vulnerability to even relatively minor shocks.

The interconnected nature of the temporal dimensions of food insecurity has profound implications for food security interventions. First, it is inadequate only to return chronically food insecure people who are affected by a food shock to their *status quo ante* situation. Second, for chronically food insecure people subsisting in a low-level poverty trap, a food crisis has no defined beginning and end, so the precise timing of the transition from one type of intervention to another is unclear – a 'continuum' approach might be

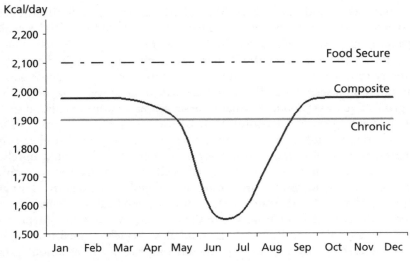

Source: Devereux, 2006a

Figure 7.2 *'Composite' food insecurity. Three states are indicated: (1) food
security – a level of food consumption that is always above minimum
subsistence (here set at 2100kcal/person/day); (2) chronic food insecurity – a
level of food consumption that is always inadequate – that is, less than
2100kcal (here set at 1900kcal/person/day); (3) 'composite food insecurity' –
a new concept which captures the reality that many poor people face, of
having inadequate food for much of the time (living with moderate hunger),
but occasionally falling much further below the subsistence threshold,
towards critically inadequate (life-threateningly low) food consumption levels*

more appropriate. Third, since there is a high probability that people who
receive food aid will require food aid again in the near future, the most effective strategy for tackling food insecurity is to address both 'chronic' and
'transitory' variants simultaneously, by tackling the underlying causes of
vulnerability, not just its symptoms.

Extremes and predictability

Means, extremes and variance

Some food systems have evolved to cope with the majority of their current
exposure to climatic variability, but there will always be extreme and unexpected events for which it is not worth investing in preparation because they
occur so rarely. Many effects of GEC will indeed be first felt through the
impact of changes in weather extremes (Howden et al, 2007; IPCC, 2007b;
Tubiello et al, 2007). Moreover, if the frequency of the extremes changes rapidly, then the food system may not evolve to adjust to the range of new
conditions quickly enough (Howden et al, 2007). The effects of extremes can

thus be felt either because the variability is remaining the same but the mean is shifting (e.g. a drier or hotter climate with the same distribution but more events outside the adapted envelope); or through an actual increase in variability whether or not the mean changes (see Figure 7.3). For example, most farming societies are adapted to cope with perhaps a one in ten year drought, building up sufficient reserves of grain or money to survive through such an event. However, where GEC is increasing the frequency of these extremes, either the local farming systems or the wider state support apparatus may not be adapting fast enough to avoid major impacts. This process can be easily visualized in terms of Figure 7.1.

Changes in socio-economic factors (e.g. in world markets, population flows or transport) can also show both forms of change over time. In the context of food systems, greater variability implies the need for greater reserves to cope with shortages that might be triggered by more variable rainfall, but also by factors such as slower food distribution systems. Globalization of trade has the potential to cause both decreases and increases in variability for food systems in this regard.

The idea of system processes woven throughout the discussions in this chapter implicitly acknowledges multiple interactive pathways of change. Environmental feedbacks and social responses to one change or driver can trigger further changes that reverberate throughout social–ecological systems, that in turn can give rise to extreme (or unpredictable) perturbations. Food system price shocks is one such example of this.

Food system price shocks

During the food price crisis (see Chapter 1), factors such as declines in cereal stocks and unilateral trade actions by individual countries restricted supply in the market and increased food prices. For example, world wheat stocks-to-use ratios declined from over 40 per cent in 1970 to 20 per cent in 2007 – below the oil crisis level. Corn stocks-to-use ratios declined from their 45 per cent peak in the 1980s to about 12 per cent in 2007, a level also previously only seen during the world oil crisis. These trends reflect deliberate policy reforms which sought to reduce the costs associated with maintaining large cereals stocks, adopting more of a 'just-in-time' approach to managing food inventories – some of which were encouraged by global institutions like the World Bank in their structural adjustment programmes. There were also increasing levels of private capital invested in grain markets (as well as other commodity markets) in search of portfolio diversification, and as a response to periods of poor performance within the world's financial markets. The financial market crash of 2008, in particular, led many analysts to re-evaluate the role of speculative pressures on agricultural markets, and coincided with a relaxation in food prices, although this could also have been due to the decline in fuel prices that was happening around the same time.

Unfavourable macroeconomic developments (such as the dollar devaluation) can furthermore lead some countries and consumers to re-evaluate and reposition their assets in a way that might create asset bubbles and speculative pressures. Market speculation can also drive 'bubbles' in markets due to

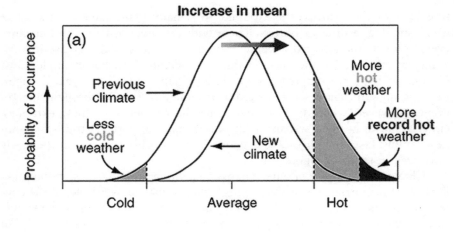

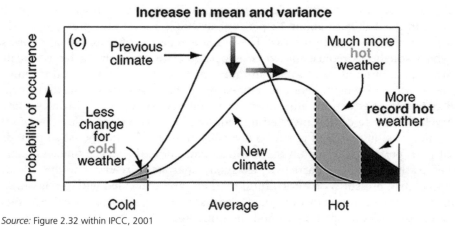

Source: Figure 2.32 within IPCC, 2001

Figure 7.3 *The effect on extreme temperatures when (a) the mean temperature increases, (b) the variance increases, and (c) when both the mean and variance increase for a normal distribution of temperature*

expectations about short- to medium-term trends; however, these can reverse rapidly due to economic conditions and fast-changing market information. This type of activity explains the spikes that develop in some markets, even contrary to the indicators provided by the supply and demand fundamentals that usually determine price formation (von Braun et al, 2008).

The upward pressure on key commodity prices worldwide that has been a focus of recent literature (Abbot et al, 2008; Evans, 2008) can be accounted for by a diverse range of underlying factors or 'drivers of change' that demonstrate the growing strength of linkages between agricultural, energy and asset markets in an increasingly globalized economy.

The factors driving the long-term trends in food supply and demand that have contributed towards a tightening of global food markets during the past decade, and the relative importance of these in explaining the sharp price increases during 2007–8, have been explored in a number of papers (FAO, 2008; Trostle, 2008). As noted previously, these trends are driven by environmental and socio-economic changes, as well as by agricultural and energy policy, including those encouraging biofuel production from agricultural feedstocks. The demand for biofuels, such as ethanol and biodiesel, tends to be strong when fossil-based fuel prices are high, and when national fuel policies push for increased levels of blending to reduce the cost of fuel imports. This has been the case in a number of countries around the world, and is a major determinant of the rapid expansion of biofuel production which has been observed in the past six years. The trend of steadily declining real prices of food commodities that prevailed during the years following the energy crisis spike of the 1970s and through the 1980s and 1990s is likely a cause of less emphasis being placed on productivity-enhancing crop research, with the growing assurance that cheap food could always be obtained from the market. This is leading some countries, like those in the Middle East and North Africa region, to face greater vulnerability and exposure to volatile food markets as they become bigger net importers of food and have to face higher import bills and shocks to their macro-economies.

Well-functioning global food markets are key to enhancing the accessibility of food, as are efficient and low-transaction regional and local markets that allow consumers to purchase food at affordable and stable prices. These conditions prevail in many parts of the world, for most commodities. However, they were not met in a number of key food markets during the food price spike of 2007–8. The political decisions on the part of some countries in South-East Asia to restrict the exports of staples like rice are considered to be major factors in the price dynamics that evolved in regional and global rice markets. The economic 'ripple' effects of imposing unilateral trade actions such as export bans export taxes, and other types of trade distortions tend to provide disincentives to local-level producers. This means they fail to respond to the signal of higher prices in a way that raises their production and offsets the price increases with supply expansion. In turn, this prolongs the price spike and reduces the speed at which the system can return to equilibrium.

In the face of the food price fluctuations of the recent past, a number of national policies are reconsidering buffer stock programmes. These were slowly abandoned and discouraged in the 1990s during a climate of increasing economic liberalization and decentralization of state control of food systems to the private sector. India and China, in particular, are looking at these policies with an eye to ensuring that the cyclical patterns of hunger which occurred in their pre-industrialized past do not recur. This is significant given the positions these countries occupy in the world food economy as major players on international markets for key cereals and staples. As an alternative to the past buffer stock regimes, some analysts have proposed a combination of financial and physical reserves to provide greater stability to global commodity markets, and to reduce the likelihood of speculative price 'bubbles' occurring (von Braun and Torero, 2009). Others have argued for an approach that avoids the management of reserves, and focuses instead on providing financing and guarantees of procurement transactions to poorer countries that would otherwise be rationed out of the market in a period of high volatility and uncertainty (Sarris, 2009). The vigorous debate over the best type of institutional arrangement that can deal with increased supply uncertainty illustrates the growing attention placed on enhancing both stability and accessibility of food supplies within an increasingly globalized food system.

Change and reversibility

Drivers of vulnerability in food systems can exert a 'linear' and continuous impact, but may also reach thresholds in their functioning and interactions which can lead to major changes in food systems.

Tipping points

Tipping points are conceptualized in earth system science as thresholds in key earth system processes, beyond which positive feedbacks trigger further change (Lenton et al, 2008), or which result in major system reorganization in which the system shifts from one major state to another (IPCC, 2007b). The likelihood of major shifts in ecosystems, characterized by tipping points and extremes, has been argued to increase when ecosystem response diversity is compromised by human actions such as the removal of whole functional groups of species or whole trophic levels within an ecosystem (Folke et al, 2004). This is one of the ways in which ecosystems become compromised in their ability to provide ecosystem services; in other words they become vulnerable.

In food systems conceptualized as coupled social–ecological systems (SES; see Chapter 5), changes in either subsystem can trigger tipping points in the coupled system. Thresholds of relevance here may emerge in various ways, but have particular significance if they (1) are long-term and hard to reverse, and (2) occur at a scale that has global ramifications. One or both of these conditions may be met in such systems in various ways.

Many documented cases that meet condition (1) are at landscape level, and thus do not meet condition (2). However, there is growing evidence of examples with a higher level impact, often involving feedbacks through both human and environmental subsystems. The loss of pollination 'services' at a highly integrated level, for example, would have catastrophic effects on human food systems (Kremen et al, 2007). Further, non-linear change might occur at the local or landscape level, but trigger large-scale feedbacks through human subsystems, particularly where there is a deep dependence on biophysical resources as in the case of subsistence livelihoods.

The points discussed above present an exclusive focus on the *negative* effects of system transformations. However, this provides a skewed picture of possible global futures under GEC, as shocks and stresses and their impacts, including major reorganization of ecosystems, social systems and coupled systems, may create opportunities for adaptation. Indeed, in systems theory, it is recognized that a state of reorganization is a state of reassembly to new, relatively stable system configurations (Garmestani et al, 2009).

The marginalization of 'zones' of food production through various mechanisms, including environmental conditions, exemplifies the kind of changes that will herald – or necessitate – reorganization in food systems.

Climate and the marginalization of production systems and regions

The impact of GEC on the production – or supply – component of food systems is often the most readily recognized impact. The term 'climatic marginalization' is used here to refer to processes whereby changes in climate constrain or limit the viability of the food production systems that are vital to food security and the sustenance of livelihoods. In other words, the suitability of a particular crop production system, and/or the biophysical factors enabling production, are altered due to changing climatic conditions such that the system or geographic area becomes unviable in terms of cost of inputs versus outputs and livelihood benefits. Farming systems may be pushed towards climatic marginalization through factors including temperature increases, changes in rainfall, sea-level rises, or changes in solar radiation in combination with atmospheric pollution. Climatic marginalization can thus be felt at a range of levels: at the micro-level, such as the household whose fields are drying out, and at macro-level, where it might be coastal communities whose fields disappear under water when sea-levels rise (Misselhorn, 2005; FAO, 2008; Sanchez et al, 2009).

The literature suggests that failures to cope with weather anomalies occur in areas where existing farming systems become unsuited to new climatic conditions, or where individuals use farming practices that are unsuitable for new climatic regimes; and/or where institutions fail to create incentives to produce crops that can tolerate new climatic conditions (Fraser, 2007). Altered climatic conditions due to GEC is a fundamental driver of food system vulnerability (Howden et al, 2007). As temperatures increase, potential crop productivity will undergo various changes across regions and crop types, and in high-risk areas there may be few alternative varieties to which farmers can turn (Burke et al, 2009). Integrated assessment models have shown that climate change effects on temperature and rainfall will, at least initially, have positive yield effects in cooler climates, but will decrease cereal yields in low latitude regions, the geographical location of most developing countries (Easterling et al, 2007).

Specifically, developing countries are predicted to have a 9 to 21 per cent decline in overall agricultural productivity due to global warming by 2050, while industrialized countries will face a 6 per cent decline to an 8 per cent increase, depending on the offsetting effects that additional atmospheric carbon dioxide could have on rates of photosynthesis and long-growing seasons (Cline, 2007; Mercado et al, 2009). As a result of these differentials in predicted production capabilities, some regions will benefit from increases in yield, while others will need to import an increasing amount of food to meet demand. Fischer et al (2005) estimate that cereal imports will increase in developing countries by 10 to 40 per cent by 2080. While agriculture may benefit in some areas, ground-level ozone is argued to be already reducing crop yields by up to 35 per cent in parts of the USA, India and China (Felzer et al, 2007). In some areas, plant growth functions may arguably be compromised to the extent that no new farming measures would be equal to the adaptation required.

Climate change is also predicted to have a negative impact on a wider range of ecosystem services, in addition to food production, which insert into many links of the food system. Climatic marginalization of food systems is thus not limited to cropping; for example, rising temperatures can increase the demand for irrigation which affects the water cycle, and sea-level rise may not only submerge farmland, but also lead to saltwater intrusion into freshwater systems and habitat losses. In addition, warmer water temperatures can result in poisonous algal blooms (Aggarwal et al, 2004; Allison et al, 2009) that curb the capacity of coastal and lake systems to provide safe water for drinking, irrigation or aquaculture.

Cyclical climatic conditions – not necessarily coupled with climate change per se – may also stress food production systems. An illustration of this occurs in East Asia, where cyclical climate patterns (that may operate at the scale of decades) have pushed the borders of vegetation zones across much of that region (Yang et al, 2005). As this alters ecosystems, current practices for natural pollination (Thomas et al, 2004), pest control (Gregory et al, 2009) and intercropping are being forced to change. Globally, tropical farming systems often exist at the extreme edge of the climate envelopes that food crops can tolerate.

The persistence of farming practices unsuited to new climate regimes and/or the inability of farming practices to meet the challenges of changing climates represents a situation to which people are vulnerable but very limited in their ability to adapt. This illustrates the kinds of 'limits to adaptation' discussed in Chapter 8 that food systems may reach in the face of GEC, and describes a mechanism of vulnerability partly determined by constraints in adaptive capacity. Where farmers or pastoralists have been able to move freely over vast land areas in response to climate, or where local markets are functioning well, adaptive farming practices may be more readily adopted (Yang et al, 2007; van Aalst et al, 2008; Barrett et al, 2009). However, there are a number of issues at play. In some cases, farmers' capacity to cope with weather-related stress may be compromised due to insecure or undefined land tenure or a lack of capital (Sen, 1981; Cotula et al, 2009). In addition, changes in the seasonality of rainfall may narrow the window for much needed double cropping patterns, particularly where subsistence agriculture plays an important role for local food security (e.g. Rosell and Holmer, 2007; Devereux, 2009). In extreme cases, livelihoods may be permanently lost and have to be located elsewhere. This would be the case for low-lying coastal areas that are flooded and submerged by sea-level rise (Adger, 1999; Aggarwal et al, 2004).

Countries with a poor institutional capacity to respond to weather-related crises with timely measures are particularly vulnerable to adverse weather (Devereux, 2009; Fraser and Stringer, 2009), such as cold and dry springs that have caused recurring crop failures in Ukraine during the 2000s (USDA, 2004), as well as to climate change. In some areas, simple proactive institutional interventions have been able to mitigate potential marginalization. For example, after a subsidized seeds and fertilizer programme in 2005, Malawi was able to turn from being food aid recipient to maize exporter in 2007 (Sanchez et al, 2009). Slow climate changes, like temperature increase, in theory allow the best opportunities for governments to take action. This might be to open up new land areas or introduce new crops or improved varieties (Yang et al, 2007; Simelton et al, 2009). Improved varieties can also be designed to tolerate a particular stress (e.g. drought, pest or temperature). However, this often means that, while they are more productive, they only tolerate narrow climatic ranges. In the face of increased weather variability commensurate with climate change, the outcomes of stress-tolerant varieties will become more uncertain.

Crop changes in response to shifts in demand, or to GEC, can also have unforeseen or poorly thought through consequences. Such responses exemplify 'negative feedbacks'; impacts that give rise to unintended negative environmental and/or food system impacts, at times causing a cycle of intensifying vulnerability. For example, as result of a transition to high-yielding 'thirsty' crops, major grain producers like Mexico, the USA, China, India and the Middle East have used up groundwater and aquifer water for agriculture faster than it is being replenished (Brown, 2005). Over-extraction of water for irrigation can lead to water-shortages and can also encourage tapping of deeper water resources which has other knock-on environmental impacts (Aggarwal et al, 2004; Ma et al, 2006).

Changes in food demand

Environmental change processes interact with key social change processes. Concern over the current and potentially future harmful consequences of GEC for food systems is stimulated by several narratives (Ericksen, 2008), one of which is particularly powerful: projected changes in demographics and consumption patterns have led many to worry that some populations will be unable to access sufficient food in the coming 50 to 100 years, particularly in the absence of adequate policy responses (von Braun, 2005). The issue of demographic change has tremendous consequences for the world's food needs and for the nutritional side of food security. Just over 200 years ago in 1798, Thomas Malthus famously predicted that populations would outrun food production, as the ecological 'carrying capacity' of the Earth was exceeded by human demand, resulting in mass starvation (Malthus, 1817; cited in Runge et al, 2003). Neo-Malthusian arguments (Ehrlich, 1968; Brown, 1973; Hardin, 1977) gained ascendancy in the 1970s during high world food prices. Although these scenarios have yet to fully materialize, many parts of the world may still face a neo-Malthusian threat, primarily because people are numerous, poor and isolated (Runge et al, 2003). When poverty is tied to rapid population growth rates as it generally is, the risk of widespread hunger is ever present.

Although land availability is neither the only nor always the most important aspect to be considered for achieving food security, it remains a crucial issue (UN, 2001a), particularly for developing countries. It is estimated that an additional 70 million hectares of cropland worldwide will be needed to meet food demands in developing countries over the next 30 years, with seven countries in Latin America and sub-Saharan Africa providing most of the land potential (FAO, 2006). The International Food Policy Research Institute predicts that future production growth will stem largely from yield improvements rather than area expansion, as has been found in past assessments of global agricultural futures such as the Millennium Ecosystem Assessment (MA) (2005). However, yield growth rates for major grains have been declining in the last decades (World Bank, 2008), and have dropped by roughly 50 per cent since their highs during the 1960s and late 1970s. One of the causes of this decline may be a fall in the growth of public agricultural research and development spending (World Bank, 2008). As the USA and other developed regions have shifted their research focus to reflect consumer preferences for processed, organic and humane products, the diffusion of more relevant yield-enhancing technology in developing countries has slowed (Alston and Pardey, 2006). Many developing countries that are generously land-endowed find it easier to convert forest and other land cover for agricultural production than to disseminate yield-enhancing technologies.

While at the global level forecasts suggest that future production gains will be sufficient to keep pace with increased population and rising demand, at the regional level they all indicate persistent and in some cases worsening food insecurity and malnutrition, particularly in sub-Saharan Africa (FAO, 2000, 2007). It is significant that all projected population increases will occur in the developing world, and much of it will be concentrated in urban areas (UN, 2001a). Six countries account for half of this annual growth: India for 21 per cent; China for 12 per cent; Pakistan for 5 per cent; Nigeria for 4 per cent; Bangladesh for 4 per cent; and Indonesia for 3 per cent. As a group, the 49 countries with the lowest per capita incomes will triple in size. While there is a large variation in the predictions, the combined effects of population growth, lower yields and increasing reliance on trade policy for food imports due to varying yields across the globe, could leave between an additional 5 and 170 million additional people malnourished in 2080, with up to 75 per cent of the total in Africa depending on the projection scenario (Schmidhuber and Tubiello, 2007). Economic and other development policies – especially policies pertaining to agricultural research and technology – will be critical in influencing future human well-being.

The demand for food, as well as the pattern of farming practices and agricultural land use worldwide, is shaped not only by population growth in the developing world, but also by other demographic trends globally that have a profound impact on existing food systems and their vulnerability. In particular, rising incomes in the developed world, lower incomes in developing regions, increasing urbanization globally, migration patterns, ageing populations, and the HIV and AIDS pandemic (Runge et al, 2003), change the economic behaviour of consumers in terms of their dietary demand for food, energy products, and the quality and concentration of markets and supply chains. These factors also shape crop production capacity and cropping patterns, thus having an impact on both the demand and production elements of food systems. Such socio-economic changes are illustrative of the multiple stressors that give rise to food system vulnerability, which include but are not limited to those arising in the biophysical environment.

Rising income and urbanization increase the demand for livestock products and highly processed foods, and decreased demand for staples (Rosegrant et al, 2001). This places additional pressure on land resources for pasture and coarse grain markets for feed, including maize. It is predicted that by 2020 the demand for beef, meat, poultry, pork and milk will at least double from 1993 levels (Delgado et al, 1999). The trend of increasing urbanization all over the world could also create enormous problems of adequate food supply, especially for the growing number of mega-cities (Stamoulis and Zezza, 2003). In this century, food insecurity will increasingly become an urban problem as more than 57 per cent of people in developing countries are expected to live in cities by 2030 (UN, 1998).

Migration patterns are a strong force in shaping food systems. A review of recent literature on population growth, migration and the rural environment has provided numerous examples in which migration of farmers to the agricultural frontiers has resulted in tropical deforestation or the desiccation of land in dryland areas (UN, 2001a). Out-migration of farmers from rural areas may have locally positive effects of reducing pressures on rural resources, or offer opportunities of remittances from employment, but may also negatively affect local agricultural labour availability. In many areas, migration is taking place from rural to urban areas, due to perceptions that agriculture can no longer support livelihoods, and processes of de-agrarianization in which rural dwellers orientate their identities, occupations and income-earning activities away from those that are strictly agricultural-based (Bryceson, 2006). In Africa, urbanization is not necessarily associated with an increase in GDP (Mugisha and Zulu, 2004), and rapid urbanization without commensurate economic growth can contribute not only to food insecurity in urban areas, but also to higher levels of other health risks including exposure to HIV (Mugisha and Zulu, 2004; Amuyunzu-Nyamongo et al, 2007; Hunter, 2007; Richards, 2007; and see Chapter 20).

Human health factors also have a powerful impact on food systems through mechanisms such as migration, shifts in the labour force and demographic structure. HIV and AIDS are having a profound impact on food systems; from the local level of the community and household, through to the regional level. One-third of the global total of HIV-infected people lives in southern Africa, where HIV prevalence in 2007 exceeded 15 per cent in seven countries (Botswana, Lesotho, Namibia, South Africa, Swaziland, Zambia and Zimbabwe) (UNAIDS, 2008). Another third lives elsewhere in sub-Saharan Africa, mainly in the central and eastern regions. Ironically, despite the devastating impact of the AIDS epidemic, the populations of the most affected countries are expected to be larger by mid-century than today (UN, 2001b). For the nine most affected countries in Africa with HIV prevalence at or above 14 per cent, the population is projected to increase from 115 million in 2000 to 196 million in 2050.

At the household and community level, the coexistence and interaction between acute food insecurity and high HIV prevalence often precipitates a downward spiral towards destitution and famine (de Waal and Whiteside, 2003). Not only does AIDS interact with and worsen livelihood shocks and stressors, but it selectively undermines the very strategies that historically were employed to respond to such shocks (Drimie and Casale, 2009).

Population age distribution changes in many parts of the world also drive change in food systems. In many places changes are due in part to the demographic transition; as incomes expand, families around the world choose to have fewer children, opting instead to invest more in the children they have since these are likely to survive to adulthood (Caldwell et al, 2006). Improvements in health care and nutrition also play a key role in age distribution (Barnett and Whiteside, 2006). In regions severely affected by AIDS, particularly southern and eastern Africa, population age distribution is severely affected, with the central section of the population pyramid being 'squeezed', and the most economically active section of the population being those most affected by the pandemic (Barnett and Whiteside, 2006). This transition means that in some regions fewer economically active people have to care for an increasing number of elderly people. It will also have tremendous consequences for social security networks, many of which contribute to food security systems (Stamoulis and Zezza, 2003).

Changes in ecosystem services

The above discussions on changes in supply and demand in food systems emphasize that the direct provision of food and water are ecosystem services that are integral to sustainable food systems. Ecosystem services, however, encompass a wide range of tangible and intangible benefits that people derive from natural living and non-living environments (Biggs et al, 2004; Scholes and Biggs, 2004; IPCC, 2007a). Four categories of services have been identified (IPCC, 2007a):

1 provisioning services: products including food, fibre and medicinal products;
2 supporting services, such as primary and secondary production and biodiversity that provide the basic support for higher-level services;
3 regulating services, including carbon sequestration, climate and water regulation, protection from natural hazards, water and air purification, and disease and pest regulation;

4   cultural services, which satisfy human spiritual and aesthetic needs – for example, 'sense of place' (Raymond et al, 2009).

Recent science assessments show that the supporting primary, and secondary, production capacity and biodiversity of ecosystems are in many places being compromised by environmental changes (MA, 2005). The ecosystem service of sufficient food production to support people – a cornerstone of a viable global food system – is intrinsically linked to global water cycles, the carbon energy cycle and climatic conditions (Khan and Hanjra, 2009).

Biodiversity is often explicitly or implicitly recognized as underpinning the ability of natural environments to sustain ecosystem services. Species diversity is, for instance, thought to increase the stability of agricultural practices; biodiversity loss is seen as threatening ecosystems, agricultural sustainability and food security (MA, 2005; Turner et al, 2007). Currently, three crops alone – rice, wheat and maize – account for about 60 per cent of the calories globally derived from plants (Thrupp, 2000). Although there are many varieties within these crops that respond differently to climate change, this is nevertheless an indication of limited global crop diversity. The active pursuit of farming systems that promote agrobiodiversity has been reviewed as yielding a suite of ecosystem service as well as food security benefits (Thrupp, 2000; Reidsma and Ewert, 2008). These include such wide-ranging advantages as the viability of agricultural systems, ecosystem cultural services, increased nutritional value of foods, insect and disease management, soil water retention, and crop science and technology benefits (Thrupp, 2000).

Other than the well-documented stresses on food production imposed by current and projected climate change (e.g. IPCC, 2007b; Reidsma and Ewert, 2008), changes in factors such as production system technologies and methods, crop choices, demographics and global markets have direct and indirect impacts on the ability of our natural environments to sustain the provision of food. The doubling of the world's population from 3 to 6 billion between 1960 and 2000 (Ringler, 2008) has led to commensurate increases in agricultural production, evident in expansion and agricultural intensification of croplands, plantations and pastures. The resulting enormous energy and water footprints on the natural environment are calling into question the sustainability of the food provisioning ecosystem service (Khan and Hanjra, 2009). Intensification practices – characterized by increased inputs such as pest controls, fertilizers and germplasm – have led to changes in the nitrogen cycle and increased concentrations of atmospheric gases (Gregory and Ingram, 2003). The rising demand for food, together with development and increasing shares of land for housing, industry and infrastructure, also puts pressure on ecosystem services by pushing agricultural production on to increasingly marginal lands that typically have low soil fertility, restricted land tenure and low economic return (Cotula et al, 2009). As much of this land is already situated in climatic conditions that are unfavourable for agriculture, there is a risk that agriculture will become even more marginal as a result of climate changes.

The World Bank expects dwindling water supplies to become a major factor inhibiting agricultural growth (Martin et al, 2006). Crop irrigation in the world's two most populous countries, India and China, has allowed them to become self-sufficient in grain, but both depend on the same water sources: potentially less reliable monsoons, retreating Himalayan glaciers and dwindling groundwater supplies. Many large water basins, including the Yellow River and Ganges, are expected to pump relatively less water for irrigation over the next 20 years due to unfavourable competition from other sectors. As a result, irrigated cereal yields in water-scarce basins are expected to decline between 11 and 22 per cent in 2025 over 1995 levels (Rosegrant et al, 2005). International water management strategies are clearly critical for using water more efficiently in crop production (Qadir et al, 2003). Water is also critical for many other ecosystem services. By 2015, 40 per cent of the world's population, about 3 billion people, are expected to live in countries where it is difficult or impossible to get enough water to satisfy basic needs. Less than 1 per cent of the world's water supply is fresh. Over the last 70 years water demand has increased six-fold, and about half of the world's available fresh water is now used for agriculture, households or industry.

It is known that food systems are best described by linked social and ecological systems and their interactions, and that assessing the vulnerability of food systems to GEC has to accommodate this integrated perspective (Ericksen, 2008). Changes in ecosystem services not only influence food systems through the primary supporting and provisioning functions described above, but also by all the higher-level functions that shape food systems through social mechanisms. It is clear that an array of human activities and choices, and environmental conditions and responses, affect both changes in ecosystem services and how these changes make food systems vulnerable.

Spatial-scale dynamics

The linkages embodied in the interaction between local land-use change and global demand for biofuels (and the feedstock crops that produce them) illustrate the local-to-global relationships that characterize today's increasingly globalized food and energy system (and conditions vulnerability). (Scales and levels are discussed in Chapter 2.) In the light of increasingly volatile food and fuel prices, many countries are looking for ways to ensure their future food security, while at the same time providing for their longer-term energy security by reducing their dependence on imported fossil fuels. These actions, explicitly designed to reduce vulnerabilities in the face of rising oil and commodity prices, can, in turn, become drivers of vulnerability in distant places. For example, the growth in maize-based ethanol in the USA led to significant changes in export volumes to the world market, as well as to the prices for maize and closely related coarse grains that are important for food and feed uses. In this case, policy decisions regarding land use and energy in the USA had significant repercussions with the USA's primary trading partners around the globe, many of whom are dependent on US maize imports for meeting food needs. Even in cases where energy feedstocks are not directly competing with food consumption, the resulting land-use pattern changes can induce a 'knock-on' effect that has consequences for either the crop or livestock sector in the form of higher demands, reduced land area or higher input costs. While second-generation biofuel technologies are now in development and are expected to reduce the direct competition between land-for-food and land-for-fuel, the case of global biofuel expansion has illustrated how globalization enables a decision taken in one location to have unexpected and disproportionate influence on the domestic markets and policies of distant places (Cash et al, 2006).

The vigorous discussion on 'indirect land use' consequences of current bioenergy policies centres around the contention that the theoretical gains that biofuels would otherwise have (in avoided fossil-fuel consumption) could be offset by the induced loss of carbon sequestration if land is cleared to make way for feedstock production. These offsetting and countervailing land-use changes could occur in the same region where the biofuels are being produced and consumed – or in far-away regions where the feedstocks are produced and whose expansion is driven by price signals received through the international markets for food, feed and fuel (e.g. Melillo et al, 2009).

Vulnerability has traditionally been conceived in place-specific terms, recognizing the geographically specific nature of direct exposure to biophysical hazards, and the particular social and institutional factors that define the sensitivity and capacities of vulnerable systems. Less often discussed is what might be called 'local–local' linkages in the manifestation of vulnerability. Here, the losses and harm experienced in one geographic location is not independent from that at another location; moreover, the vulnerabilities of both places have counteractive or synergistic interactions with one another. Such linkages can be thought of as 'teleconnections': impacts and responses in one location affect the vulnerability of another through multiple mechanisms, including biophysical linkages, tightly coupled markets and the movement of people, information and resources (Adger et al, 2009; Eakin et al, 2009).

The importance of local–local linkages in vulnerability is particularly evident in food systems. Farmers are often acutely aware of these linkages, knowing that the viability of their harvests is in part dependent on the viability of the harvests of their competitors. Small-scale vegetable farmers in central Mexico, for example, can identify by name the specific villages in neighbouring states where frosts, floods and droughts have direct implications for the prices they receive from buyers. The losses of their competitors provide small windows of opportunity that can make significant livelihood differences in an otherwise bad year for production (Eakin, 2006). Local–local vulnerabilities can also be more distantly connected (see Box 7.1).

The movement of people and plants intensified by globalization is also contributing to local–local vulnerabilities. One controversial example is the introduction of genetically modified (GM) material into native corn varieties in Mexico. Although the planting of GM seed is not yet permitted in Mexico, traces of such material have been discovered in the local maize seed stocks in Mexico (Dyer et al, 2009). Food insecurity is a prominent feature of many rural communities where traditional varieties of maize are most commonly sown. Households supplement their harvests and diets with publicly subsidized corn grain, some of it imported from the USA where GM corn is now

---

Box 7.1 *The interdependent vulnerabilities of Mexican and Vietnamese coffee-producing households*

Eakin et al (2009) illustrate the interdependent vulnerabilities of Mexican and Vietnamese coffee-producing households to market shocks, climate hazards and environmental degradation. The dramatic expansion of coffee farming in response to market liberalization in Vietnam produced a ripple effect, eventually contributing to increased vulnerability for both Mexican and Vietnamese farmers. As production expanded in Vietnam, producers there became vulnerable to processes exacerbated by their own success: the vagaries of the coffee market (Dang Van Ha and Shively, 2008) and to environmental degradation caused by irrigation expansion and deforestation to plant coffee plantations (D'haeze et al, 2005). Thus although poverty levels declined in the primary coffee production region from 70 to 52 per cent between 1993 and 1998, further reductions in poverty in the area stagnated after 1998, despite significant progress in reducing poverty at the national level in the same period (Glewwe et al, 2004).

At the same time, as the global market became flooded with coffee, farmers in the highlands of Mexico suddenly found themselves facing historically low prices, while simultaneously struggling to maintain production in a context of declining labour and increased losses to climate hazards (Eakin et al, 2006). In this context, the livelihood security – and hence food security – of households in the two countries are linked not only by fluctuations in world coffee prices, but also by the land-use responses of farm households, the material flow of coffee beans, and the global flow of information and knowledge.

---

the norm. In the face of high rates of international migration the transport of seed is not uncommon. In this case, the efforts of households to assure their livelihood security through migration and resourceful use of available seed in formal and informal markets, in the long term, may threaten the genetic and cultural value associated with local maize varieties.

Tightly coupled international commodity markets are perhaps the strongest driver of local–local vulnerabilities. Vulnerability is of particular concern for commodities where export production is geographically concentrated, volumes traded are relatively small compared to demand, and demand is constant or increasing. The international market for rice is one example. Only 7 per cent of rice is traded internationally, yet demand for rice is growing internationally (von Braun and Soledad Bos, 2005). Most rice produced is heavily protected, and the cultural and political significance of the crop forces exporting countries to satisfy domestic food security needs first when stocks are low and prices high. The media widely reported that hoarding of rice stocks in Thailand and Vietnam in 2007–8 contributed to rising food insecurity and political unrest in Haiti, Pakistan, Cambodia and Bangladesh among other countries. Once again, the vulnerability of populations in geographically disparate locations is intricately tied through concentrated global markets and the interaction of domestic policy interventions in these markets. The result is that in some cases, land-use conversion, rising input costs and food needs in one location contribute to hunger and instability elsewhere (von Braun, 2007; Oxfam, 2008).

Conclusion

The question of 'vulnerability to what' in the context of food systems signals the need to consider to which stressors a system is exposed, its sensitivity or likely response, and its capacity to adapt or change while maintaining or attaining robustness in the form of food security. The question of 'vulnerability to what' requires an understanding not only of the drivers of vulnerability, but also of their dynamic interactions and synergisms, and of how these are shaped over time and through space; this encapsulates the challenge of addressing multiple stressors in the context of GEC.

Addressing the title question of this chapter has been framed by a reflection on, first, the varying temporal scale over which drivers of change function, in which consideration is given to the short- and long-term impacts of climate changes and climate variability, as well as the short- and long-term interaction between states of food insecurity. The interconnected nature of the temporal dimensions of food insecurity has profound implications for food systems and for policies and programmes seeking to strengthen them.

Second, perturbations in food systems that are 'abnormal' – extreme or unpredictable – are traced, primarily through the lens of food price shocks, which in turn illustrates the complexities of an increasingly globalized food system.

Third, the idea of reversibility and 'linear' and non-linear change in food systems draws attention to tipping points and thresholds in earth systems science, the marginalization of production systems, and alterations to ecosystems services. Not all system responses are linear, which makes for potentially rapid and unpredictable changes in food systems or their components. There is thus a possibility of unanticipated outcomes of food system exposures to stresses, whether in terms of the nature or the speed of change. This potential for 'surprises' is explored further in Chapter 20. The challenge of systematically integrating the full spectrum of food system elements, as well as the multiple stressors with which they interact, is highlighted in the discussion on ecosystem services. Ecosystems need to be recognized for their full range of connectivity with social systems – not just food provisioning – and this in turn dictates a need for better scientific methods of analysis to negotiate coupled systems.

Finally, spatial and scalar interactions between system changes and between vulnerabilities in the context of food systems are traced. Intersections within and between food systems across space and time necessitate, for example, the development of mechanisms to address transitory crises, such as hunger and food shortages, without undermining the mechanisms that will prevent chronic food insecurity or avert future food crises. Similarly, long-term structural vulnerabilities cannot be ignored in the pursuit of robust food systems.

Note

1   'Knowledge system' in this chapter is used to refer to the knowledge and perceptions generated, framed and held by a particular social group; such groups might include indigenous groups, civil society, minority social groups, politicians and scientists.

References

Abbot, P. C., C. Hurt and W. E. Tyner (2008) What's Driving Food Prices? *Issue Report*. Oak Brook, Farm Foundation

Adger, W. N. (1999) 'Social vulnerability to climate change and extremes in coastal Vietnam', *World Development*, 27, 2, 249–69

Adger, W. N., H. Eakin and A. Winkels (2009) 'Nested and teleconnected vulnerabilities to environmental change', *Frontiers in Ecology and the Environment*, 7, 3, 150–57

Aggarwal, P. K., P. K. Joshi, J. S. I. Ingram and R. K. Gupta (2004) 'Adapting food systems of the Indo-Gangetic plains to global environmental change: Key information needs to improve policy formulation', *Environmental Science & Policy*, 7, 487–98

Allison, E. H., A. L. Perry, M. C. Badjeck, W. N. Adger, K. Brown, D. Conway, A. S. Halls, G. M. Pilling, J. D. Reynolds, N. L. Andrew and N. K. Dulvy (2009) 'Vulnerability of national economies to the impacts of climate change on fisheries', *Fish and Fisheries*, 10, 2, 173–96

Alston, J. and P. G. Pardey (2006) Developing-country perspectives on agricultural R&D: New pressures for self-reliance? In Pardey, P. G., J. M. Alston and R. R. Piggott (eds) *Agricultural R&D in the Developing World: Too Little, too Late?*, Washington, DC, International Food Policy Research Institute

Amuyunzu-Nyamongo, M., L. Okeng'O, A. Wagura and E. Mwenzwa (2007) 'Putting on a brave face: The experiences of women living with HIV and AIDS in informal settlements of Nairobi, Kenya', *AIDS Care*, 19, 25–34

Bankoff, G., G. Frerks and D. Hilhorst (eds) (2004) *Mapping Vulnerability: Disasters, Development and People*, London, Earthscan

Barnett, T. and A. Whiteside (2006) *AIDS in the 21st century: Disease and Globalization*, London, Palgrave Macmillan

Barrett, C. and D. Sahn (2001) *Food Policy in Crisis Management*, Ithaca, Cornell University Press

Barrett, C. B., R. Bell, E. C. Lentz and D. G. Maxwell (2009) 'Market information and food insecurity response analysis', *Food Security*, 1, 2, 151–68

Biggs, R., E. Bohensky, P. Desanker, C. Fabricius, T. Lynam, A. Misselhorn, C. Musvoto, M. Mutale, B. Reyers, R. J. Scholes, S. Shikongo and A. S. van Jaarsveld (2004) Nature Supporting People: The Southern African Millennium Ecosystem Assessment Integrated Report. Pretoria, Council for Scientific and Industrial Research

Bradbury, M. (1998) 'Normalizing the crisis in Africa', *The Journal of Humanitarian Assistance*, http://jha.ac/1998/02/04/normalising-the-crisis-in-africa/ (accessed: undated)

Brown, L. R. (1973) 'The next crisis? Food', *Foreign Policy*, 13, 3–13

Brown, L. R. (2005) *Outgrowing the Earth: The Food Security Challenge in an Age of Falling Water Tables and Rising Temperatures*, London, Earthscan

Bryceson, D. F. (2006) Vulnerability and viability of East and Southern Africa's apex cities. In Bryceson, D. F. and D. Potts (eds) *African Urban Economies*, New York, Palgrave Macmillan

Burke, M. B., D. B. Lobell and L. Guarino (2009) 'Shifts in African crop climates by 2050, and the implications for crop improvement and genetic resources conservation', *Global Environmental Change,* 19, 3, 317–25

Caldwell, J. C., B. K. Caldwell, P. Caldwell, P. F. McDonald and T. Schindlmayr (2006) *Demographic Transition Theory,* Dordrecht, Springer

Carter, M., P. Little, T. Mogues and W. Negatu (2005) 'Shocks, sensitivity and resilience: Tracking the economic impacts of environmental disaster on assets in Ethiopia and Honduras', *Development and Comp Systems,* http://129.3.20.41/eps/dev/papers/0511/0511029.pdf (accessed: undated)

Cash, D. W., W. Adger, F. Berkes, P. Garden, L. Lebel, P. Olsson, L. Pritchard and O. Young (2006) 'Scale and cross-scale dynamics: governance and information in a multilevel world', *Ecology and Society,* 11, 2

Cline, W. R. (2007) *Global Warning and Agriculture: Impact Estimates by Country,* Washington, DC, Centre for Global Development

Coghlan, A. (2002) 'GM row delays food aid', *New Scientist,* 175, 2354, 4

Cotula, L., S. Vermeulen, R. Leonard and J. Keeley (2009) Land grab or development opportunity? Agricultural investment and international land deals in Africa. London/Rome, IIED/FAO/IFAD

D'haeze, D., D. Raes, J. Deckers, T. A. Phong and H. V. Loi (2005) 'Groundwater extraction for irrigation Coffea canephora in Ea Tul watershed, Vietnam – a risk evaluation', *Agricultural Water Management,* 73, 1, 1–19

Dang Van Ha and G. Shively (2008) 'Coffee boom, coffee bust, and smallholder response in Vietnam's central highlands', *Review of Development Economics,* 12, 2, 312–26

Darcy, J. and C. Hofmann (2003) *According to Need? Needs Assessment and Decision-Making in the Humanitarian Sector,* London, Overseas Development Institute

de Waal, A. and A. Whiteside (2003) 'New variant famine: AIDS and food crisis in southern Africa', *The Lancet,* 362, 9391, 1234–37

Delgado, C., M. Rosegrant, H. Steinfeld, S. Ehui and C. Courbois (1999) Livestock to 2020: The next food revolution. *Food, Agriculture and the Environment Discussion Paper 28,* Rome, FAO

Devereux, S. (2006a) Identification of methods and tools for emergency assessments to distinguish between chronic and transitory food insecurity. *Desk Review for the 'Strengthening Emergency Needs Assessment Capacity' (SENAC) project,* Rome, World Food Programme

Devereux, S. (2006b) Vulnerable Livelihoods in Somali Region, Ethiopia. *IDS Research Report 57,* Brighton, Institute of Development Studies

Devereux, S. (2009) 'Why does famine persist in Africa?', *Food Security,* 1, 1, 25–35

Devereux, S. and Z. Tiba (2007) Chapter seven: Malawi's first famine: 2001–2. In Devereux, S. (ed) *The New Famines,* London, Routledge

DFID (2004) Scoping Study Towards DFIDSA's Hunger and Vulnerability Programme (accessed: 19 October 2005). Southern African Regional Poverty Network (SARPN) for DFID

Drimie, S. and M. Casale (2009) 'Multiple stressors in Southern Africa: the link between HIV/AIDS, food insecurity, poverty and children's vulnerability now and in the future', *AIDS Care: Psychological and Socio-medical Aspects of AIDS/HIV,* 21, supp 1, 28–33

Dyer, G. A., J. A. Serratos-Hernánadez, H. R. Perales, P. Gepts, A. Piñeyro-Nelson, A. Chávez, N. Salinas-Arreortua, A. Yúnez-Naude, J. E. Taylor and E. R. Alvarez-Buylla (2009) 'Dispersal of transgenes through maize seed systems in Mexico', *PLoS ONE,* 4, 5, e5734

Eakin, H. (2006) *Weathering Risk in Rural Mexico: Climatic, Institutional and Economic Change*, Tucson, The University of Arizona Press

Eakin, H., C. Tucker and E. Castellanos (2006) 'Responding to the coffee crisis: A pilot study of farmers' adaptations in Mexico, Guatemala and Honduras', *The Geographical Journal*, 172, 2, 156–71

Eakin, H., A. Winkels and J. Sendzimir (2009) 'Nested vulnerability: Exploring cross-scale linkages and vulnerability tele-connections in Mexican and Vietnamese coffee systems', *Environmental Science and Policy*, 12, 4, 398–412

Easterling, W. E., P. K. Agrawal, P. Batima, K. Brander, L. Erda, M. Howden, A. Kirilenko, J. Morton, J.-F. Soussana, J. Schmidhuber and F. Tubiello (2007) Food, fibre, and forest products. In Parry, M. L., O. F. Canziani, J. P. Palutikof, P. J. van der Linden and C. E. Hanson (eds) *Climate Change 2007: Impacts, Adaptation and Vulnerability. Contribution of Working Group II to the Fourth Assessment Report of the Intergovernmental Panel on Climate Change*, Cambridge, Cambridge University Press

Ehrlich, P. (1968) *The Population Bomb*, New York, Ballantine Books

Ericksen, P. J. (2008) 'What is the vulnerability of a food system to global environmental change?', *Ecology and Society*, 13, 2

Evans, A. (2008) Rising Food Prices: Drivers and Implications for Development. *Briefing Paper 08/02*, London, Chatham House

FAO (2000) Agriculture: Towards 2015/2030. *Technical Interim Report*, Rome, FAO

FAO (2006) *World agriculture: Towards 2015/30*, Rome, FAO

FAO (2007) *The State of Food Insecurity and Agriculture*, Rome, FAO

FAO (2008) *The State of Food and Agriculture in Asia and the Pacific Region 2008*, Bangkok, FAO Regional Office for Asia and the Pacific

Felzer, B. S., T. Cronin, J. M. Reilly, J. M. Melillo and X. Wang (2007) 'Impacts of ozone on trees and crops', *Comptes Rendus Geosciences*, 339, 11–12, 784–98

Fischer, G., M. Shah, F. N. Tubiello and H. van Velhuizen (2005) 'Socio-economic and climate change impacts on agriculture: an integrated assessment, 1990–2080', *Philosophical Transactions of the Royal Society Biological Sciences*, 360, 2067–83

FIVIMS (2002) *Understanding Food Insecurity and Vulnerability: Tools and Tips*, Rome, FAO FIVIMS

Folke, C., S. Carpenter, B. Walker, M. Scheffer, T. Elmqvist, L. Gunderson and C. S. Holling (2004) 'Regime shifts, resilience and biodiversity in ecosystems management', *Annual Review of Ecology, Evolution, and Systematics*, 35, 1, 557–81

Fraser, E. D. G. (2007) 'Travelling in antique lands: using past famines to develop an adaptability/resilience framework to identify food systems vulnerable to climate change', *Climatic Change*, 83, 495–514

Fraser, E. D. G. and L. C. Stringer (2009) 'Explaining agricultural collapse: Macro-forces, micro-crises and the emergence of land use vulnerability in southern Romania', *Global Environmental Change*, 19, 1, 45–53

Garmestani, A. S., C. R. Allen and L. Gunderson (2009) 'Panarchy: Discontinuities reveal similarities in dynamic system structure of ecological and social systems', *Ecology and Society*, 14, 1

Glewwe, P., N. Agrawal and D. Dollar (eds) (2004) *Economic Growth, Poverty, and Household Welfare in Vietnam*, Washington, DC, The World Bank

Gregory, P. J. and J. S. I. Ingram (2003) *Global Environmental Change and Future Crop Production*, 11th Australian Agronomy Conference, Geelong, Victoria

Gregory, P. J., S. N. Johnson, A. C. Newton and J. S. I. Ingram (2009) 'Integrating pests and pathogens into the climate change/food security debate', *J. Exp. Bot.*, 60, 10, 2827–38

Haddad, L. and T. Frankenberger (2003) *Integrating Relief and Development to Accelerate Reductions in Food Insecurity in Shock-Prone Areas*, Washington, DC, International Food Policy Research Institute

Hardin, G. (1977) *The Limits of Altruism: An Ecologist's View of Survival*, Bloomington, Indiana, Indiana University Press

Howden, S. M., J. F. Soussana, F. N. Tubiello, N. Chhetri, M. Dunlop and H. Meinke (2007) 'Climate change and food security special feature: Adapting agriculture to climate change', *Proceedings of the National Academy of Sciences of the United States of America*, 104, 19691–96

Hunter, M. (2007) 'The changing political economy of sex in South Africa: The significance of unemployment and inequalities to the scale of the AIDS pandemic', *Social Science & Medicine*, 64, 3, 689–700

IPCC (2001) IPCC, 2001: Climate Change 2001: The Scientific Basis. Contribution of Working Group I to the Third Assessment Report of the Intergovernmental Panel on Climate Change. In Houghton, J. T., Y. Ding, D. J. Griggs, M. Noguer, P. J. van der Linden, X. Dai, K. Maskell and C.A. Johnson (eds) Cambridge, United Kingdom and New York, NY, Cambridge University Press

IPCC (2007a) Climate Change 2007: Impacts, adaptation and vulnerability. In Parry, M., O. Canziani, J. Palutikof, P. van der Linden and C. Hanson (eds) *Contribution of Working Group II to the Fourth Assessment Report of the Intergovernmental Panel on Climate Change*, Cambridge, Cambridge University Press

IPCC (2007b) Climate Change 2007: The physical science basis. In Solomon, S., D. Qin, M. Manning, Z. Chen, M. Marquis, K. B. Averyt, M. Tignor and H. L. Miller (eds) *Contribution of Working Group I to the Fourth Assessment Report of the Intergovernmental Panel on Climate Change*. Cambridge, United Kingdom, and New York, United States, Cambridge University Press

Khan, S. and M. A. Hanjra (2009) 'Footprints of water and energy inputs in food production – Global perspectives', *Food Policy*, 34, 2, 130–40

Kremen, C., N. M. Williams, M. A. Aizen, B. Gemmill-Herren, G. LeBuhn, R. Minckley, L. Packer, S. G. Potts, T. Roulston, I. Steffan-Dewenter, D. P. Vázquez, R. Winfree, L. Adams, E. E. Crone, S. S. Greenleaf, T. H. Keitt, A. M. Klein, J. Regetz and T. H. Ricketts (2007) 'Pollination and other ecosystem services produced by mobile organisms: a conceptual framework for the effects of land-use change', *Ecology Letters*, 10, 299–314

Leichenko, R. M. and K. L. O'Brien (2008) *Environmental Change and Globalization: Double Exposures*, Oxford, Oxford University Press

Lenton, T. M., H. Held, E. Kriegler, J. W. Hall, W. Lucht, S. Rahmstorf and H. J. Schellnhuber (2008) 'Tipping elements in the Earth's climate system', *Proceedings of the National Academy of Sciences of the United States of America*, 105, 1786–93

MA (2005) *Ecosystems and Human Well-being: Synthesis*, Washington, DC, Island Press

Ma, J., A. Y. Hoekstra, H. Wang, A. K. Chapagain and D. Wang (2006) 'Virtual versus real water transfers within China', *Philosophical Transactions of the Royal Society B: Biological Sciences*, 361, 1469, 835–42

Martin, P., M. Abella and C. Kuptsch (2006) *Managing Labor Migration in the Twenty-First Century*, New Haven, Yale University Press

Melillo, J. M., A. C. Gurgel, D. W. Kicklighter, J. M. Reilly, T. W. Cronin, B. S. Felzer, S. Paltsev, C. A. Schlosser, A. P. Sokolov and X. Wang (2009) Unintended Environmental Consequences of a Global Biofuels Program. *MIT Joint Program on the Science and Policy of Global Change*, Cambridge, MA, MIT Press

Mercado, L. M., N. Bellouin, S. Sitch, O. Boucher, C. Huntingford, M. Wild and P. M. Cox (2009) 'Impact of changes in diffuse radiation on the global land carbon sink', *Nature*, 458, 7241, 1014–17

Misselhorn, A. A. (2005) 'What drives food insecurity in southern Africa? A meta-analysis of household economy studies', *Global Environmental Change*, 15, 33–43

Mugisha, F. and E. M. Zulu (2004) 'The influence of alcohol, drugs and substance abuse on sexual relationships and perception of risk to HIV infection among adolescents in the informal settlements of Nairobi', *Journal of Youth Studies*, 7, 3, 279–93

O'Brien, K. L. and R. M. Leichenko (2000) 'Double exposure: Assessing the impacts of climate change within the context of economic globalization', *Global Environmental Change*, 10, 3, 221–32

Oxfam (2008) Double-edged prices: Lessons from the food price crisis: 10 actions developing countries should take. *Briefing paper 121*, Oxford, Oxfam International.

Prendiville, N. (2003) Nutrition and food security information systems in crisis-prone countries. *International Workshop on Food Security in Complex Emergencies*, Tivoli, Italy, FAO

Qadir, M., T. M. Boers, S. Schubert, A. Ghafoor and G. Murtaza (2003) 'Agricultural water management in water-starved countries: challenges and opportunities', *Agricultural Water Management*, 62, 3, 165–85

Raymond, C. M., B. A. Bryan, D. H. MacDonald, A. Cast, S. Strathearn, A. Grandgirard and T. Kalivas (2009) 'Mapping community values for natural capital and ecosystem services', *Ecological Economics*, 68, 5, 1301–15

Reidsma, P. and F. Ewert (2008) 'Regional farm diversity can reduce vulnerability of food production to climate change', *Conservation Ecology (now Ecology and Society)*, 13, 1

Richards, R. B. O'Leary and K. Mutsonziwa (2007) 'Measuring quality of life in informal settlements in South Africa', *Social Indicators Research*, 81, 2, 375

Ringler, C. (2008) 'The millennium ecosystem assessment: Tradeoffs between food security and the environment', *Turkish Journal of Agriculture and Forestry*, 32, 3, 147–58

Rosegrant, M. W., M. S. Paisner, S. Meijer and J. Witcover (2001) *Global Food Projections to 2020: Emerging Trends and Alternative Futures*, Washington, DC, International Food Policy Research Institute

Rosegrant, M. W., S. A. Cline, W. Li, T. B. Sulser and R. Valmonte-Santos (2005) Looking Ahead: Long-Term Prospects for Africa's Agricultural Development and Food Security 2020. *Discussion Paper No. 41*, Washington, DC, International Food Policy Research Institute

Rosell, S. and B. Holmer (2007) 'Rainfall change and its implications for Belg harvest in South Wollo, Ethiopia', *Geografiska Annaler: Series A, Physical Geography*, 89, 287–99

Runge, C., B. Senauer, P. G. Pardey and M. W. Rosegrant (2003) *Ending Hunger in our Lifetime: Food Security and Globalization*, Baltimore, John Hopkins University Press

Salama, P., F. Assefa, L. Talley, P. Spiegel, A. van der Veen and C. A. Gotway (2001) 'Malnutrition, measles, mortality, and the humanitarian response during a famine in Ethiopia', *JAMA*, 286, 5, 563–71

Sanchez, P. A., G. L. Denning and G. Nziguheba (2009) 'The African green revolution moves forward', *Food Security*, 1, 37–44

Sarris, A. (2009) Evolving Structure of World Agricultural Trade and Requirements for New World Trade Rules. *Expert Meeting on How to Feed the World in 2050*, Rome, FAO

Schmidhuber, J. and F. N. Tubiello (2007) 'Global food security under climate change', *Proceedings of the National Academy of Sciences of the United States of America*, 104, 19703–8

Scholes, R. J. and R. Biggs (eds) (2004) *Ecosystem Services in Southern Africa: A Regional Assessment*, Pretoria, Council for Scientific and Industrial Research

Sen, A. (1981) *Poverty and Famines. An Essay on Entitlement and Deprivation*, Oxford, Clarendon Press

Sharp, K. and S. Devereux (2004) 'Destitution in Wollo (Ethiopia): chronic poverty as a crisis of household and community livelihoods', *Journal of Human Development*, 5, 227–47

Simelton, E., E. D. G. Fraser, M. Termansen, P. M. Forster and A. J. Dougill (2009) 'Typologies of crop-drought vulnerability: an empirical analysis of the socio-economic factors that influence the sensitivity and resilience to drought of three major food crops in China (1961–2001)', *Environmental Science & Policy*, 12, 4, 438–52

Stamoulis, K. and A. Zezza (2003) A conceptual framework for national agricultural, rural development, and food security strategies and policies. *ESA Working Paper Number 03–17*, Rome, Food and Agriculture Organization

Taylor, A. and J. Seaman (2004) Targeting Food Aid in Emergencies. *Emergency Nutrition Network (ENN): Special Supplement*, London, Save the Children UK

Thomas, C. D., A. Cameron, R. E. Green, M. Bakkenes, L. J. Beaumont, Y. C. Collingham, B. F. N. Erasmus, M. F. de Siqueira, A. Grainger, L. Hannah, L. Hughes, B. Huntley, A. S. van Jaarsveld, G. F. Midgley, L. Miles, M. A. Ortega-Huerta, A. Townsend Peterson, O. L. Phillips and S. E. Williams (2004) 'Extinction risk from climate change', *Nature*, 427, 6970, 145–48

Thrupp, L. A. (2000) 'Linking agricultural biodiversity and food security: The valuable role of sustainable agriculture', *International Affairs*, 76, 2, 265–81

Trostle, R. (2008) Global Agricultural Supply and Demand: Factors Contributing to the Recent Increase in Food Commodity Prices. *Economic Research Service WRS-0801*, US Department of Agriculture

Tubiello, F. N., J.-F. Soussana and S. M. Howden (2007) 'Climate Change and Food Security Special Feature: Crop and pasture response to climate change', *PNAS*, 104, 19686–90

Turner, W. R., K. Brandon, T. M. Brooks, R. Costanza, G. A. B. da Fonseca and R. Portela (2007) 'Global conservation of biodiversity and ecosystem services', *BioScience*, 57, 10, 868–73

UN (1998) *World Population Growth from Year 0 to 2050*, New York, United Nations Population Division

UN (2001a) *World Population Monitoring: Population, Environment and Development*, New York, Department of Economic and Social Affairs, Population Division, United Nations

UN (2001b) *World Population Prospects: The 2000 Revision: Highlights*, New York, Population Division, Department of Economic and Social Affairs, United Nations

UNAIDS (2008) *Report on the Global AIDS Epidemic*, Geneva, UNAIDS

USDA (2004) Ukraine: Average Harvest Prospects For Winter Grains, Production Estimates and Crop Assessment Division, Foreign Agricultural Service, www.fas.usda.gov/pecad2/highlights/2004/05/Ukraine%2004%20Trip%20Report/index.htm (accessed: 12 December 2008)

van Aalst, M. K., T. Cannon and I. Burton (2008) 'Community level adaptation to climate change: The potential role of participatory community risk assessment', *Global Environmental Change*, 18, 1, 165–79 von Braun, J. (2005) The world food situation: an overview. *CGIAR Annual General Meeting, 2005*, Marrakech, International Food Policy Research Institute

von Braun, J. (2007) *The world food situation: new driving forces and required actions*, IFPRI's Biannual Overview of the World Food Situation, CGIAR Annual General Meeting, Beijing

von Braun, J. and M. Soledad Bos (2005) The changing economics and politics of rice: implications for food security, globalization and environmental sustainability. In Toriyama, K., K. L. Heong and B. Hardy (eds) *Rice is Life: Scientific Perspectives for the 21st Century. Proceedings of the World Rice Research Conference, Tokyo and Tsukuba, Japan, 4–7 November 2004*, International Rice Research Institute

von Braun, J. and M. Torero (2009) Implementing Physical and Virtual Food Reserves to Protect the Poor and Prevent Market Failure. *Policy Brief*, Washington, DC, International Food Policy Research Institute

von Braun, J., A. Ahmed, K. Asenso-Okyere, S. Fan, A. Gulati, J. Hoddinott, R. Pandya-Lorch, M. W. Rosegrant, M. Ruel, M. Torero, T. van Rheenen and K. von Grebmer (2008) High food prices: The what, who, and how of proposed policy actions. *Policy Brief*, Washington, DC, International Food Policy Research Institute

Watson, F. (2007) Chapter twelve: Why are there no longer "war famines" in contemporary Europe? Bosnia besieged, 1992–1995. In Devereux, S. (ed) *The New Famines*, London, Routledge

WFP (2004) Emergency Needs Assessment. *WFP/EB.1/2004/4-*. Rome, World Food Programme

World Bank (1986) *Poverty and Hunger: Issues and Options for Food Security in Developing Countries*, Washington, DC, World Bank

World Bank (2008) *World Development Report 2008: Agriculture for Development*, Washington, DC, World Bank

Yang, J., Y. Ding, R. Chen and L. Liu (2005) 'Fluctuations of the semi-arid zone in China, and consequences for society', *Climatic Change*, 72, 1, 171–88

Yang, X., E. Lin, S. Ma, H. Ju, L. Guo, W. Xiong, Y. Li and Y. Xu (2007) 'Adaptation of agriculture to warming in Northeast China', *Climatic Change*, 84, 1, 45–58

8
Adapting Food Systems

Polly Ericksen, Beth Stewart, Siri Eriksen, Petra Tschakert, Rachel Sabates-Wheeler, Jim Hansen and Philip Thornton

Why adapt food systems?

Given the prevalence of food insecurity worldwide and the contribution of food systems to global environmental change (GEC), modern food systems need to be adapted to enhance food security and minimize negative environmental feedbacks. In addition, the GEC impacts (both positive and negative) on food systems and food security heighten the need for action to adapt to future change over the long term. Although adaptation in ecosystems has long been studied, and numerous researchers have documented adaptive rural livelihood strategies to manage environmental and other change, the concentration on planned adaptation of complex, globalized food systems to unprecedented and GEC is relatively new. As with studies of vulnerability, multiple approaches have led to a variety of conclusions. Chiefly some studies have been more actor-oriented rather than systems-focused (Nelson et al, 2007), which has led to differences in both conceptual frameworks and evaluation of 'successful' adaptation. Some have viewed adaptation as about reducing current risks while others have focused more on managing future and highly uncertain change (Eakin et al, 2009). Most researchers now recognize that high adaptive capacity of actors or individual units will not lead to pro-active adaptation strategies alone, as higher level institutional and policy reforms are also needed (Adger et al, 2007). Adaptation is an ongoing learning process (Armitage et al, 2008), which may ultimately lead to transformations of undesirable systems (Walker et al, 2004; Lebel et al, 2006).

This chapter will discuss the following issues with respect to adapting food systems to GEC. First, current approaches to adaptation are summarized, with a discussion of their merits as well as shortcomings. A number of examples are included from cases of ongoing adaptation by communities as well as with the assistance of researchers. Second, consideration is given to how thinking about adaptation of dynamic food systems raises conceptual as well as practical issues, including tradeoffs among adaptation priorities and outcomes, uncertainty, and the governance challenges in global food chains. This chapter is not intended to be a catalogue of all adaptation options and interventions. Rather, the aim is to raise key issues for future research and analysis. A few key messages the chapter will highlight are: lessons from current discourse and practice; challenges for scaling up to the required level of effort; high potential for maladaptation in food systems; and promising new directions for implementing adaptation in food systems.

Framing adaptation

An adaptation is an action or strategy intended to buffer against harm, particularly in the face of or after large shocks, such as economic or climatic. Adaptation also occurs in response to new stresses and movement into a new environment, and may occur to take advantage of new opportunities. With the acceptance that GEC, especially climate change, is inevitable and progressive, the discourse around adaptation is trying to motivate planned or anticipatory adaptation, to prevent negative impacts and to plan for sustainable development in the face of future uncertain change. As with discussions of vulnerability, social framings differ in origin and emphasis from ecosystem approaches. Different approaches to adapting food systems leads to differences in notions of what is feasible and what should be a policy priority (Nelson et al, 2007; Eakin et al, 2009). Nelson et al (2007) say that approaches focused on systems versus those focused on actors differ in terms of attention given to processes of decision-making and governance, versus maintaining response capacity throughout a system. In addition, systems-oriented approaches have focused more on future issues, not only current change and stressors. Eakin et al (2009) expand on this to highlight the tradeoffs between actor-focused approaches, which may compromise future system resilience, and resilience approaches, which may sacrifice a few vulnerable agents but the system overall will persist.

Social framings

Adaptation to environmental and other economic or social shocks has a tradition of study in the rural livelihoods and food security communities, as well as more recently in the climate change community. The focus is usually protecting livelihood strategies or maintaining some level of household food security. At the micro level, studies highlight the importance of the adaptive capacity of individuals, households or communities. Adaptive capacity is a function of assets or capitals in the broadest sense, including physical, financial, social and human (Ford et al, 2006). Field studies indicate that people adapt to many stressors concurrently, particularly economic in the case of farmers or to maintain food security (Eakin, 2005; Eriksen et al, 2005). Thus for rural households, for example, both migration and diversification out of agriculture can be considered adaptation strategies, the long-term impacts of which are complex. Although often both strategies provide crucial remittance or off-farm income that supports on-farm agricultural activities and helps households to ensure food security (Ellis, 2000), not all households diversify or migrate as part of a wealth-accumulating strategy. Sometimes these diversifications are survival strategies at best (Dercon, 2005); at worst they may actually put people at greater risk in the future or expose them to a new risk (e.g. migration in the context of HIV/AIDS). Also, climate variability is often considered a normal stress, and people focus on other social and economic constraints instead (Reid and Vogel, 2006).

An important distinction should be made between coping and adaptation. Coping with a shock allows survival, and perhaps protection of short-term food security or income, but often wears down assets which will be needed in future (Cromwell and Slater, 2004). This is particularly true in the case of coping with repeated shocks such as drought, or in the face of chronic disabling illness such as HIV/AIDS (see Chapter 7). Adaptation is best considered as modifications in behaviour or strategies that enable people to continue to develop in the face of change over the long run; adaptation is therefore the more useful goal when thinking about progressive and long-term GEC (Berkes and Jolly, 2001).

It is increasingly recognized that people also need an enabling institutional and policy environment to successfully adapt in the longer term (Adger et al, 2007). Adaptive capacity alone is insufficient for long-term and proactive adaptation at society-wide scales. For example, much of the research on sustainable livelihoods finds that lack of institutional support is a significant barrier to local adaptive capacity to diversify livelihoods for positive wealth accumulation (Ellis and Freeman, 2004). Eakin (2005) finds that the institutional and policy changes in Mexico shape the context in which farmers are able to make choices, and often economic issues override climatic stresses. Thus although adaptation ultimately takes place at the household and community level, without proactive intervention by governments, system-wide and long-term adaptation will not occur. As discussed later in the chapter, this introduces issues pertaining to power dynamics and political will or choices.

Ecosystem framings

Ecologists have long studied how ecosystems adapt to change. The discussion of temporal and spatial dynamics in Chapter 7 draws on many of these studies. While a full discussion of the evolution of understanding about adaptation in dynamic ecosystems is beyond the scope of this chapter, resilience theory explains that in dynamic ecosystems disturbance happens regularly, and resilience is a measure of how the ecosystem adapts to such disturbance. Resilient systems rely on adaptive capacity to maintain function and identity in spite of disturbance (Carpenter et al, 2001; Folke et al, 2004). As food systems rely upon key ecosystem services such as climate regulation, clean water, disease regulation, etc., loss of resilience in desirable ecosystems that maintain these services is a concern.

Current approaches to adaptation for agriculture or food security

As the level at which choices are made and enabled is so important in food systems and decisions about adaptation, in this section a distinction is made between local and higher-level adaptations. As adaptation is a broad and emerging topic, this chapter cannot address all relevant issues. The chapters in Part V pick up on the issue of adaptation more fully, as they deal with challenges for food systems under GEC in the coming decades. The critical political dimensions and conflicts are addressed in Chapters 17 and 18, as is the key and powerful role of private sector actors and the need to rethink food systems governance. Chapter 19 addresses tradeoffs between increasing food production and maintaining ecosystem services, while Chapter 20 discusses managing food systems under uncertainty and surprise.

Local adaptation
Rural households protecting livelihoods

Of course farmers and rural households have been adapting to climate variability and environmental change for centuries, often without outside interventions. Such local adaptation offers many useful lessons for strategies to adapt to future GEC. Traditional agricultural coping and food security strategies include migratory pastoralism, water harvesting techniques, cultivating a range of crops with different sensitivities to rainfall and temperature variability, storing food post-harvest to last through the year, soil and water conservation, moderating seasonal hunting patterns, etc. (Leach and Mearns, 1996; Scoones, 1996; Berkes and Jolly, 2001; Ford et al, 2006). They are based on intimate knowledge of local ecological and social systems. Many argue that local level or 'community based adaptation' must be at the forefront of adaptation plans, especially in developing countries, as ultimately adaptation actions are local and higher level policies must support these. The case of West African farmers (Box 8.1) illustrates some key issues.

In relation to long-term and progressive GEC, rain-fed agricultural systems in semi-arid and sub-humid West Africa have experienced higher variability in rainfall patterns over the last decade, mainly in the form of delayed onsets of rains, more and more pronounced dry spells during the cropping season, and more extreme rainfall events towards the end of the rainy season. Since the large majority of rural populations depend directly on agriculture as a main source of their livelihood, climatic changes adversely affect not only agricultural production and productivity but also household food security. Typical coping strategies to reduce negative impacts on production include experimenting with alternative planting dates and new crops such as drought-tolerant varieties, replanting, and cultivation at higher elevation in the case of severe rainfall events and flooding. As food production fails or is seriously curtailed, labouring for food and cash as well as consumption of unusual food such as wild plants and insects are not uncommon (e.g. Becker, 2000). Food storage is often the most effective pro-active coping mechanism; more capital-intensive technical strategies such as irrigation remain exceedingly rare in this part of Africa.

Less well understood are adaptive strategies to maintain food security in the face of environmental changes that go beyond agronomic and technical improvements in production. Such strategies relate to the exchange and distribution of food, two other critical aspects of the food system, and include innovative arrangements that link social actors in time and space, through

---

Box 8.1 *Proactive local adaptation*

Michael Mortimore and colleagues have highlighted the long tradition of successful adaptation by smallholder farmers in sub-Saharan Africa in the face of challenging climatic and environmental conditions. Mortimore and Harris (2004), examining the performance of smallholder farming systems at the national, district and farm level, highlight the achievements of smallholders in maintaining soil fertility and sustaining production, contradicting the dominant narrative of soil degradation. In another example, looking past solely food production, Mortimore (2005) breaks well-adapted food systems, what he calls 'success stories', into four components: more sustainable ecosystem management; increasing investment on- and off-farm; stable or improving output or output value per hectare; and evidence of improving incomes and welfare. Examining three cases in West Africa (the Kano close-settled zone, Nigeria; the Diourbel region, Senegal; and the Maradi department, Niger), Mortimore concludes that, although not all of these components are evident in every case, the sustained achievements of smallholders are evident. Mortimore (2005) further identifies key areas where policy support has yielded 'success' in the past, including promoting agricultural product markets; physical infrastructure; institutional infrastructures; knowledge management, supporting the spread of farmers' own expertise; investment incentives; and income diversification incentives:

> By providing evidence of real achievements and internal potentials, success stories can therefore point the way toward laying a new foundation for evidence-led policies for dryland development: Rather than aiming to transform 'inappropriate' local practices, such policies instead aim to build on local experience, suggesting a more organic model for development. (Mortimore, 2005)

---

both geographical and moral ties. Focusing on adaptive capacities and creative solutions of vulnerable and potentially food-insecure populations can provide vital opportunities to identify and reinforce networks of agency that transcend the limits of individual fields and other locales of production.

Evidence from Senegal suggests that social processes of household food allocation within and between communities are not only complex but also highly dynamic, mirroring times of climatic, economic and socio-political stress (Tschakert, 2007). While 80–100 per cent of households in a sample of vulnerable populations in the Old Peanut Basin exchanged seed and harvest gifts within their own community, independent of their resource endowment, 60–75 per cent also engaged in reciprocal arrangements beyond their village boundaries. Similarly, 50–65 per cent of families exchanged processed food and meals, a few with more than 15 other households in their community. Some of these moral ties explicitly addressed a concrete food crisis in 2005 that was triggered by extensive rainfalls well into the harvesting period of millet and groundnuts, the main staple and cash crops. Other exchanges represented longer-term, strategic investments so as to fortify existing social capital in anticipation of future shocks and vulnerabilities.

In Ghana, local–local processes of food exchange are becoming more and more frequent among migrants who have left their rural homes, drastically altered by environmental degradation and increasingly unpredictable rainfalls, in the northern parts of the country. Despite remarkable national increases in the production of core staples such as cassava, yams, plantains and maize over the last 15 years, regional unevenness in food security remains (Luginaah et al, 2009). Not surprisingly, large numbers of men and women seek employment in the agricultural heartland in the west central regions (mainly Brong Ahafo), and mining and urban centres in the south. The 2003 welfare indicators, for instance, reported 31 per cent of all rural households in the Upper West Region and 47 per cent in the Upper East Region as having difficulties meeting their food needs, compared to the national average of 13 per cent (Ghana Statistical Service, 2003). A study in two slums of Accra, the capital, revealed that one-fourth of all interviewed migrants (n=73) sends food to family members who stayed behind, while one-third receives food from home (Tschakert and Tutu, 2008). For more than half of the participants, food gifts also constituted a critical type of assistance during the migration process.

On a more integrated spatial level, linkages between farm migrants and their family members who stay behind in increasingly 'pathological homes' (Tschakert and Tutu, 2008) have become a crucial lifeline through which agricultural surpluses are channelled back as domestic 'food aid' to sustain the survival of entire communities in the Upper North (Luginaah et al, 2009). Contrary to well-known seasonal and circular migration movements during the northern dry season, this recent pattern of food remittances can be seen as a permanent adaptive arrangement. Those who can afford to engage in relatively expensive sharecropping agreements commit to sending truckloads of dried foodstuffs (mainly maize and yams) to less mobile family members back home to reduce their precarious food security.

While the example of Ghanaian migrants illustrates how creative self-organization among two vulnerable populations can enhance food security and livelihood resilience across spatial levels, such local adaptation may eventually require higher-level support. In the absence of a national food policy that provides for populations at risk beyond times of acute crises (such as the 2007 floods), collectively organized transportation systems that reduce individual costs of food shipments as well as government-funded road infrastructure are needed to effectively sustain such informal actor networks.

External livelihood protection interventions

Often economic or climatic shocks trigger agricultural production shortfalls or food insecurity, particularly in poor communities. The disaster relief community, composed of international donors and non-governmental organizations (NGOs), has traditionally intervened with safety net or livelihood recovery interventions, usually in-kind (food or seed) or cash transfers. Again, the extensive debate about the most effective type of safety net is outside the scope of this chapter, but it is highly relevant to considerations of adapting food systems to progressive GEC. A lot of innovative research and practice has demonstrated that when carefully designed, immediate interventions are useful for helping families to get back on their feet or to maintain their farming livelihoods after a shock (Devereux, 2001; Maxwell, 2002). However, both the Food and Agriculture Organization (FAO), with its emphasis on a 'twin track' for agricultural growth and poverty alleviation, and the World Bank, with its emerging social risk management agenda, recognize that social protection is an important policy agenda for developing countries facing chronic poverty and food insecurity, as well as low agricultural productivity and sensitivity to environmental stress.

The dominant framing of social protection is as a policy agenda for responding to income and consumption vulnerability, where vulnerability is a function of the frequency and nature of shocks and the capacity to cope with these shocks. Often these programmes are designed to specifically target the 'most vulnerable', which is a more difficult task than it may seem (see Chapter 7). As Devereux (2001) and others argue (Barrett and Maxwell, 2005), social protection policies ideally provide ex-ante risk reduction, in contrast to ex-post strategies such as food aid. This more proactive social protection requires higher level support for adaptation.

Higher level adaptation

Future, broad-scale and large-impact environmental change requires anticipatory action; hence many in the development and GEC communities are focused on proactive, planned adaptation, and are concerned about how to build adaptive capacity, as well as provide a sufficiently enabling environment to bring about adaptation on the broad scale necessary (Bapna et al, 2009; Padgham, 2009). Progress on increasing agricultural productivity and enhancing food security has been made in a number of areas, as discussed below. Not all of these interventions have been integrated with the official discourse around adaptation to GEC, as the agricultural development and food security communities and the GEC communities have been slow to collaborate.

Agro-technologies

There is a history of research-assisted technical adaptation in food systems. Not surprisingly, the most work has been done in the area of agricultural technology, primarily breeding crops for drought or heat tolerance or developing best practices for flood management. Recent publications such as *Millions Fed* (Spielman and Pandya-Lorch, 2009) highlight significant success stories. Table 8.1 summarizes cases where research and farmer innovation has succeeded in adapting agricultural systems to an environmental stress. Although these may not all be adaptations to progressive environmental change, they do illustrate cases where efforts by researchers, NGOs and farmers succeeded in enhancing production.

Table 8.1 *Examples of innovation to increase agricultural production*

Type of innovation	Adapting to	Example of where it worked
Sorghum and millet hybridization	Arid lands and limited rainfall	India where yields have roughly doubled since 1960 (Pray and Nagarajan, 2009)
Development of vaccine and mass vaccination of cattle against rinderpest	Rinderpest, a contagious disease characterized by necrosis and erosion in the digestive tract	Africa, Asia and Europe where rinderpest has now been eliminated despite afflicting cattle for thousands of years (Roeder and Rich, 2009)
Rice hybridization	Population growth and resultant reductions in arable land per person	China, where average rice yield rose from 3.4 tonnes per hectare in 1978 to 6.7 tonnes per hectare in 2008, allowing an additional 60 million people to be fed, while total acreage reduced (Li et al, 2009)
Shallow tubewells and boro rice development	Water scarcity leading to limited cultivatable land and reliance on fertilizer	Bangladesh where production increased from 23 million tonnes in 1989–90 to 43 million tonnes in 2007–8 (Hossain, 2009)
Introduction of natural predators (a South American wasp) to the mealybug pest	The mealybug pest, which threatened cassava cultivation	Across Africa where mealybug population declined substantially after introduction of the wasps (Nweke, 2009)
Tilapia (tropical finfish) hybridization	Reacting to opportunity (large areas suitable for aquaculture) as well as stresses (dwindling marine stocks as a result of over-exploitation)	Philippines where 280,000 people are estimated to benefit from direct or indirect employment in the tilapia industry (Yosef, 2009)
Water harvesting through mechanized construction of traditional micro-catchment ridges	Water scarcity	Syria where it has improved survival rates of plants on livestock grazing land (CGIAR, 2009)
Soil-less agriculture or floating gardens	Flooding and enabling growing in harsh conditions	Bangladesh (Haq et al, 2005)

Climate information

A second area of promising adaptation interventions involve the development and use of climate information. One of these is the use of seasonal forecasts. Interactions between the global atmosphere and the more slowly varying ocean and land surfaces, such as those associated with the El Niño/Southern Oscillation (ENSO) in the Pacific, provide a degree of predictability months in advance in several agriculturally important regions. Current capability to produce and disseminate seasonal forecasts resulted from investments such as the TOGA programme (1985–94), which advanced understanding of ENSO and developed an ocean-observing system (McPhaden et al, 1998); establishment of the International Research Institute for Climate Prediction (1996); and regional climate outlook forums (RCOFs) initiated in Africa and Latin America in 1997. A study that demonstrated a strong link between ENSO-related Pacific sea surface temperatures and Zimbabwe maize yields sparked interest in applying seasonal prediction to agriculture and food security challenges (Cane et al, 1994). Seasonal forecasts should, in principle, enable farmers to adopt technology, replenish soil nutrients and intensify production when climatic conditions are favourable; and to protect against the long-term consequences of adverse conditions.

As with many proposed adaptation interventions, the rationale for promoting seasonal climate prediction as a planned adaptation relies more on intuitive argument than on empirical evidence. First, to the degree that climate change impacts are amplifications of the substantial challenges that climate variability already imposes on agriculture, efforts to adapt to the impact of climate change must address current climate risk. Second, supporting adaptive management in response to forecast information should foster capacity to adapt more quickly to change at all time-scales, reducing lags associated with reactive autonomous adaptation. Third, periodic interactions with agricultural stakeholders around seasonal forecasts would provide an opportunity to also examine climate observations for evidence of recent change. Finally, seasonal climate prediction offers a means for agricultural communities to deviate from costly precautionary ex-ante risk management strategies by identifying and capitalizing on relatively good seasons even in the face of an adverse long-term climatic trend.

A flurry of pilot-scale research with developing country farmers, following the highly publicized 1997 El Niño, provides useful insights on the communication and use of climate information more generally. These studies typically show strong interest in using forecast information, and identify a range of potential responses (e.g. Ngugi, 2002; Phillips, 2003; Tarhule and Lamb, 2003; Ziervogel, 2004). They also reveal widespread constraints, including communication failures associated with access to information, and mismatch between farmers' needs (summarized in Hansen, 2002) and the scale, content, format or accuracy of available information (O'Brien et al, 2000; Ingram et al, 2002; Archer, 2003; Phillips, 2003; Ziervogel, 2004).

There are promising opportunities to improve seasonal forecasts as an adaptation intervention. The most obvious is to address the known communication failures. Accumulating evidence shows that it is feasible to address the spatial and temporal scales of operational forecasts by downscaling locally (Gong et al, 2003; Moron et al, 2006; Robertson et al, 2008), presenting predictive information about rainfall frequency (Moron et al, 2006; Moron et al, 2007) and risk of damaging dry spells (Sun et al, 2007; Robertson et al, 2008), and translating forecasts into agricultural production terms (Cane et al, 1994; Hansen et al, 2004a; Hansen et al, 2006). Giving the agricultural sector greater ownership of the process and more influence over the design of the information products would arguably improve institutional mechanisms such as the RCOFs (Cash et al, 2006; Hansen et al, 2007). A few pilot studies provide useful guidance on improving communication of probabilistic forecast information (Hansen et al, 2004b; Patt et al, 2005; Hansen et al, 2007; Marx et al, 2007; Roncoli et al, 2009), and demonstrate that smallholder farmers can use the information effectively when communication is improved (Huda et al, 2004; Patt et al, 2005; Meinke et al, 2006; Roncoli et al, 2009).

A second opportunity is to extend forecast information to other actors in the food system, such as providers of credit and production inputs, whose actions greatly influence farmers' flexibility to respond to advance information. In general, input supply chains need information at a longer lead time than farmers if they are to adjust supply to changing demand, but should benefit from the greater predictability that exists at aggregate scales.

Finally, much of the impact of climate information on vulnerability is likely to come through enabling more effective adaptive response to change, and through reducing the disincentive effect that climate risk has on adoption of innovation and market participation. This potential can be realized only to the degree that climate information services are integrated within a comprehensive approach to adaptation, and mainstreamed into agricultural development.

Policy reform for food security

At national and regional levels, some interventions in the policy domain are promising. Many of these are aimed at protecting access to food. Although not always successful, mandated grain reserves have a long history, and are still used by many countries to ensure domestic food security by maintaining an accessible supply of food at regulated prices. In Bangladesh and India, for example, grain is still distributed through public mechanisms such as Food for Work. One of the more controversial ideas to emerge from the recent high-level food security discussions has been the idea of a virtual grain reserve. The controversy arises in part because national grain reserves have been subject to mismanagement and provided perverse incentives to production and trade. However, von Braun and others argue that these issues could be overcome with a virtual reserve that relied on futures markets and flexible financing monitored by the World Food Programme (IFPRI, 2008).

Domestic market interventions have also been studied and analysed, in response to economic research which emphasized the importance of market failures in limiting food security and agricultural growth (Jayne et al, 2002; Dorward et al, 2003). As with grain reserves, domestic price controls are also highly controversial, with most economic analysts favouring liberalized markets that allow for trade (Dorosh, 2009). In spite of these studies and efforts, there are still gaps in the understanding of policies needed to better manage climate risk and markets, particularly in increasingly globalized food systems. The issue of how much capacity and policy leverage many developing country nations have to respond to food price shocks is hotly contested (Paarlberg, 2002; Birdsall, 2008).

Towards a framing for complex, globalized food systems

GEC involves unfolding processes that will bring impacts that are not fully understood and predictable. In addition, these changes will interact with social, economic and political changes. Although there is a long history of adapting food systems to environmental and socio-economic change, most current planning and policy for adaptation will be insufficient for adapting to future GEC. Additionally, community or household adaptation needs long-term support. Many adaptations to enhance food security have led to environmental negatives; and food security is asymmetrically distributed around the world, not just between countries but also within countries, often (but not only) due to differences in income and wealth. The processes which have produced the negative food system outcomes are increasingly connected. The current evidence of food insecurity and environmental degradation suggests that maladaptation may already be occurring. For food systems, maladaptation would be any intervention or strategy that results in more losses of ecosystem services, a decline in food security or increases in inequitable food security outcomes.

Food security at the local level is embedded in higher level national and international processes such as markets and trade networks. Technical solutions require institutional support and ongoing collaboration, as explained in the case of seasonal forecasts. Many responses to food insecurity are short term (e.g. safety nets or food aid relief). Box 8.2 highlights the need for better social protection policy. National-level responses to the food price crisis of 2008 created international problems, as they exacerbated the high prices when governments such as India and Vietnam (inter alia) imposed export bans on rice (Lustig, 2009). The shortcomings in much of the discussion about adaptation options for food security are limited both by which food system activities and outcomes are considered, as well as a focus on only one level of governance (e.g. local) or a small set of actors (e.g. farmers and NGOs).

The multiple actors engaged in food systems have different objectives, and hence their adaptation priorities and motivations to adapt to environmental change will be different (Thompson and Scoones, 2009); see also Chapter 17. But for many national and international policy-makers, one of the main goals of adaptation in food systems is to enhance food security and also protect desirable ecosystem services without contributing to more negative GEC. It is also important to note that for many households as well as nations, food systems are an important income earning strategy, and achieving food security is still a development goal internationally and for many nations. So the adaptation of food systems to environmental change has to be compatible with these development goals. For economic objectives to be aligned with food security and environmental goals, the private sector must be engaged and in some cases their leadership should be recognized. This is further discussed in Chapter 18.

There are examples of how to intentionally promote or protect ecosystem services through agri-food systems, allowing communities to become more food secure while also reducing the environmental impacts of these systems.

Box 8.2 *How should social protection incorporate seasonality?*

If seasonality re-emerges as an overwhelming and increasingly unpredictable phenomenon, social protection programmes will need to be reframed. Clearly, seasonal episodes can act as regular shocks and serve to undermine livelihoods. Where households are already living on the margins, repeated seasonal negative events can send them into destitution. The inability of poor or disadvantaged (excluded) people to recover from cyclical stresses leads to catastrophic problems (e.g. poverty or hunger traps) (see Chapter 7).

Income, consumption and assets are crucial in helping to overcome poverty and minimize livelihood shocks; however, 'resilience' is more complex than household income and asset portfolios. It comprises a complex function of existing behaviour, reflected in livelihood profiles that themselves represent long-term or structural adaptation to predictable shocks and stresses; crisis response behaviour (such as the ability to rely on formal and informal insurance and networks in times of crisis); and external (policy) responses to a predicted and actual crisis. Provision of consumption, income and asset insurance is only a partial response to vulnerability. An expanded view of social protection must incorporate responses to both chronic and structural vulnerability. Thus, an outcome of the effect of a seasonal episode, such as increased malnutrition, chronic poverty or increased food insecurity, depends on a complex array of factors that all need to be considered in policy formulation for risk and vulnerability reduction.

While seasons are very distinct in 'northern' countries, consumption is largely seasonal, reflecting the ability of the more affluent to access food from around the globe all year. Thus, the notion that changes in climate are the primary cause of downside seasonal impacts is highly questionable; instead, inequitable distribution of wealth and the politics of access and distribution should be seen as the fundamental drivers of the negative impacts of seasonality. Seasonality per se is not the problem. It is the interaction of limited opportunities, disadvantage and weak fall-back positions that leads to negative outcomes. Seasonality simply reveals these underlying or structural problems.

To minimize seasonal blind spots, social protection policy must recognize that: (i) seasonality is about the impacts on livelihoods of regular, cyclical fluctuations; (ii) seasonality represents a predictable stress, and as such needs to be built into risk management or social programmes in such a way that the programme is long-run, sustainable and appropriated timed; (iii) seasonal effects are highly complicated, reflecting the way that different spheres of livelihoods are intimately connected and dependent upon each other; and (iv) the way in which the impacts of seasonality are felt are location- and group-specific, but also mediated by systems of distribution and access. Putting considerations of access and distribution at the heart of a seasonality analysis leads to a differentiation of the impact of seasonality by location, wealth, gender and a range of other factors. Recognizing the varied impacts across groups and socio-economic status also implies the need to acknowledge the interconnectedness of prosperous seasonal living in one location or group, with poverty-entrenched seasonal livelihoods in another location or group.

The 'Bright Spots' project of the Comprehensive Assessment on Water Management in Agriculture highlights success stories. The project examined the linkages between land and water degradation and agricultural productivity, gathering examples of cases of 'food production that makes preeminent use of nature's goods and services while not permanently damaging these assets' (Noble et al, 2006), focusing on the small scale; on individuals, households and small communities. Table 8.2 outlines the cases where yield has been increased along with ecosystem benefits: 'the benefits are loosely arranged by scale of impact, such that the first are primarily factors contributing to the social-ecological resilience of communities..., while others become more important for increasing the resilience of ecosystems at regional scales' (Bossio et al, 2008, p209).

Framing adaptation in coupled social–ecological systems

As discussed in Chapter 5, the concept of coupled social–ecological systems introduces potentially appealing concepts for thinking about adaptation in food systems. This framework seeks to improve human management of ecosystems by explaining dynamics and change as the norm, and fostering systems that can cope with uncertainties and surprises such as climate change or world-wide market shocks affecting commodity prices (Carpenter et al, 2001; Gunderson, 2003).

However, the implications of managing food systems with this conceptual framework have not been fully explored. In particular, the role of social actors, from public to private sector as well as from local up to international levels, has not been sufficiently investigated.

Managing food systems for resilience?

Semi-arid northern Kenya is largely inhabited by nomadic pastoralists who are marginalized politically and economically. These regions have experienced recurrent droughts in the past ten years; the drought of 2008–9 was one of the worst on record. Droughts trigger food insecurity, as pastoralists have to rely on grains and purchased foods during droughts but often cannot sell livestock since cash markets collapse when everyone tries to sell off dying animals. In many ways, the explanation of these recurrent and escalating food insecurity crises 'fits' with a resilience approach. Droughts are the fast and short shocks that cause 'surprise' or a negative outcome; although they occur regularly, they are responded to each time as an emergency rather than a normal part of a cycle. The slower underlying variables that actually explain why food insecurity results from a drought are chronic poverty and marginalization, combined with the slow degradation of the natural resource base near settlements, the inability to access better grazing areas because of insecurity, and ongoing sedentarization of pastoralists in towns (particularly for food relief distribution).

The adaptive capacity of pastoralists to manage during droughts has been slowly eroded through a gradual loss of assets (livestock) with recurrent droughts from which they cannot recover. Their social networks are stretched too thin, and often food aid relief is insufficient. Furthermore, government and NGO assistance is unable (or unavailable) to support this eroded adaptive

Table 8.2 *Summary of selected ecosystem benefits beyond increased production of food derived in bright spot case studies (adapted from Bossio et al, 2008)*

Local-level: —Resource-use efficiency↑— —Ecological Footprint↓— —Environmental Pollution↓—

Landscape-level: —Ecosystem Services↑—

Ecosystem Benefit / Bright Spot case study	Soil Quality	Water Productivity	Low External Inputs	Integrated Pest Management	Water Cycling	Biodiversity	Carbon Sequestration	Social Capital
Huang-Huai-Hai river plain	Y	Y						
Bukhara Shirkat, Ikrom farm and Shermat farm	Y	Y						Y
Water harvesting in northern Ethiopia		Y		Y	Y			
System of rice intensification, Madagascar	Y	Y	Y	Y				
Bonganyilli-Dugu-Song agrodiversity project	Y	Y	Y					
Rio do Campo watershed, no till	Y	Y	Y	Y			Y	Y
Adarsha watershed	Y	Y		Y	Y		Y	Y
Powerguda watershed	Y	Y	Y	Y	Y	Y	Y	Y
Quesungual slash and mulch agroforestry system	Y	Y	Y	Y	Y	Y	Y	Y
Farmer networks in North-East Thailand	Y	Y	Y	Y	Y	Y	Y	Y

capacity. In this sense, social thresholds have been breached. Thus the persistent food insecurity has become an alternative state with significant consequences; the damage that results to human and livestock health with each food crisis is long-lasting.

A resilience approach also helps to explain how the external (government and NGO) response to food insecurity in these pastoralist communities has failed in the past. For most of the 1990s, the early warning system and food aid response system were limited by poor information, institutional weaknesses and lack of trust and coordination among the NGOs and donors working on relief responses. This can be explained as a problem of cross-level and cross-scale interactions among actors. The responses are most often food aid, a response of last resort that is more focused on saving lives rather than livelihoods. Numerous evaluations continue to highlight the lack of interventions that protect livelihoods by protecting livestock (Morton, 2001; Aklilu and Wekesa, 2002). Responses aimed at supporting livelihoods are still plagued by lack of long-term contingency planning and regular veterinary services, combined with insufficient infrastructure (i.e. issues related to chronic poverty and marginalization). The growing number of sedentarized pastoralists points to the need for greater support for alternative livelihoods.

This example illustrates that resilient states may be undesirable; the concept of managing for resilience may be useful. As Walker et al (2004) explain it, managers can erode the resilience of an undesirable system deliberately, transforming it to a different state. Developing the capacity to bring about such transformations within food systems is an open area for research. The idealized concepts of adaptive management and governance will be extremely challenging to achieve in food systems. Recent analyses of efforts to promote transformation (Lebel et al, 2006; Olsson et al, 2006) propose the importance of adaptive governance for managing system change. Olsson et al suggest that transformation requires a careful phase of preparation, which relies heavily on building knowledge and social networks. Lebel et al, recognizing the power dimensions in who decides when to transform a system, suggest that adaptive governance requires participation, polycentric institutions to manage across scales and levels, and strong accountability of authorities. In food systems, this will require significant engagement with the private sector actors controlling distribution and retail mechanisms.

Challenges to adapting food systems to GEC

Adaptations beyond food production

A limitation of many examples of efforts to build more resilient food systems is their overemphasis placed on food production, neglecting other elements of food systems vital in achieving food security. Some examples covering more of the food system have been outlined by IFPRI's Millions Fed project. Cunningham (2009) documents the replacement of ad hoc milk production, marketing and selling with a more organized system from production to consumption in India from the 1970s to 1990s. This was achieved through technological and infrastructural advances, as well as the creation of a national milk grid incorporating village dairy cooperatives, district and regional cooperative unions and state marketing federations. By the early 21st century 11 million households were employed by the cooperatives, milk production increased from 42 to 67 million litres per day and per capita consumption of milk by dairy farmers increased from 290 to 339g between 1988 and 1989 and 1995 and 1996 (Cunningham, 2009). This development model is now being followed in other countries and for other commodities.

Kaminski et al (2009) highlight the impact of cotton reforms in Burkina Faso (an easing in of market liberalization while strengthening local institutions and building capacity) on rural livelihoods (taking the impacts past solely production). Although cotton is not a food crop, increased employment and earnings in this sector leading to higher farm incomes has improved food security as by 2006; 70 per cent of those involved in cotton production reported to be food secure, compared to 40–45 per cent in 1996 (Kaminski et al, 2009).

Many more studies of this type need to be drawn into the current discourse on adaptation to enhance food security in the face of GEC. This is particularly true for traditional research on agricultural development and food insecurity.

Feedbacks and maladaptation

Political economic analyses of global food systems (such as McMichael, 2007) point to the deeply rooted nature of inequity in food security status, resource control and profits between rich and poor countries; a result of colonial history as well as existing trade policies and barriers. This is exacerbated by disparity in the policy and governance capabilities between countries (Birdsall, 2008). GEC will in many cases increase these asymmetries. Recognizing the interconnected nature of food systems across scales and levels means that adaptation cannot be treated as an isolated change in agronomic technology or local practices. Loss of key ecosystem services (especially climate regulation) have already resulted from failure to manage cross-scale and cross-level interactions such as greenhouse gas emissions, nutrient loading in water systems, population pressure and soil erosion (Folke, 2006; Walker et al, 2006). Availability and affordability of food between locations are becomingly increasingly interconnected. If the discussion of adaptation is broadened to the whole food system, the high potential for maladaptation emerges as a primary concern. One example is described in Box 8.3.

Much GEC results from the interactions in coupled social–ecological systems, often as the result of unanticipated feedbacks. Managing adaptation to enhance ecosystem services must consider these feedbacks, which implies the need to constantly assess the links between human well-being (food security, health and natural resource-based livelihoods) and ecosystem services (De Fries et al, 2004; Carpenter et al, 2009). To what extent will food security be enhanced at the expense of water quality, biodiversity and climate regulation, and at what point will climate change and modified nutrient cycles overwhelm the ability of food systems to provide food security are major questions, taken up in Chapters 19 and 20.

Gichuki and Molden (2008) examine the cross-scale externalities which can arise from adaptation projects such as the cases analysed in the 'Bright Spots' projects mentioned above. They show that externalities can be both positive and negative. For example, in the Athi River basin in Kenya's Machakos district, while soil and water conservation led to reduced reservoir

> **Box 8.3** *Agricultural subsidies and maladaptation in the Mississippi basin*
>
> The Mississippi basin extends across 48 per cent of US land, housing a $100 billion agricultural economy (Malakoff, 1998), including the cultural heartland of US farming, the American corn belt. For years this area has contributed to keeping people and livestock fed across the country and the globe. Yet the food systems stemming from these areas, largely shaped by federal agricultural policies, are having a number of unforeseen negative outcomes with widespread consequences.
>
> Agriculture in the Mississippi basin has been profoundly shaped by US agricultural policies, particularly the US farm bill, which outlines governmental spending on commodity programmes, including subsidies for the production of a number of different crops. Supported by agricultural subsidies, farmers in the Mississippi basin have been steadily increasing yields. This has been achieved by more intensive farm management, techniques such as increased inputs of pesticides and fertilizer, higher planting densities and increased irrigation (Foley et al, 2004). Yet with these practices comes an unforeseen outcome: the extensive environmental degradation that is increasingly becoming evident in the region and beyond. Soil erosion, destruction of habitats and declines in biodiversity have all been associated with the subsidized industrial system (Eubanks II, 2009). In addition, industrial agricultural practices impact on important nutrient cycles inside and outside of the region (carbon, nitrogen and phosphorous). As explained in Chapter 1, the increased addition of nutrients in food production is having negative effects on aquatic ecosystems. In an extreme example of this, nitrate runoff from farmland in the Mississippi basin is resulting in the emergence of an aquatic 'dead zone' in the Gulf of Mexico, where hypoxia is causing massive fisheries decline (Turner and Rabalais, 1991, 1994; Rabalais et al, 1996).

sedimentation, increased evapotranspiration led to reduced catchment water yield. They also cite the example from Barbier (2003), where upstream benefits from irrigation in the Hadejia-Jama'are floodplain, northern Nigeria, were associated with economic losses as a result of reduced flows into the floodplains. The tradeoffs associated with the positive upstream impacts included increased cost of domestic water, and economic losses to vegetable farmers due to drops in groundwater.

The negative feedbacks within food systems to both food security as well as ecosystems suggest that to adapt them, they may need to be transformed. Resilience theory proposes that as disturbance offers opportunities for innovation or transformation, these opportunities can be managed to bring about more desirable system configurations (Walker et al, 2004). For example, rangelands can be deliberately managed to contain more desirable grassy species for cattle, while bad management can result in those same rangelands 'degrading' into a largely shrubby state. Thus the challenge is to manage resilience with intent, which is chiefly a factor of the adaptive capacity within the system. It is up to social actors, the system managers, to create a system with a new structure if the current one is undesirable. In complex food systems, it is unclear who the key actors are and how power is shared among them; as discussed in Chapter 18, the emergence of private sector and non-state actors is rapidly altering food system governance structures. Deliberation and negotiation about 'desirable' adaptation strategies will be critical in determining whether the outcomes of such adaptation increase or decrease food security and ecosystem services. Sundkvist et al (2005) build upon these ideas in their critique of what they see as unsustainable dependence of western European food systems on imported foods. This distancing between consumers and producers allows feedbacks, especially to ecosystem services, to be ignored. The solution they propose is to either shorten this distance, or to introduce institutions to 'translate' or manage the feedbacks so that consumers and producers are forced to confront them. However, this may face a number of deeply ingrained social barriers (see Table 8.3).

An emerging concern is with the many limits or barriers to adaptation. A barrier is an obstacle to further adaptation which can be overcome; a limit is an absolute barrier which cannot be surpassed (Hulme et al, 2007). Both concepts suggest measurable thresholds, but in social systems such thresholds are socially, culturally and politically determined. Actors will often not agree on whether a threshold has been surpassed, nor if it matters enough to warrant adaptation (Adger et al, 2009). This may often happen because individuals have differing incentives, as well as immediate short-term pressures that may override longer-term concerns. For example, although it may be clear that an ecosystem service such as water quality is threatened, the economic benefits of using pesticides to enhance crop yields are often the main criteria for farmer decisions. Although overconsumption of highly processed convenience foods puts people at risk of obesity or diabetes, their low price and convenience often override these risks. Thus overcoming the social and political barriers to adaptation is a challenging area of future research.

Institutional and policy limits

As also suggested through the discussion in this chapter, many of the limits to adaptation in food systems lie squarely in the institutional and policy domain. These include the following limits:

- Institutional capacity to plan pro-actively – examples here draw from the food aid and disaster response communities, demonstrating that lack of coordination and trust among local, national and international agencies and governments results in chronic failure to have more forward-looking relief responses, as discussed in the example from northern Kenya.
- Inability to formulate integrated food policy – examples here draw on failures of 'joined up food policy' in the UK and USA, so there are still crop subsidy programmes at odds with nutritional recommendations; other policies emphasize intensive production at odds with ecosystem service degradation. Again the behaviour and influence of large marketing and retailing companies on food sourcing, processing and marketing pose challenges to the development and implementation of coherent food policy.

Table 8.3 *Potential social barriers to food system adaptation*

Values	Adaptation is inherently linked to values. For example, are people anthropocentric or ecocentric, and to what extent are they discounting or valuing the future? How much do they value social justice or environmental justice? Consumers often make the choice between fair trade and organic or low food miles and nutritious foods with only limited knowledge of the complexities of food systems, and generalizations as to what is 'good'. The differences in values throughout civil society add to the complexity of food system governance.
Desires for prestige	Food is a central part to everyone's lives and associations are often drawn between what people eat and social class. The desire for prestige, or social status, can shape food consumption; for example, the association between meat-eating and wealth in China or the link between organic food consumption and status in California ('yuppie chow' as Guthman (2003) calls it). Often what is driven by a desire for social standing distracts from needs for more socially and environmentally just food systems.
Societal norms	People's behaviour is often confined by social norms or cultural habits to which many feel a need to conform. These norms can stand in opposition to the creation of more adapted food systems. For example, the tradition of eating meat with most meals in the UK ('meat and two veg' society) could be seen as a major barrier to reducing meat consumption and associated environmental ills. Associated to this is the extent to which behaviour fits in with wider culture. For example, many would argue that the culture of food waste (a third of the food bought in the UK is thrown away) is linked to a wider culture of waste and undervaluation of resources.
Not knowing how to leverage food systems	It is hard for consumers (and to a lesser extent farmers and policy-makers) to know where the leverage points through which to influence change in food systems are located, and which are the pressure points which would have most impact on mitigating the impacts of food systems and adapting current food systems to future GEC. Most consumers just feel that they can influence through consumption, which many would argue is a product of present neoliberal ideas that the markets can solve all.
Perceptions of risk to future climate change	Much of the discourse surrounding alternative food movements in Europe and the USA centres around mitigation, reducing the impacts of food systems, rather than focusing on the need for food system adaptation (which is often seen as something that will need to take place elsewhere). As such, perceptions of risk to future climate change may be underestimated. This takes away some of the urgency of change in food systems, making it seem more like a charitable thing to be done.

- Limits to market-based or consumerist solutions in modern food systems – examples here draw on critiques of ethical trade initiatives and local food projects which, while they are successful in the case of some commodities (coffee, chocolate, fruit) and in some areas (e.g. middle-class communities that support lots of community supported agriculture), fall short of wide-scale food system transformation.
- Inability to reform global trade rules and manage impacts of market failures for food commodities – the main example here is the failure to finalize

the Doha round, and the market difficulties contributing to the 2008 price shocks.

Given these limits, more fundamental transformations of rural livelihood and food systems – for instance, through diversification out of agriculture – are essentially inconceivable, at least for the time being. Despite an abundance of more or less successful short-term and mostly reactive coping strategies to climatic and other environmental changes, there are hardly any incentives that encourage truly innovative experiments. For the vast majority of rural populations, costs associated with such experiments are too high and, lacking sound safety nets, the risk of failure too great. Low diversity in assets and entitlements, a key characteristic of vulnerability in food systems (Ericksen, 2008), prevents resource-poor social actors from developing diversified livelihood portfolios that could be implemented once an opportunity arises. Finally, alternative strategies of accessing food are hampered by inadequate climate information and projections, and a widespread perception that changes in rainfall patterns are a temporary anomaly rather than a lasting reality. These barriers undermine awareness-building that a more radical change in the production, distribution and consumption of food may eventually become indispensable.

Politics: A barrier or a limit?

Adaptation is a fundamentally political process. First, cross-scale and cross-level interactions in complex food systems imply that although vulnerability may be experienced locally, the causes of vulnerability may be generated by processes or decisions located elsewhere, that is either in a different locality or at a national or global level (Ziervogel et al, 2006; O'Brien and Leichenko, 2008; Eriksen and Silva, 2009). Such processes may involve land tenure change, conflict, policy reforms, demographic change or economic restructuring, as observed for southern Africa (Eriksen and Watson, 2009). Hence, if adaptation involves the reduction of vulnerability, adaptation must mean addressing some of these processes. Confronting global and national processes are often a matter of national or even international politics. If vulnerability among a particular group is generated in part by economic liberalization, such as among farmers in Mexico (Eakin, 2006), addressing this cause of vulnerability means challenging national policy choices and possibly altering the balance of emphasis from goals of free trade and commercial production to priorities of poverty reduction and sustainable smallholder agriculture. The policy goal of restructuring agriculture towards fewer larger farms may contribute to undermining ecological diversity and climate-adapted ways of production, for example. Since adaptation is not only a local measure, but part and parcel of societal transformations, contesting current national development may be part of adaptation. The ability to negotiate local adaptation interests in the face of economic liberalization and national development priorities may differ both between countries, such as Norway and Mozambique, and within countries depending on the social and political context (Eriksen, 2009).

Demographic change, economic liberalization and conflict exemplify processes that may be international in scope; hence adaptation also involves international politics. Addressing the way economic liberalization generates vulnerability, for example, impinges on rules set out in international trade agreements. Where incorporating climate and vulnerability considerations may contradict principles of free trade (such as supporting local niche production in the face of competition from internationally traded foods), the extent to which the principle of free trade is to override other concerns may need to be renegotiated.

Second, and related to the above, adaptation is determined by the multiple and competing objectives of different actors involved in food systems, and the ability of different actors to negotiate these objectives. Since causes of vulnerability in one place or among one group may be generated by actions elsewhere or by another group, there are often conflicts of interest regarding adaptation. What may be desirable actions in terms of adaptation in one place may disadvantage a different group or area. This means that some adaptation efforts may be constrained by strong vested interests by the powerful, or that other adaptation efforts may actually increase the vulnerability of other groups. Expanding irrigation to reduce reliance on rain-fed crops in one area may well reduce access to water of people in another area, and more fundamentally affect land rights. Protection of a watershed may deprive groups of drought grazing rights critical for managing climatic variability and change. A given adaptation is not necessarily 'benign' for everyone (Eriksen and O'Brien, 2007). Eriksen and Lind (2009) showed that adaptation is instead strongly contested, both in local political processes as well as in national processes. Reduction of violent conflict may be the most effective way of reducing vulnerability of many populations to drought; however, there are often political or economic reasons for conflict and insecurity in the first place and hence interests that are threatened by efforts to promote peace. Adaptation, in terms of trade and new income sources made possible by peaceful circumstances, may threaten existing power and economic configurations. Hence there are observations that conflict is used deliberately to resist shifting of power bases or hinder local adaptation mechanisms.

It can never be assumed that adaptation is apolitical or divorced from the social and political context. Addressing the causes of vulnerability necessarily means reconfiguring social and political processes, or at the very least considering how particular practical measures (such as altering agricultural activities in response to climate change) may be biased towards certain interests over others, undermining the interests of some groups, shifting power balances, and potentially affecting the way that conflicting interests are negotiated. It is particularly important to consider whether the interests of those most vulnerable are safeguarded – in terms of resource rights, political position and ability to continue and strengthen current livelihood and adaptation strategies – through interventions, decisions or policies aimed at strengthening adaptation.

Biophysical limits

As discussed in Chapters 1 and 7, the need for increased food production is urgent. This raises the issue of technical limits to agricultural production. The consensus is that the genetic productivity limits of livestock or crop species have not yet been reached, although obviously, the technical limits vary by geographical and ecological context. In already intensive agricultural production systems, discussions about limits are couched in terms of minimizing ecosystem feedbacks (Tilman et al, 2002; Godfray et al, 2010). Chapter 19 examines these issues in more detail.

In many parts of the developing world, and particularly in sub-Saharan Africa, yield gaps in cereal and livestock production are still considerable. There is no doubt that increased productivity can be achieved, but targeting of technological interventions needs to be much more nuanced and focused. In highly intensive systems, there is much scope for increasing efficiencies in the use of resources and minimizing deleterious impacts on soil, water and air quality, for example (Steinfeld et al, 2006). There are also many opportunities to sustainably increase production and household incomes in both the more extensive and the more market-orientated mixed systems. For instance, from a livestock perspective, increased reliance on the wide range of adaptive traits associated with local livestock breeds could contribute substantially to the resilience of production systems (Seré et al, 2007). This highlights the need for characterizing, conserving and utilizing the genetic resources of the developing world.

While there will be opportunities for some, many others will have to deal with considerable threats to their existing livelihood systems and food security. There is plenty of evidence to show that it is the resource-poor who are least able to adapt, particularly those households in marginal environments with few options. As land becomes increasingly risky for cropping, such households may need to consider livelihood transitions; for example, by increasing reliance on livestock rather than crops (Jones and Thornton, 2009) or moving to alternative livelihood strategies. Even in some of the more productive areas, pressures on natural resources and/or a changing climate are bound to bring about livelihood changes: the future of coffee production in East Africa is a case in point.

Technology will have a crucial role in helping food system actors to adapt in the future. There are plenty of off-the-shelf options that may be appropriate in different situations, but technical 'silver bullets' are no more going to apply in the future than they have in the past. It will be more like 'silver grapeshot': lots of different options, some of which may apply in some places, but many which will not (the recent review on conservation agriculture by Giller et al (2009) is a good example of this). It is also the case that much technology is developed by private businesses, who to date have not been sufficiently integrated into food security development strategies and pathways.

Conclusion: Managing under uncertainty

The take-home message from this chapter is that adaptation in complex food systems requires a multisectoral, multilevel approach that deals with issues far beyond improving methods for agricultural production. Furthermore, adaptation will not only be an issue for vulnerable smallholder agriculturalists or food insecure households and communities. One inherent characteristic of complex systems is the uncertainty about what future conditions and trends will be. This uncertainty is unlikely to decrease in the near future; the challenge is to manage in spite of this uncertainty – but this is an area where little work has been done, beyond the idea of 'robust' decision strategies. In conclusion, food systems are dynamic and constantly changing – and any adaptation choice made today will provide the context for future choices. Adaptation is thus a moving target, not something that can be solved with a one off 'quick fix'. These challenges are discussed in Parts III, IV and V.

References

Adger, W. N., S. Agrawala, M. M. Q. Mirza, C. Conde, K. O'Brien, J. Pulhin, R. Pulwarty, B. Smit and K. Takahashi (2007) Assessment of adaptation practices, options, constraints and capacity. In Parry, M. L., O. F. Canziani, J. P. Palutikof, P. J. van der Linden and C. E. Hanson (eds) *Climate Change 2007: Impacts, Adaptation and Vulnerability. Contribution of Working Group II to the Fourth Assessment Report of the IPCC*, Cambridge, UK, Cambridge University Press

Adger, W. N., S. Dessai, M. Goulden, M. Hulme, I. Lorenzoni, D. R. Nelson, L. O. Naess, J. Wolf and A. Wreford (2009) 'Are there social limits to adaptation?', *Climatic Change*, 93, 335–54

Aklilu, Y. and M. Wekesa (2002) *Drought, Livestock and Livelihoods: Lessons from the 1999–2001 Emergency Response in the Pastoral Sector in Kenya*, London, Overseas Development Institute

Archer, E. R. M. (2003) 'Identifying underserved end-user groups in the provision of climate information', *Bulletin of the American Meteorological Society*, 84, 11, 1525–32

Armitage, D., M. Marschke and R. Plummer (2008) 'Adaptive co-management and the paradox of learning', *Global Environmental Change*, 18, 1, 86–98

Bapna, M., H. McGray, G. Mock and L. Withey (2009) Enabling adaptation: priorities for supporting the rural poor in a changing climate. *WRI Issue Brief*, Washington, DC, World Resources Institute

Barbier, E. B. (2003) 'Upstream dams and downstream water allocation: the case of the Hadejia-Jama'are floodplain, northern-Nigeria', *Water Resources Research*, 39, 11, 1311–19

Barrett, C. B. and D. G. Maxwell (2005) *Food Aid after Fifty Years: Recasting its Role*, New York, Routledge

Becker, L. C. (2000) 'Garden money buys grain: food procurement patterns in a Malian village', *Human Ecology*, 28, 2, 219–49

Berkes, F. and D. Jolly (2001) 'Adapting to climate change: Social-ecological resilience in a Canadian Western Arctic community', *Conservation Ecology*, 5, 2

Birdsall, N. (2008) Addressing inherent asymmetries of globalization. In Von Braun, J. and E. Diaz-Bonilla (eds) *Globalization of Food and Agriculture and the Poor*, Oxford, Oxford University Press

Bossio, D., A. Noble, N. Aloysius, J. Pretty and F. Penning de Vries (2008) Ecosystem benefits of 'bright' spots. In Bossio, D. and K. Geheb (eds) *Conserving Land, Protecting Water*, Wallingford, CAB International

Cane, M. A., G. Eshel and R. W. Buckland (1994) 'Forecasting Zimbabwean maize yield using eastern equatorial Pacific sea surface temperature', *Nature*, 370, 6486, 204–5

Carpenter, S., B. Walker, J. M. Anderies and N. Abel (2001) 'From metaphor to measurement: resilience of what to what?', *Ecosystems*, 4, 765–81

Carpenter, S. R., H. A. Mooney, J. Agard, D. Capistrano, R. S. De Fries, S. Díaz, T. Dietz, A. K. Duraiappah, A. Oteng-Yeboah, H. M. Pereira, C. Perrings, W. V. Reid, J. Sarukhan, R. J. Scholes and A. Whyte (2009) 'Science for managing ecosystem services: Beyond the Millennium Ecosystem Assessment', *Proceedings of the National Academy of Sciences of the United States of America*, 106, 5, 1305–12

Cash, D. W., J. C. Borck and A. G. Patt (2006) 'Countering the loading-dock approach to linking science and decision making: Comparative analysis of El Nino/Southern Oscillation (ENSO) Forecasting Systems', *Science Technology Human Values*, 31, 4, 465–94

CGIAR (2009) *Global Climate Change: Can Agriculture Cope?*, Washington, DC, CGIAR

Cromwell, E. and R. Slater (2004) *Food Security and Social Protection*, London, ODI/DfID

Cunningham, K. (2009) Connecting the milk grid: smallholder dairy in India. In Pandya-Lorch, D. J. S. a. R. (ed) *Millions Fed: Proven Successes in Agricultural Development*, Washington, International Food Policy Research Institute

De Fries, R. S., J. A. Foley and G. P. Asner (2004) 'Land-use choices: balancing human needs and ecosystem function', *Frontiers in Ecology and the Environment*, 2, 5, 249–57

Dercon, S. (2005) 'Risk, poverty and vulnerability in Africa', *Journal of African Economics*, 14, 4, 483–88

Devereux, S. (2001) 'Livelihood insecurity and social protection: a re-emerging issue in rural development', *Development Policy Review*, 19, 507–19

Dorosh, P. A. (2009) 'Price stabilization, international trade and national cereal stocks: world price shocks and policy response in South Asia', *Food Security*, 1, 137–49

Dorward, A., N. Poole, J. Morrison, J. Kydd and I. Urey (2003) 'Markets, institutions and technology: Missing links in livelihoods Analysis', *Development Policy Review*, 21, 3, 319–32

Eakin, H. (2005) 'Institutional change, climate risk, and rural vulnerability: Case studies from Central Mexico', *World Development*, 33, 11, 1923–38

Eakin, H. (2006) *Weathering Risk in Rural Mexico: Climatic, Institutional and Economic Change*, Tucson, The University of Arizona Press

Eakin, H., E. L. Tompkins, D. R. Nelson and J. M. Anderies (2009) Hidden costs and disparate uncertainties: tradeoffs in approaches to climate policy. In Adger, W. N., I. Lorenzoni and K. L. O'Brien (eds) *Adapting to Climate Change: Thresholds, Values, Governance*, Cambridge, Cambridge University Press

Ellis, F. (2000) *Rural Livelihoods and Diversity in Developing Countries*, Oxford, Oxford University Press

Ellis, F. and H. A. Freeman (2004) 'Rural Livelihoods and Poverty Reduction Strategies in Four African Countries', *Journal of Development Studies*, 40, 4, 1–30

Ericksen, P. J. (2008) 'What is the vulnerability of a food system to global environmental change?', *Ecology and Society*, 13, 2

Eriksen, S. (2009) Understanding global change and multiple stressors across geographic contexts: Human security in two sites in Norway and Mozambique. *Global Environmental Change and Human Security Conference*, Oslo.

Eriksen, S. and K. O'Brien (2007) 'Vulnerability, poverty and the need for sustainable adaptation measures', *Climate Policy*, 7, 337–52

Eriksen, S. and J. Lind (2009) 'Adaptation as a political process: Adjusting to drought and conflict in Kenya's drylands', *Environmental Management*, 43, 5, 817–35

Eriksen, S. and J. A. Silva (2009) 'The vulnerability context of a savanna area in Mozambique: household drought coping strategies and responses to economic change', *Environmental Science and Policy,* 12, 33–52

Eriksen, S. E. H. and H. K. Watson (2009) 'The dynamic context of southern African savannas: investigating emerging threats and opportunities to sustainability', *Environmental Science & Policy,* 12, 1, 5–22

Eriksen, S. H., K. Brown and P. M. Kelly (2005) 'The dynamics of vulnerability: locating coping strategies in Kenya and Tanzania', *Geographical Journal,* 171, 4, 287–305

Eubanks II, W. S. (2009) *The Sustainable Farm Bill: A Proposal for Permanent Environmental Change,* Environmental Law Reporter, http://ssrn.com/abstract= 1410800 (accessed: 20 November 2009)

Foley, J. A., C. J. Kucharik, T. E. Twine, M. T. Coe and S. D. Donner (2004) Land use, land cover, and climate change across the Mississippi basin: impacts on selected land and water resources. In De Fries, R., G. Asner and R. Houghton (eds) *Ecosystems and Land Use Change,* American Geophysical Union, Geophysical Monograph Series 153

Folke, C. (2006) 'Resilience: the emergence of a perspective for social-ecological systems analyses', *Global Environmental Change,* 16, 3, 253–67

Folke, C., S. Carpenter, B. Walker, M. Scheffer, T. Elmqvist, L. Gunderson and C. S. Holling (2004) 'Regime Shifts, Resilience and Biodiversity in Ecosystems Management', *Annual Review of Ecology, Evolution, and Systematics,* 35, 1, 557–81

Ford, J. D., B. Smit and J. Wandel (2006) 'Vulnerability to climate change in the Arctic: A case study from Arctic Bay, Canada', *Global Environmental Change,* 16, 2, 145–60

Ghana Statistical Service (2003) Core Welfare Indicators Questionnaire (CWIQ) Survey, Ghana, www.statsghana.gov.gh/CoreWelfare.html (accessed: undated)

Gichuki, F. and D. Molden (2008) Bright basins – do many bright spots make a basin shine? In Gaheb, D. B. a. K. (ed) *Conserving Land, Protecting Water,* Wallingford, CAB International

Giller, K. E., E. Witter, M. Corbeels and P. Tittonell (2009) 'Conservation agriculture and smallholder farming in Africa: The heretics' view', *Field Crops Research,* 114, 1, 23–34

Godfray, H. C. J., J. R. Beddington, I. R. Crute, L. Haddad, D. Lawrence, J. F. Muir, J. Pretty, S. Robinson, S. M. Thomas and C. Toulmin (2010) 'Food security: The challenge of feeding 9 billion people', *Science,* 327, 812–18

Gong, X., A. G. Barnston and M. N. Ward (2003) 'The effect of spatial aggregation on the skill of seasonal precipitation forecasts', *Journal of Climate,* 16, 18, 3059–71

Gunderson, L. H. (2003) Adaptive dancing: interactions between social resilience and ecological crises. In Berkes, F., J. Colding and C. Folke (eds) *Navigating Social-Ecological Systems: Building Resilience for Complexity and Change,* Cambridge, Cambridge University Press

Guthman, J. (2003) 'Fast food/organic food: reflexive tastes and the making of "yuppie chow"', *Social & Cultural Geography,* 4, 1

Hansen, J., S. Marx and E. Weber (2004b) *The Role of Climate Perceptions, Expectations, and Forecasts in Farmer Decision Making: The Argentine Pampas and South Florida,* Palisades, New York, International Research Institute for Climate Prediction

Hansen, J., W. Baethgen, D. Osgood, P. Ceccato and R. K. Ngugi (2007) 'Innovations in climate risk management: Protecting and building rural livelihoods in a variable and changing climate', *Journal of Semi-Arid Tropical Agricultural Research,* 4, 1

Hansen, J. W. (2002) 'Realizing the potential benefits of climate prediction to agriculture: issues, approaches, challenges', *Agricultural Systems*, 74, 3, 309–30

Hansen, J. W., A. Potgieter and M. K. Tippett (2004a) 'Using a general circulation model to forecast regional wheat yields in northeast Australia', *Agricultural and Forest Meteorology*, 127, 1–2, 77–92

Hansen, J. W., A. Challinor, A. Ines, T. Wheeler and V. Moron (2006) 'Translating climate forecasts into agricultural terms: advances and challenges', *Climate Research*, 33, 1, 27–41

Haq, R. A. H. M., P. Ghosh and A. M. Islam (2005) Wise use of wetland for sustainable livelihood through participatory approach: A case of adapting to climate change. *Asian Wetland Symposium 2005*, Orissa, India, Bhubaneswar

Hossain, M. (2009) Pumping up production: shallow tubewells and rice in Bangladesh. In Spielman, D. J. and R. Pandya-Lorch (eds) *Millions Fed: Proven Successes in Agricultural Development*, Washington, DC, International Food Policy Research Institute

Huda, A. K. S., R. Selvaraju, T. N. Balasubramanian, V. Geethalakshmi, D. A. George and J. F. Clewett (2004) Experiences of using seasonal climate information with farmers in Tamil Nadu, India. In Huda, A. K. S. and R. G. Packham (eds) *Using Seasonal Climate Forecasting in Agriculture: A Participatory Decision-Making Approach*, Canberra, ACIAR Technical Reports, Australian Centre for International Agricultural Research

Hulme, M., W. N. Adger, S. Dessai, M. Goulden, I. Lorenzoni, D. Nelson, L. Naess, J. Wolf and A. Wreford (2007) Limits and barriers to adaptation: four propositions. *Tyndall Briefing Note*, Norwich, Tyndall Centre for Climate Change Research

IFPRI (2008) *Physical and Virtual Global Food Reserves to Protect the Poor and Prevent Market Failure*, Washington, DC, IFPRI

Ingram, K. T., M. C. Roncoli and P. H. Kirshen (2002) 'Opportunities and constraints for farmers of west Africa to use seasonal precipitation forecasts with Burkina Faso as a case study', *Agricultural Systems*, 74, 3, 331–49

Jayne, T. S., J. Govereh, A. Mwanaumo, J. K. Nyoro and A. Chapoto (2002) 'False promise or false premise? The experience of food and input market reform in eastern and southern Africa', *World Development*, 30, 11, 1967–85

Jones, P. G. and P. K. Thornton (2009) 'Croppers to livestock keepers: Livelihood transitions to 2050 in Africa due to climate change', *Environmental Science & Policy*, 12, 4, 427–37

Kaminski, J., D. Headey and T. Bernard (2009) Navigating through reforms: Cotton reforms in Burkina Faso. In Spielman, D. J. and R. Pandya-Lorch (eds) *Millions Fed: Proven Successes in Agricultural Development*, Washington, DC, International Food Policy Research Institute

Leach, M. and R. Mearns (eds) (1996) *The Lie of the Land: Challenging Received Wisdom on the African Environment (African Issues Series)*, The International African Institute in association with James Currey, Oxford, and Heinemann, Portsmouth

Lebel, L., J. M. Anderies, B. Campbell, C. Folke, S. Hatfield-Dodds, T. P. Hughes and J. Wilson (2006) 'Governance and the capacity to manage resilience in regional social-ecological Systems', *Ecology and Society*, 11

Li, J., Y. Xin and L. Yuan (2009) Pushing the yield frontier: Hybrid rice in China. In Spielman, D. J. and R. Pandya-Lorch (eds) *Millions Fed: Proven Successes in Agricultural Development*, Washington, DC, International Food Policy Research Institute

Luginaah, I., T. Weis, S. Galaa, M. K. Nkrumak, R. Benzer-Kerr and D. Bagah (2009) Environment, migration and food security in the Upper West Region of Ghana. In Luginaah, I. N. and E. K. Yanful (eds) *Environment and Health in Sub-Saharan Africa: Managing an Emerging Crisis*, Berlin, Springer

Lustig, N. (2009) Coping with rising food prices: Policy dilemmas in the developing world. *Working Papers*, Washington, DC, Center for Global Development

Malakoff, D. (1998) 'Coastal ecology: Death by suffocation in the Gulf of Mexico', *Science*, 281, 5374, 190–92

Marx, S. M., E. U. Weber, B. S. Orlove, A. Leiserowitz, D. H. Krantz, C. Roncoli and J. Phillips (2007) 'Communication and mental processes: Experiential and analytic processing of uncertain climate information', *Global Environmental Change*, 17, 1, 47–58

Maxwell, D. (2002) *The International Humanitarian Community and Famine Prevention: Lessons Learned from the Ethiopian Crisis of 1999–2000, Ending Famine in the 21st Century*, Brighton, IDS, University of Sussex

McMichael, P. (2007) Feeding the world: agriculture, development and ecology. In Leys, C. and L. Panitch (eds) *Socialist Register 2007: The Ecological Challenge*, New York, Monthly Review Press

McPhaden, M. J., A. J. Busalacchi, R. Cheney, J.-R. Donguy, K. S. Gage, D. Halpern, M. Ji, P. Julian, G. Meyers, G. T. Mitchum, P. P. Niiler, J. Picaut, R. W. Reynolds, N. Smith and K. Takeuchi (1998) 'The tropical ocean-global atmosphere observing system: A decade of progress', *Journal of Geophysical Research*, 103, C7, 14169–240

Meinke, H., R. Nelson, P. Kokic, R. Stone, R. Selvaraju and W. Baethgen (2006) 'Actionable climate knowledge: from analysis to synthesis', *Climate Research*, 33, 1, 101–10

Moron, V., A. W. Robertson and M. N. Ward (2006) 'Seasonal predictability and spatial coherence of rainfall characteristics in the tropical setting of Senegal', *Monthly Weather Review*, 134, 11, 3248–62

Moron, V., A. W. Robertson, M. N. Ward and P. Camberlin (2007) 'Spatial coherence of tropical rainfall at the regional scale', *Journal of Climate*, 20, 21, 5244–63

Mortimore, M. (2005) 'Dryland development: Success stories from West Africa', *Environment*, 47, 1, 8–21

Mortimore, M. and F. Harris (2004) 'Do farmers' achievements contradict the nutrient depletion scenarios for Africa?', *Land Use Policy*, 22, 1, 43–56

Morton, J. (ed) (2001) *Pastoralism, Drought and Planning: Lessons from Northern Kenya*, Chatham, UK, NRI

Nelson, D. R., W. N. Adger and K. Brown (2007) 'Adaptation to environmental change: Contributions of a resilience framework', *Annual Review of Environment and Resources*, 32, 1, 395–419

Ngugi, R. K. (2002) Climate forecast information: The status, needs and expectations among smallholder agro-pastoralists in Machakos District, Kenya. *IRI Technical Report 02-04*. Palisades, New York, International Research Institute for Climate Prediction, Columbia University

Noble, A. D., D. A. Bossio, F. W. T. Penning de Vries, J. Pretty and T. M. Thiyagarajan (2006) Intensifying agricultural sustainability: An analysis of impacts and drivers in the development of 'bright spots'. *Comprehensive Assessment Research Report*, Colombo, Comprehensive Assessment Secretariat

Nweke, F. (2009) Resisting viruses and bugs: Cassava in sub-Saharan Africa. In Spielman, D. J. and R. Pandya-Lorch (eds) *Millions Fed: Proven Successes in Agricultural Development*, Washington, DC, International Food Policy Research Institute

O'Brien, K. and R. Leichenko (2008) *Environmental Change and Globalization: Double Exposures,* New York, Oxford University Press

O'Brien, K., L. Sygna, L. O. Naess, R. Kingamkono and B. Hochobeb (2000) Is information enough? User responses to seasonal climate forecasts in Southern Africa. *Report 2000:03,* Oslo, Center for International Climate and Environment Research (CICERO)

Olsson, P., L. H. Gunderson, S. R. Carpenter, P. Ryan, L. Lebel, C. Folke and C. S. Holling (2006) 'Shooting the rapids: Navigating transitions to adaptive governance of social-ecological systems', *Ecology and Society,* 11, 1

Paarlberg, R. L. (2002) Governance and food security in an age of globalization. *Food, Agriculture and the Environment Discussion Papers,* Washington, DC, International Food Policy Research Institute

Padgham, J. (2009) *Agricultural Development under a Changing Climate: Opportunities and Challenges for Adaptation,* Washington, DC, World Bank, Agriculture and Rural Development and Environment Departments

Patt, A., P. Suarez and C. Gwata (2005) 'Effects of seasonal climate forecasts and participatory workshops among subsistence farmers in Zimbabwe', *Proceedings of the National Academy of Sciences of the United States of America,* 102, 35, 12623–28

Phillips, J. G. (2003) Determinants of forecast use among communal farmers in Zimbabwe. In O'Brien, J. J. and C. Vogel (eds) *Coping with Climate Variability: The Use of Seasonal Climate Forecasts in Southern Africa,* Abingdon, Ashgate Publishing

Pray, C. E. and L. Nagarajan (2009) Improving crops for arid lands: pearl millet and sorghum in India. In Spielman, D. J. and R. Pandya-Lorch (eds) *Millions Fed: Proven Successes in Agricultural Development,* Washington, DC, International Food Policy Research Institute

Rabalais, N. N., W. J. Wiseman, R. E. Turner, B. K. Sen Gupta and Q. Dortch (1996) 'Nutrient changes in the Mississippi River and system responses on the adjacent continental shelf', *Estuaries,* 19, 2B, 386–407

Reid, P. and C. Vogel (2006) 'Living and responding to multiple stressors in South Africa – Glimpses from KwaZulu-Natal', *Global Environmental Change,* 16, 2, 195–206

Robertson, A. W., V. Moron and Y. Swarinoto (2008) 'Seasonal predictability of daily rainfall statistics over Indramayu district, Indonesia', *International Journal of Climatology,* 29, 10, 1449–62

Roeder, P. and K. Rich (2009) Conquering the cattle plague: The global effort to eradicate rinderpast. In Spielman, D. J. and R. Pandya-Lorch (eds) *Millions Fed: Proven Successes in Agricultural Development,* Washington, DC, International Food Policy Research Institute

Roncoli, C., C. Jost, P. Kirshen, M. Sanon, K. Ingram, M. Woodin, L. Somé, F. Ouattara, B. Sanfo, C. Sia, P. Yaka and G. Hoogenboom (2009) 'From accessing to assessing forecasts: an end-to-end study of participatory climate forecast dissemination in Burkina Faso (West Africa)', *Climatic Change,* 92, 3, 433–60

Scoones, I. (1996) Range management science and policy: Politics, polemics and pasture in southern Africa. In Leach, M. and R. Mearns (eds) *The Lie of the Land,* Oxford/Portsmouth NH, James Currey/Heinemann for the International African Institute

Seré, C., A. van der Zijpp, G. Persley and E. Rege (2007) Dynamics of livestock production systems, drivers of change and prospects for animal genetic resources. *International Technical Conference on Animal Genetic Resources for Food and Agriculture,* Interlaken, Switzerland, FAO

Spielman, D. J. and R. Pandya-Lorch (2009) *Millions Fed: Proven Successes in Agricultural Development*, Washington, DC, International Food Policy Research Institute

Steinfeld, H., P. Gerber, T. Wassenaar, V. Castel, M. Rosales and C. de Haan (2006) *Livestock's Long Shadow: Environmental Issues and Options*, Rome, FAO

Sun, L., H. Li, M. N. Ward and D. F. Moncunill (2007) 'Climate variability and corn yields in semiarid Ceara, Brazil', *Journal of Applied Meteorology and Climatology*, 46, 2, 226–40

Sundkvist, A., R. Milestad and A. Jansson (2005) 'On the importance of tightening feedback loops for sustainable development of food systems', *Food Policy*, 30, 224–39

Tarhule, A. and P. J. Lamb (2003) 'Climate research and seasonal forecasting for West Africans: Perceptions, dissemination, and use?', *Bulletin of the American Meteorological Society*, 84, 12, 1741–59

Thompson, J. and I. Scoones (2009) 'Addressing the dynamics of agri-food systems: an emerging agenda for social science research', *Environmental Science and Policy*, 12, 386–97

Tilman, D., K. G. Cassman, P. A. Matson, R. Naylor and S. Polasky (2002) 'Agricultural sustainability and intensive production practices', *Nature*, 418, 6898, 671–77

Tschakert, P. (2007) 'Views from the vulnerable: Understanding climatic and other stressors in the Sahel', *Global Environmental Change*, 17, 3–4, 381–96

Tschakert, P. and R. Tutu (2008) Solastalgia: Environmentally-induced distress and migration due to climate change among Africa's poor, Environment, Forced Migration & Social Vulnerability – International Conference, Bonn, Germany

Turner, R. E. and N. N. Rabalais (1991) 'Changes in Mississippi River water quality this century', *BioScience*, 41, 3, 140–47

Turner, R. E. and N. N. Rabalais (1994) 'Coastal eutrophication near the Mississippi river delta', *Nature*, 368, 619–21

Walker, B., C. S. Holling, S. R. Carpenter and A. Kinzig (2004) 'Resilience, adaptability and transformability in social-ecological systems', *Ecology and Society*, 9, 2

Walker, B., L. Gunderson, A. Kinzig, C. Folke, S. Carpenter and L. Schultz (2006) 'A handful of heuristics and some propositions for understanding resilience in social-ecological systems', *Ecology and Society*, 11, 1, art 13

Yosef, S. (2009) Farming the aquatic chicken: Improved tilapia in the Philippines. In Spielman, D. J. and R. Pandya-Lorch (eds) *Millions Fed: Proven Successes in Agricultural Development*, Washington, DC, International Food Policy Research Institute

Ziervogel, G. (2004) 'Targeting seasonal climate forecasts for integration into household level decisions: the case of smallholder farmers in Lesotho', *The Geographical Journal*, 170, 1, 6–21

Ziervogel, G., S. Bharwani and T. E. Downing (2006) 'Adapting to climate variability: Pumpkins, people and policy', *Natural Resources Forum*, 30, 4, 294–305

9
Part II: Main Messages

1 Food systems can be conceptualized as coupled social–ecological systems, in which vulnerability arises from multiple stressors operating across different scales (e.g. temporal, spatial, institutional) and levels on them (e.g. micro to macro). However, this requires a deeper and richer empirical understanding of the relationships between ecosystem services and food security. It also forces society to confront the tradeoffs between key ecosystem services and social welfare outcomes.

2 Assessing the vulnerability of food systems to global environmental change requires combining social framings of vulnerability with ecological analyses of resilience. The former framings are useful for their emphasis on people, and explaining that vulnerability is embedded in social, political and economic processes at multiple levels. The latter framings have contributed an appreciation of vulnerability as a feature of system structure, and the notion that vulnerability can be evaluated as a loss of system function. Loss of resilience can cause a system to flip into a new state, with a different set of functions.

3 The evaluation of food system vulnerability requires not only the assessment of the vulnerability of individual system elements (e.g. the activities, resources and actors involved in the production and processing of food), the movement of food across space and time (e.g. trade, distribution and storage), and among food consumers, but also an understanding of how vulnerabilities are produced, exacerbated or mitigated through the synergistic or antagonistic interaction of the different food system elements and actors across different scales. The capacity of a collection of actors, activities and processes (not necessarily geographically contiguous) to deliver important ecological and social services is then vulnerable, conditioned on distinct drivers of global environmental change.

4 For food systems, answering 'vulnerability to what?' requires not only identifying what stresses and changes cause vulnerability, but also an exploration of the interactions among these stresses and changes, along and across a variety of scales, and the mechanisms by which these interactions generate vulnerability.

5 Another important step in assessing the vulnerability of food systems to global environmental change is to understand the timeframes and pace of emerging drivers, their spatial extent, and their reversibility, in the context of constant interaction. Some change processes may lead to positive opportunities for system reorganization. The challenge is to understand and identify food system processes such that opportunities are exploited and reorganization is effective for human well-being.

6 Managing food systems for resilience is an appealing idea, in principle. In practice, however, there are numerous conceptual and governance issues to resolve. Adaptive capacity is conceptualized very differently in social and ecological literature. The conditions necessary for better 'managing' food systems will be extremely difficult to achieve given the current levels of institutional capacity, political contexts and policy arrangements in food systems.

7 The potential for adaptation strategies to improve food security or ameliorate loss of ecosystem services is not guaranteed. Given the current structure of globalized food systems, and their contribution in exacerbating global environmental change, the potential for 'maladaptation' of food systems is high.

PART III

ENGAGING STAKEHOLDERS

10
The Science–Policy Interface

John Holmes, Gabriele Bammer, John Young,
Miriam Saxl and Beth Stewart

Introduction

As discussed in Parts I and II, global environmental change (GEC) and food security both involve a host of interacting issues. The GEC/food security debate is therefore highly complex, and both science and policy agendas are similarly complex. How, then, can science best help the policy process?

Despite the complexity of the issues involved, the concept of evidence-based policy-making is fairly straightforward: decisions based on a good understanding of how the relevant natural and social aspects of food systems work are more likely to lead to successful outcomes than those which are not; and research can be commissioned to generate new knowledge where currently it is inadequate from the policy-makers' or other stakeholders' point of view. The practice of evidence-based policy-making is, however, anything but straightforward, has rarely lived up to expectations and remains controversial.

As Owens et al (2006) point out, social and political research has identified the complexities of knowledge and policy processes and the shortcomings of 'the linear-rational model in which "sound science" is straightforwardly translated into policy'. They suggest that it is more helpful 'to think in terms of a continuum of influence and utility, ranging from clear and immediate impacts to long-term, subtle processes in which problem definitions and modes of thinking change'. Parsons (2002) goes further and questions whether research can provide objective answers to policy questions and whether policy-making can become a more rational process, and asserts that 'evidence-based policy-making is a missed opportunity for improving government and has only served to make the relationship between knowledge and policy-making in a democratic society more muddled rather than less confused'. Nutley (2003) takes a more moderate line and argues that 'evidence-informed' or 'evidence-aware' policy-making is a sensible aspiration, whereas 'evidence-based' overstates the case.

Various authors have examined the underlying causes of the problems experienced at the science–policy interface. They include relatively straightforward observations that scientists rarely understand policy processes and lack the language to communicate with policy-makers and vice versa (Gregrich, 2003); using science to justify decisions already made (Porter, 1996); the deliberate misinterpretation of research-based evidence by policy-makers (Rosenstock, 2002), or manipulation of evidence by scientists themselves (Lackey, 2006); over-use of research-based evidence to justify political ends by removing scope for value-based political debate (Hoppe, 1999; Weingart, 1999); and deeply rooted cultural and institutional differences between science and policy communities as summarized by Scott et al (2005) and reproduced in Table 10.1. (Some ways to overcome these issues are discussed in Chapter 11, while Chapter 14 identifies benefits to be gained from so doing.) Research by the Overseas Development Institute (ODI) and the Research and Policy in Development (RAPID) Programme points to even greater complexity in developing country contexts (Court et al, 2005) (and see Box 10.1 for an example of the hazards of policy-making that fails to take

---

Box 10.1 *Misguided policy-making in the developing world*

The paradigm promoted by many of the international finance institutions is grounded in neoclassical economics and quantitative modelling, and Omamo and Farrington (2004) highlight how agricultural policy-making in Africa, based upon a neoliberal economic reform agenda, has often proven misguided. They argue that policy so grounded is not reflective of reality and thus is hazardous. They further argue that 'market imperfections are the norm, not the exception' and that the economic reform agenda does not incorporate understandings of how and why these imperfections arise. These could include, for example, failing to incorporate an understanding of the more closed networks of information sharing (which are controlled by the more powerful actors in business), preventing the inclusion of other, often newer, actors. Another example relates to the insurance market which, given unusually high levels of risk faced by actors in African agriculture, is a particularly important buffering instrument. Yet, as Omamo and Farrington highlight, the market works imperfectly with unusually high premiums and some things simply uninsurable against, such as political meddling. Despite knowledge of such imperfections, this is generally not being incorporated into policy-making. And, as they further point out, '(t)he major task of including the poor more fully in markets will continue to be neglected for as long as policy advice remains dominated by unrealistic assumptions concerning the structure, conduct and performance of markets'.

Additionally, again according to Omamo and Farrington (2004), this economic reform agenda often overlooks the 'how' of implementation, particularly constraints to implementation such as weak infrastructure, corruption and power relations. As such, chasing after ideals can blind one to the realities of policy implementation. They argue that policy-making will have to be reconstructed around these observations, making use of more qualitative methodology in addition to, currently dominant, quantitative methodology.

Table 10.1 *Cultural and institutional factors affecting science–policy relationships (from Scott et al, 2005)*

Factor/barrier	Science	Policy
Aims	New and valid knowledge, rigorous research, publication in academic literature, additional research funding.	Respond to political pressures, balance interests and values in policy-making, cope with events, and manage resources.
Timescale	Slow, need to conduct research and publish. Timescale often in years.	Fast, need to react to events, difficult to see years ahead.
People	Specialized, narrow focus, disciplinary, stay in field.	Generalists, broad view, move often. Don't want to get 'bogged down' in detail. May have little understanding of science.
Communication	Mainly to other academics via academic publications and conferences, some respond to consultations, wary of media.	Mainly with other policy-makers, don't read research publications, increasingly consult when making policy, strongly affected by media.
Success measures	Publication, new funding, discovery of new knowledge.	Make problems go away, avoid crisis, marshal evidence to support policies.
Evidence	Based in experimentation, data collection, monitoring and disciplinary traditions.	Not just science but also relating to economic, political and other interests and priorities. What is feasible?
Quality control	Central to science. Conservative; via peer review/publication. Often a narrow focus on the scientific aspects; little role for 'relevance' criteria.	Via policy machine; timetables often too fast for peer review; consultants' research often not reviewed externally. Have to 'make do' with available evidence; think scientists too cautious.
Synthesis	Few incentives – who conducts it? A lot of work; may not lead to new insights; can be hard to publish.	Want scientific consensus and 'strong science'. Set up mechanisms to achieve synthesis.
View of interaction	Wary, jealously guard independence and autonomy, takes time, skills and resources they may not have.	Willing to interact with researchers if useful. But information overload and little time, so need to target efforts.
View of 'relevance'	Want control of research agendas, suspicion of 'applied' research, lower reputational rewards than 'pure' science.	Why else do research? Frustration at irrelevance of much research and vagueness of researchers.
Research problems	Often set within disciplines, can be narrow and/or methodological.	Want a focus on real-world problems – surely these are interesting for research?
Interdisciplinarity	Institutionally difficult; often organized in disciplinary units, few incentives or resources for interdisciplinary working, often less highly regarded and harder to publish. Barriers from different disciplinary world views, assumptions, methods.	A problem focus requires interdisciplinarity. Frustration at the lack on interdisciplinary research. Critical of researchers' policy prescriptions for lacking realism. But also face problems in being interdisciplinary themselves.

such complexity into account), but surprisingly, a greater willingness of policy-makers to seek policy advice from researchers, than in developed countries (Jones et al, 2008).

In this chapter two strands of literature that provide useful thinking about the research–policy nexus are reviewed. In the first the implications of post-normal science and the Cynefin framework (now a part of the suite of tools and frameworks developed by the Cognitive Edge initiative, www.cognitive-edge.com) are examined. These raise important conceptual issues, but are still in early stages in terms of having widely accepted practical outcomes. The second strand involves an examination of a sample of the political science literature on theories of policy-making. It shows how different theories can illuminate different aspects of the complex policy-making process, as well as how they can guide practical action. The chapter concludes by briefly reviewing the emergence of a new category of workers in this space – namely knowledge brokers and boundary organizations.

While this chapter focuses on the issues around linking science and policy-making, the opening up of modern policy-making processes (in, for example, the European Commission) requires consideration of the challenges arising from engaging with a broader range of stakeholders on the associated science. Given the need for different epistemic communities to share their different understandings of 'science' and its implications for the policy decision, further difficulties consequently arise (Wynne, 1992).

The evolving context

The last decade has seen an increasing emphasis by governments on accessing science to inform policy-making and on securing the engagement of scientists with the policy-making process (Holmes and Clark, 2008). This is particularly so in GEC issues given the high priority most governments are now placing on climate change. However, challenges to the effective operation of the science–policy interface are exacerbated as issues become more contested, policy-making boundaries become more blurred and system complexity increases. Consequently, it becomes more difficult for science to deliver robust knowledge in this more challenging context. This, unfortunately, is the direction in which the context for science–policy interactions is travelling, reflecting the increasingly global nature of the issues that must be addressed, the complexity of the interacting factors that cannot be ignored and the competing interests that must be resolved. And science is being asked to inform decision-making at these higher, and more highly integrated, levels, not just to provide information on the workings of component parts. This is now also particularly true for food security issues, for which concepts of food systems (rather than just agriculture) as the underpinning of food security are gaining increased traction (see, for example, Box 10.2 for details of the UK Food Security Assessment).

The implications for science are discussed by Ravetz (1999) in his reflections on the concept of 'post-normal science', described in an earlier paper by Funtowicz and Ravetz (1993). The conditions for post-normal science arise in

---

Box 10.2 *Looking past production: The UK Food Security Assessment*

The UK Food Security Assessment (UKFSA) (Defra, 2009) provides an example of putting into practice the realization that food security is dependent on more than just agricultural production. The UK's Department for Environment, Food and Rural Affairs (Defra) defines UK food security as 'ensuring the availability of, and access to, affordable, safe and nutritious food sufficient for an active lifestyle, for all, at all times', bringing together the concepts of sustainability and food security to develop a wider understanding of their connectivity. Equally, in other aspects of the UKFSA, Defra has developed a more rigorous assessment of current and future food security in the UK, realizing that the complexity of UK food systems must be embraced in comprehensive policy-making. The assessment is divided into six themes (global availability, global resource sustainability, UK availability and access, UK food-chain resilience, household food security, and safety and confidence), taking the debate past solely agricultural production to a discussion of the complex array of factors at play.

Defra has used a balanced scorecard approach to develop a number of indicators for each of these themes, allowing them to assess the state of play now and in the past and predict likely positions in the future (a summary of these assessments can be seen at www.defra.gov.uk/foodrin/security/assessment.htm). For example, the assessment highlights diversity in the UK food supply as fundamental to the theme of UK availability and access, spreading risks and keeping prices competitive. Defra has selected as indicators of the health of this diversity the EU's share of imports into the UK, diversity of the fruit and vegetable supply, EU-wide production capacity, UK production capacity, potential of UK agriculture in extremes, number and diversity of entry ports into the UK, flexibility of ports in handling sea-borne imports and port concentration for non-indigenous foods. Through breaking down diversity in this manner, the assessment has enabled a more complex understanding of the challenges and risks to UK availability and access. This more sophisticated understanding of UK food security sets the food system in the context of political, technological, economic, demographic and environmental change. The assessment, and the way in which it portrays complexity, is now being used to communicate more effectively the complexity of the elements making up UK food security, to assess changes in the different dimensions of food security and to highlight areas in need of further research.

GEC debates (and in the present context, food security), in which 'typically facts are uncertain, values in dispute, stakes high, and decisions urgent'. Such issues in the context of food security may include any of the determinants relating to food availability, food access and/or food utilization (Chapter 2). The previous distinction between 'hard', objective scientific facts and 'soft', subjective value judgements is inverted: hard policy decisions must be made when scientific inputs are irremediably soft. A key guiding principle in this post-normal context is the need to focus on the quality of the decision-making process and that the quality of the science must be defined more comprehensively than in the traditional research setting.

It is important therefore to distinguish between different kinds of problem domain and to develop appropriate strategies to tackle them. In order to help structure thinking about how to address different types of problem domain, the Cynefin framework has been developed, which differentiates between known, knowable, complex and chaotic domains as illustrated in Table 10.2 (Kurtz and Snowden, 2003; Snowden and Boone, 2007; Shaxson, 2009). In the 'known' and 'knowable' domains on the right-hand side of the table, science plays a fairly traditional role exploring cause–effect relationships and generating the knowledge used by the expert, scientific advisors on whom decision-makers rely; much of the traditional GEC and agriculture research is found in these domains. In contrast, in the 'complex' and 'chaotic' domains on the left-hand side of the table (more the 'home' of GEC/food systems research), patterns may emerge but are difficult (if not impossible) to predict, or there may be no visible relationships between cause and effect as in the chaotic domain. The most relevant science in this left-hand side of the table is that of complexity theory (which studies the emergence of patterns due to the interactions of many agents) and the decision model is to create 'probes' to make patterns more visible before action is taken.

Post-normal science and the Cynefin framework are complementary concepts, and both place increased emphasis on accessing a broader range of perspectives as complexity, controversy and uncertainty increase. Ravetz and Funtowicz (1999) call for extended peer communities and a plurality of knowledge. An emphasis on the quality of process requires an open dialogue between stakeholders and recognition that people other than scientists can reformulate problems and identify solutions that may elude the accredited experts (Ravetz, 1999). For issues of food security such stakeholders may include the broader science community, funding agencies, interest groups such as national/regional policy agencies and non-governmental organizations (NGOs), and individuals and communities vulnerable to global environmental change (Brklacich et al, 2007; and see Chapter 11).

Table 10.2 *Structuring policy problems (from Shaxson (2009),*
which was adapted from Hisschemöller and Hoppé (2001),
Snowden and Boone (2007), and Shaxson (2008))

		Is there consensus and clarity in the policy question?...	
		No	Yes
...and clarity about the relevant knowledge?	No	• Unstructured problem • Policy issues are 'complex' • Domain of emergence	• Moderately structured problem • Policy issues are 'knowable' • Domain of experts
	Yes	• Badly structured problem • Policy issues are 'chaotic' • Domain of rapid response	• Well-structured problem • Policy issues are 'known' • Domain of best practice

In relation to the Cynefin framework, Kurtz and Snowden (2003) point to the need to gain multiple perspectives on the nature of the system in order to move towards greater consensus and clarity in the policy realm (right to left in the table). Similarly, Snowden and Boone (2007) indicate that levels and breadth of interaction and communication need to be increased.

ODI's RAPID framework identifies a set of key factors that researchers need to understand, as well as a range of practical actions and how to do them in order to navigate the political processes, develop the most effective evidence and collaborate with other key actors if they wish to maximize their impact on policy and practice (Young and Court, 2004). These are reproduced in Table 10.3.

Political science theory approaches

Political science offers several theories of policy-making, but these are rarely drawn on when thinking about the science–policy interface (Ritter and Bammer, in press). Three theories, which illustrate three different avenues for research to provide support, are briefly reviewed. These are to see policy-making as a technical–rational cycle; a response to pressures exerted by different interest groups; and an entrepreneurial activity involving seizing opportunities as they arise. It is important to see these as complementary rather than competing theories, with each illustrating a different aspect of the policy process. While there may be some limitations in this approach for some countries, the general concepts should be widely applicable.

Policy-making as a technical–rational cycle

When an issue or problem comes onto the policy-makers' agenda, the policy process starts with an examination of existing policy and identification of new options for dealing with the problem; in other words, with policy analysis. This leads to consideration of possible ways of intervening, for example, by changing laws or regulations, by increasing or decreasing certain types of taxation, by introducing an education programme or by changing available services. These interventions are often referred to as policy instruments or levers. Policy-makers will also assess the likely impact of proposed policy changes through consultation with affected parties. In addition to researchers, consultation generally involves various stakeholders, who may be businesses, consumer groups or service providers.

There will also be a process of coordination between relevant government departments. For GEC/food security issues, these could include agriculture, environment and treasury ministries, but the coordination may also be more widespread. For example, in the Caribbean, the importance of foreign exchange earnings for buying food on the world market is very important, so the tourism sector would be an important stakeholder. In countries highly dependent on food aid, stakeholders would include those international institutions and bilateral assistance agencies with a long history of food aid to that country.

Based on all these inputs, a decision will be made and an implementation plan established. There may be a specific evaluation plan, but more commonly

Table 10.3 *The RAPID framework to maximize the impact of research on policy and practice (from Young and Court, 2004)*

	What researchers need to know	What researchers need to do	How to do it
Political context	• Who are the policy-makers? • Is there policy-maker demand for new ideas? • What are the sources/strengths of resistance? • What is the policy-making process? • What are the opportunities and timing for input into formal processes?	• Get to know the policy-makers, their agendas and their constraints. • Identify potential supporters and opponents. • Keep an eye on the horizon and prepare for opportunities in regular policy processes. • Look out for – and react to – unexpected policy windows.	• Work with the policy-makers. • Seek commissions. • Line up research programmes with high profile policy events. • Reserve resources to be able to move quickly to respond to policy windows. • Allow sufficient time and resources. • Align with other stakeholders to influence policy-makers.
Evidence	• What is the current theory? • What are the prevailing narratives? • How divergent is the new evidence? • What sort of evidence will convince policy-makers?	• Establish credibility over the long term. • Provide practical solutions to problems. • Establish legitimacy. • Build a convincing case and present clear policy options. • Package new ideas in familiar theory or narratives. • Communicate effectively.	• Build up programmes of high-quality work. • Action-research and pilot projects to demonstrate benefits of new approaches. • Use participatory approaches to help with legitimacy and implementation. • Clear strategy for communication from the start. • Face-to-face communication.
Links	• Who are the key stakeholders? • What links and networks exist between them? • Who are the intermediaries, and do they have influence? • Whose side are they on?	• Get to know the other stakeholders. • Establish a presence in existing networks. • Build coalitions with like-minded stakeholders. • Build new policy networks.	• Partnerships between researchers, policy-makers and policy end-users. • Identify key networkers and salesmen. • Use informal contacts.
External influences	• Who are main international actors in the policy process? • What influence do they have? • What are their aid priorities? • What are their research priorities and mechanisms? • What are the policies of the donors funding the research?	• Get to know the donors, their priorities and constraints. • Identify potential supporters, key individuals and networks. • Establish credibility. • Keep an eye on donor policy and look out for policy windows.	• Develop extensive background on donor policies. • Orient communications to suit donor priorities and language. • Cooperate with donors and seek commissions. • Contact (regularly) key individuals.

governments rely on the stakeholders to alert them to problems, which may then be attended to in a new cycle of policy-making. An illustration of such a policy cycle is provided by Bridgman and Davis (2004), and reproduced in Figure 10.1.

Policy-making as a response to pressures exerted by different interest groups

Anyone who has been involved or interested in policy-making knows that such a technical–rational view de-emphasizes the critically important political aspects. Another way of looking at policy-making is to see 'public policy as the outcome of the pressures of society's many and diverse interest groups' (Fenna, 2004) (see Box 10.3 for the example of policy surrounding genetically modified foods).

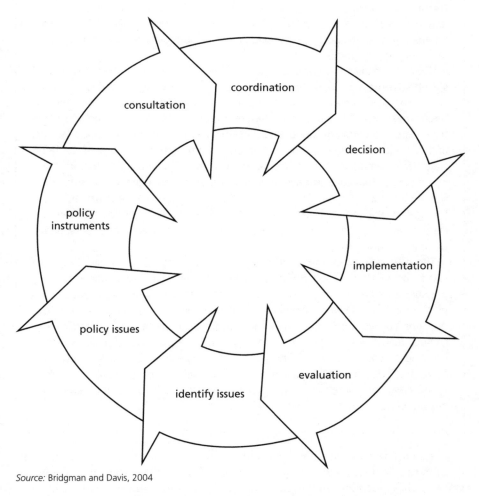

Source: Bridgman and Davis, 2004

Figure 10.1 *Key elements of a 'policy cycle'*

Box 10.3 *EU food policy as a response to stakeholder pressure*

The wake of a number of food safety scandals such as bovine spongiform encephalopathy (BSE) and dioxin has seen European Union (EU) food policy-making post-2000 attempting to become more authoritative and legitimate, trying to create a new food safety and quality framework (as Marsden (2008) discusses). At the same time as trying to satisfy public/consumer pressure, EU regulators are also trying to balance corporate, state and scientific interests. The last ten years has seen the release of a Food Safety White Paper, the establishment of the European Food Safety Authority (EFSA), and the expanding of the remit of the Health and Consumer Protection Directorate (DG SANCO). The previously retailer-led regulation has consequently adapted, now sitting alongside public sector agencies with a stronger emphasis on 'independent' scientific mitigation (DG SANCO) and assessment (EFSA) of food risks. So the 'interrelationships between the State, the private sector and the public are rearranged, and their responsibilities are redistributed' (Marsden, 2008, p196), as more retailer-led assurance schemes appear alongside states taking more responsibility and trying to guarantee minimum standards.

Genetic modification (GM) provides a specific example of the balancing of stakeholder pressure in EU food policy-making. Levidow et al (2005) outline the competing policy agendas in EU GM policy-making arenas, emphasizing the different meanings of precaution for different groups. The different agendas come from agrobiotechnology opponents (including environmentalists, small farmers and cautious consumers), the European Commission itself (aiming to bring together stakeholders in evidence-based policy-making) and the agrobiotechnology industry (highlighting environmental and economic benefits of GM). 2004 saw a new regulatory era for GM in the EU with three new pieces of legislation brought in. Prior to that, according to Marsden (2008), GM's roots in the perception of risk and scepticism of science led to excessive and unworkable piecemeal legislation.

The new regulation, involving a more comprehensive process, allows certain GM feeds, seeds and food in different circumstances. The new regulations follow pressure from the private sector to allow certain products, but the private sector's pivotal role extends past lobbying to implementation of regulation as a result of pressure applied from different interest groups, in particular concerned consumers. Specifically the private sector takes main responsibility for an expanded labelling scheme and traceability systems to verify labels. Consumers now have more (but not complete) information on genetically modified organisms in their foods, creating a more transparent system. As such, the interests of different parties are playing out in the policy-making arena.

Stakeholders are not passive entities waiting for a call for input into policy. Instead, stakeholders are often strong and pro-active advocates seeking to shape policy agendas, highlight policy failures and counteract the influence of other stakeholders with competing interests. A good example is the role of science in shaping the GEC/food security debate in, for instance, the vulnerability of the food system to climate shocks (see Chapter 7). Stakeholders often increase their power to influence policy by forming coalitions with others who seek the same specific outcomes on particular problems. Sometimes policy reflects the overwhelming dominance of one coalition. At other times there are competing coalitions and the resulting policy is a compromise rather than an outright win for any side. In any case, any given policy can be seen as representing the balance between different advocacy coalitions. Looking at policy in this way highlights that policy change occurs in three different ways: when an external perturbation occurs upsetting the balance between existing advocacy coalitions; when a new advocacy coalition gains power; or when an existing powerful advocacy coalition changes its beliefs (Sabatier, 1988).

Policy-making as an entrepreneurial activity involving the seizing of opportunities

Much policy change occurs incrementally through adjustments to existing policies – in an ongoing repetition of the cycle illustrated in Figure 10.1. But from time to time major policy changes occur. Typically the window of opportunity is opened by an unexpected event (Kingdon, 2003). It may be an incident, such as a particularly severe drought, or the publication of statistics showing that a particular problem has become urgent and serious. But such a trigger event on its own is not enough. Two additional sets of circumstances are required. First, the event needs to occur in the right political context, which is influenced by the national mood, the organization of political forces and the position of influential interest groups. For example, major policy change is more likely to occur when the electorate is deeply and obviously concerned about the issue, soon after a government has been elected rather than just before an election and when the influential interest groups are aligned or when the positions of powerful opposing groups can be managed.

The other requisite set of circumstances involves the practicalities of the policy setting process. Considerations here include other problems on the policy agenda and whether the new issue can effectively override them, whether taking action is technically feasible and whether there are any considerations that mitigate against public acceptance of the problem or the proposed solution.

How and to whom to provide research in support of policy formulation

There are three main strategies for providing research support to the policy process, which can be discussed in relation to the three theories outlined above. These are communication, advocacy and engagement.

Much has been written about how to effectively communicate research findings to policy-makers, from providing succinct summaries, to effective use of the media, to providing individual briefings to policy-makers (Brownson et al, 2006). Communication as a strategy is generally most effective for the technical–rational aspects of policy-making. Communication focuses on clearly describing the research findings, their limitations and their applicability to the policy problems. But there is less information in the literature about when it is most advantageous, for example, to target politicians directly rather than bureaucrats or key players in interest groups. There is also little information about the level at which targeting will be most effective – for example, a senior minister, up-and-coming backbencher or opposition politician. This is probably because it is highly variable and dependent on the particularly political situation in a country at a point in time (as well as on personalities). It may be hard to develop general principles about what level to target.

The second strategy – advocacy – is tied to the aspect of policy-making that involves responding to pressure from different interest groups. Researchers who act as advocates attempt to put pressure on policy-makers to ensure a particular outcome. Effective advocacy requires more than good communication skills. Among other things, it requires more attention to be paid to powerful ways of framing the issue, analysis of the strategies that will be used by opponents and how to counteract them, and decisions about whether or not to team with various other interest groups seeking the same outcomes (Chapman, 2001, 2007). Researchers undertaking advocacy may target their efforts privately at a small number of policy-makers or may seek to gain widespread public support through the media. They may seek to work continuously for policy change or may wait for a window of opportunity to open. As outlined earlier, the advocacy process is inevitably political and there are dangers, as well as advantages, for researchers when their work becomes politicized. Researchers may be subjected to personal attacks, funding may be cut and they can even lose their jobs.

A third approach that researchers may seek to employ is engagement. This aims at a problem-solving approach to a policy issue, where researchers and policy-makers work together. This approach also sits most comfortably in a technical–rational framework, but gives researchers a greater role than just providing information. It also puts researchers in a good position to be responsive when a policy window opens. It can be useful to think of engagement as fitting into one of two broad models: policy-maker-initiated and researcher-initiated.

Engagement initiated by policy-makers generally places selected researchers in the role of trusted advisors, with a direct channel of communication. Such an 'insider-confidant' role may be partisan or non-partisan. For example, shadow ministers may seek out researchers to help develop new policy platforms to take to upcoming elections and, if successful, the new government may continue to seek advice on implementing that policy. An example of non-partisan engagement is when researchers are invited to join committees providing policy advice and when those committees continue to function across successive governments. This kind of support is generally provided by more senior researchers at high levels of policy-making. But the necessary relationships may be built and some research support provided at more junior levels. For example, new PhD graduates who join government bureaucracies may seek advice from former supervisors or members of the same student cohort when helping to develop policy in particular areas.

Researcher-initiated engagement generally seeks to involve policy-makers in the design of research projects to ensure that the questions being addressed have policy relevance. The researchers may seek ongoing involvement by inviting policy-makers to be on advisory committees that provide input into the conduct of the research and the analysis and interpretation of the findings, as well as the way in which the findings will be disseminated.

The role of knowledge brokers and boundary organizations

The preceding discussion has pointed to the complexity of the issues that face us – of which GEC and food security is a prime example – and the challenges of enabling research to inform complex policy-making processes. Knowledge brokers (or interpreters) provide a critical part of the response to these challenges. They have an important role to play in bridging between research and policy and in facilitating interactions between research, policy and stakeholder communities (van Kammen et al, 2006; Lomas, 2007). Boundary organizations (Cash, 2001; Guston, 2001; see below and Box 10.4) fulfil a similar bridging role. The need for them arises from the different motivations, drivers and cultures of the research and policy communities and the time pressures faced by researchers and policy-makers.

Knowledge brokers

As Bielak et al (2008) put it:

> Knowledge brokering is usually applied in an attempt to help knowledge exchange work better for the benefit of all parties. It involves bringing people together, helping to build links, identifying gaps and needs and sharing ideas. It also includes assisting groups to understand each others' abilities and needs, and guiding people to sources of knowledge. This may include summarizing and synthesizing research and policy into easily understood formats and translating policy problems into researchable questions.
>
> Knowledge brokers help to ensure relevance: i.e., that research is answering the right questions and that policy stakeholders are engaged in the inquiry process and have some ownership of its outputs. They can also influence the research process by providing opportunities for stakeholders to get involved in a meaningful way.

In relation to complex systems, Brand and Karvonen (2007) use the term 'weaving' to describe knowledge brokering as follows:

> a problem of organized complexity consists of patterns that can be understood – albeit not by a sole individual. Instead, organized complexity necessitates the pooling of understandings and knowledges to develop a shared asset base. It acknowledges that all types of experts are needed, as well as individuals who can

weave these strands of thought together to construct the whole.
This does not imply that the 'weavers' know the whole, *but they
should be able to identify potential linkages and facilitate their
co-discovery.*

Knowledge brokers may be individuals or organizations, for example:

- in-house science advisers, potentially embedded in policy teams but maintaining contacts with the research community;
- dedicated agencies and research institutes: close contact and sustained interaction over time enables these bodies to have a good understanding of the policy process and issues;
- advisory committees whose independence can be helpful when decisions and environmental standards are challenged;
- consultants who can play an important intermediary role with regulated industries and local authorities; and
- professional, industrial and commercial bodies and associations.

Skills and capacities identified as important to be an effective interpreter are as follows (Holmes and Savgard, 2008):

- being a good mediator, able to produce a well-balanced synthesis;
- having a good sense of different arguments;
- having good social skills;
- being open and accessible to experts and with a good network of contacts;
- able to synthesize information into a structure which is meaningful;
- being familiar with the world of research and also aware of policy issues;
- able to put yourself in the shoes of the policy-makers and stakeholders;
- having breadth as well as depth: needing to take a broader view of your research field than is normal and having exposure to the international context; and
- able to see the forest, not just the trees and able to say what things mean in practice.

Concerns have been expressed that there are insufficient people and organizations with these skills, and consequently a more systematic approach should be taken to developing them. But how can a greater number of such science–policy interpreters be nurtured? The specific skill sets of the scientists themselves can be developed through role-playing and capacity-building aimed at written and oral communication development and through regular science-policy dialogues, and/or by developing a cohort of professional mediators.

The UK Department for International Development has recently introduced a number of initiatives to strengthen the use of research-based evidence in its own policy-making, including recruiting 'evidence brokers' to package and synthesize research evidence so it is usable by policy-makers, and to draw their attention to neglected issues and perspectives in policy debates; employing 15 researchers as senior research fellows to work part-time in the

Department; commissioning systematic reviews of the research evidence on a range of topical policy questions; strengthening the communications team in the central research department; and establishing a research communication facility in Africa to promote the wider uptake of the research that the Department funds.

Boundary organizations

As described, the science–policy/knowledge–action interface is widely accepted to be a difficult land to navigate, a site of competing knowledge claims, suffering severe communication problems as a result of the different languages, norms and cultures found on either side of the complex and blurred boundary between 'science' and 'politics'. In recent decades, an array of new institutions has emerged, straddling this divide, and seeking to ease the process of taking scientific findings to their application in policy and to stabilize the boundary between the different disciplines, thereby avoiding the politicization of science and vice versa, as described by Guston (2001). In 1999, Guston developed the theoretical concept of the 'boundary organization' to describe these organizations that 'lie between politics and science' (Guston, 1999).

Miller (2001) defines boundary organizations as 'institutions that are neither labs nor conventional political organizations [which] are increasingly prevalent features of institutional landscape of modern society and play key roles in managing the interactions between science and politics, economics and culture'. Their work is multifaceted (Carr and Wilkinson, 2005), but boundary organizations fulfil a number of distinct roles: they involve actors from both sides of the boundaries and mediate between science and politics; they exist between, but with responsibility and accountability to both sides of the boundary; and they create and use boundary objects (Guston, 1999, 2000; Cash, 2001). Such organizations seek to aid communication and engender mutual respect of each other's ways of knowing, by providing a forum for effective stakeholder interaction (Carr and Wilkinson, 2005).

Boundary objects, as defined by Guston (2001) (and based on the very well articulated concepts of Star and Griesemer, 1989), 'sit between two different social worlds, such as science and non-science, and they can be used by individuals within each for specific purposes without losing their identity'. Examples of boundary objects include assessment reports, maps and computer models. Cash et al (2003) describe boundary organizations as managing the interplay and tradeoffs between the three qualities that make scientific information more effective in influencing policy: credibility, salience and legitimacy. The three functions which lead to an increase in these qualities and contribute most to boundary management are communication, translation and mediation. Examples of boundary organizations include 'expert advisory committees, scientific assessments, research management agencies, consensus conferences and so on' (Miller, 2001). Specific examples include Research and Development within the Consultative Group on International Agricultural Research (CGIAR) system, El Nino Southern Oscillation forecasting (Cash et al, 2003) and the International Research Institute for Climate Prediction (Agrawala et al, 2001). Box 10.4 summarizes a case study which explores the work of one boundary organization, FEWS NET (the Famine Early Warning System).

Box 10.4 *Boundary organization case study: FEWS NET*

The Famine Early Warning System (FEWS NET; www.fews.net) was initiated in 1985, in response to the deadly famine that rocked Sudan and Ethiopia in 1984–85. Funded by the US Agency for International Development (USAID), and working in collaboration with international, regional and national partners, it seeks to provide early warning and vulnerability information on emerging and evolving food security issues. Professionals in Africa, Central America, Haiti, Afghanistan and the USA monitor and analyse relevant information and data in terms of its impacts on livelihoods and markets to identify potential threats to food security. Once issues are identified, FEWS NET helps decision-makers to act to mitigate food insecurity using a suite of communications and decision-support products (see www.fews.net/ml/en/product for a full list of products). Among other things, the 'boundary objects' it creates include a monthly food security update for 25 countries, regular food security outlooks and alerts, as well as brief-ings and support to contingency and response planning efforts. To support analysis, and programme and policy development, more in-depth studies are also available, for example, in areas such as livelihoods and markets, providing addi-tional useful information.

According to John Scicchitano, USAID Programme Manager for FEWS NET (per-sonal communication, 28 October 2009), these boundary objects have been developed using lessons learnt from the 25 years that the network has been active, and products have evolved with feedback from food security policy-makers, as well as others; for example, economists suggested the importance of integrating markets and trade into the FEWS NET models. Brown and Choularton (2008) describe the role of FEWS NET as translating observations into useful infor-mation, combining remote sensing data and livelihoods to better inform decisions; Scicchitano also draws attention to this role, describing the main form of science-policy mediation performed by the network as being that of language translation, making information more 'decisionable' and thereby more salient. FEWS NET also takes an active role in efforts to strengthen early warning and food security networks, including developing capacity, developing policy-useful infor-mation, and building consensus around food security problems and solutions.

FEWS NET's partners provide it with credibility and legitimacy; NASA (National Aeronautics and Space Administration), NOAA (National Oceanic and Atmospheric Administration) and USGS (United States Geological Survey) are scientific implementation partners, collecting satellite and remote sensing data; the United States Department of Agriculture's Foreign Agricultural Service is also a partner. The current prime contractor (a contract offered to a private company by USAID every five years) is the international development consulting firm Chemonics International, Inc. (www.chemonics.com); it has staff in Africa, pres-ence in the field, and also runs the Washington, DC office. According to the FEWS

NET website, it is this 'coordinated network of leading scientific institutions and private enterprise that provide unique technical and managerial talent and field presence, making FEWS NET an essential information system' (www.fews.net/contact, accessed 28.10.2009). Scicchitano states: 'FEWS NET is the point where discussions start' (personal communication, 28 October 2009). FEWS NET information is used by a large number of different organizations and individuals, a further demonstration of its credibility; according to Scicchitano, it is used by policy-makers, the main food system actors, those making decisions concerning development needs, and is the main decision support to USAID; its recent move to the office of USAID's Food for Peace allows key personnel to report directly to those making decisions concerning aid (Brown and Chourlarton, 2008). FEWS NET is accountable to those on both side of the boundary, a key feature of any boundary organization.

When asked whether FEWS NET is a boundary organization, Scicchitano suggested that the term is not actively used, but that the three roles of the boundary organization as defined by Cash (2001) and Guston (1999, 2000), describe 'exactly what [FEWS NET] is trying to do'; FEWS NET is there to mediate (a primary role of the network, often mediating between governments, as well as translating between scientists and policy-makers); it is accountable to principals on both sides of the boundary, and creates 'boundary objects galore' which makes the scientific information more salient and applicable. It is a highly regarded and successful organization, and perceived by many to have saved countless lives and to be a reliable source of ready-to-use information.

Conclusions

This chapter has examined some of the complexities of the policy-making process and the consequent challenges to securing effective inputs from science. Communication across the dynamic and shifting interface between science and policy-making can be enhanced through the deployment of skilled individuals and organizations – knowledge brokers and boundary organizations – whose role is to facilitate mutual understanding and interpretation of knowledge needs and significance. For countries taking strategic decisions on their research capacity, investment in such individuals and organizations should be a high priority.

References

Agrawala, S., K. Broad and D. H. Guston (2001) 'Integrating climate forecasts and societal decision making: Challenges to an emergent boundary organization', *Science Technology Human Values*, 26, 4, 454–77

Bielak, A. T., A. Campbell, S. Pope, K. Schaefer and L. Shaxson (2008) From science communications to knowledge brokering: the shift from science push to policy pull. In Cheng, D., M. Claessens, T. Gascoigne, J. Metcalfe, B. Schiele and S. Shi (eds) *Communicating Science in Social Contexts: New Models, new Practices*, Dordrecht, Springer

Brand, R. and A. Karvonen (2007) 'The ecosystem of expertise: Complementary knowledges for sustainable development', *Sustainability: Science, Practice and Policy*, 3, 1

Bridgman, P. and G. Davis (2004) *The Australian Policy Handbook*, Sydney, Allen and Unwin

Brklacich, M., I. F. Brown, E. J. D. Campos, A. Krusche, A. Lavell, L. Kam-biu, J. J. Jiménez-Osornio, S. Reynes-Knoche and C. Wood (2007) Stakeholders and global environmental change science. In Tiessen, H., M. Brklacich, G. Breulmann and R. S. C. Menezes (eds) *Communicating Global Change Science to Society: An Assessment and Case Studies*, Washington, DC, Island Press

Brown, M. E. and R. Chourlarton (2008) *Famine Early Warning Systems And Remote Sensing Data*, New York, Springer-Verlag

Brownson, R. C., C. Royer, R. Ewing and T. D. McBride (2006) 'Researchers and policymakers: travelers in parallel universes', *American Journal of Preventive Medicine*, 30, 2, 164–72

Carr, A. and R. Wilkinson (2005) 'Beyond participation: Boundary organizations as a new space for farmers and scientists to interact', *Society and Natural Resources*, 18, 255–65

Cash, D. (2001) 'In order to aid in diffusing useful and practical information: Agricultural extension and boundary organizations', *Science, Technology and Human Values*, 26, 4, 431–53

Cash, D. W., W. C. Clark, F. Alcock, N. M. Dickson, N. Eckley, D. H. Guston, J. Jäger and R. B. Mitchell (2003) 'Knowledge systems for sustainable development', *Proceedings of the National Academy of Sciences of the United States of America*, 100, 14, 8086–91

Chapman, S. (2001) 'Advocacy in public health: roles and challenges', *International Journal of Epidemiology*, 30, 1226–32

Chapman, S. (2007) *Public Health Advocacy and Tobacco Control: Making Smoking History*, Oxford, Blackwell

Court, J., I. Hovland and J. Young (eds) (2005) *Bridging Research and Policy in Development: Evidence and the Change Process*, London, ITDG Publishing

Defra (2009) *UK Food Security Assessment: Our Approach*, London, Department for Environment Food and Rural Affairs

Fenna, A. (2004) *Australian Public Policy*, Frenchs Forest, New South Wales, Pearson Longman

Funtowicz, S. and J. Ravetz (1993) 'Science for the post-normal age', *Futures*, 25, 739–55

Gregrich, R. J. (2003) 'A note to researchers: communicating science to policy makers and practitioners', *Journal of Substance Abuse Treatment*, 25, 3, 233–37

Guston, D. H. (1999) 'Stabilizing the boundary between US politics and science: The role of the office of technology transfer as a boundary organization', *Social Studies of Science*, 29, 1, 87–111

Guston, D. H. (2000) *Between Politics and Science: Assuring the Integrity and Productivity of Research*, New York, Cambridge University Press

Guston, D. H. (2001) 'Boundary organizations in environmental policy and science: An introduction', *Sci. Technol. Hum. Values*, 26, 4, 399–408

Hisschemöller, M. and R. Hoppé (2001) Coping with intractable controversies: The case for problem structuring in policy design and analysis. In Hisschemöller, M., R. Hoppé, W. Dunn and J. Ravetz (eds) *Knowledge, Power and Participation in Environmental Policy Analysis*, Maidenhead, UK, Transaction Publishers

Holmes, J. and R. Clark (2008) 'Enhancing the use of science in environmental policy making and regulation', *Environmental Science and Policy*, 11, 702–11

Holmes, J. and J. Savgard (2008) Dissemination and implementation of environmental research. *Report 5681*, Stockholm, Swedish Environmental Protection Agency

Hoppe, R. (1999) 'Policy analysis, science and politics: from "speaking truth to power" to "making sense together"', *Science and Public Policy,* 26, 3, 201–10

Jones, N., H. Jones and C. Walsh (2008) Political Science? Strengthening science-policy dialogue in developing countries. *Working Paper 294*, London, Overseas Development Institute

Kingdon, J. W. (2003) *Agendas, Alternatives, and Public Policy,* New York, Longman

Kurtz, C. and D. Snowden (2003) 'The new dynamics of strategy: sense-making in a complex and complicated world', *IBM Systems Journal,* 42, 3, 462–83

Lackey, R. (2006) 'Science, scientists and policy advocacy', *Conservation Biology,* 21, 1, 12–17

Levidow, L., S. Carr and D. Wield (2005) 'European Union regulation of agri-biotechnology: Precautionary links between science, expertise and policy', *Science and Public Policy,* 32, 261–76

Lomas, J. (2007) 'The in-between world of knowledge brokering', *British Medical Journal,* 334, 129–32

Marsden, T. (2008) 'Agri-food contestations in rural space: GM in its regulatory context', *Geoforum,* 39, 1, 191–203

Miller, C. (2001) 'Hybrid management: Boundary organizations, science policy, and environmental governance in the climate regime', *Science Technology Human Values,* 26, 4, 478–500

Nutley, S. (2003) Bridging the policy/research divide: reflections and lessons from the UK. *National Institute of Governance Conference: Facing the Future: Engaging stakeholders and citizens in developing public policy,* Canberra

Omamo, S. W. and J. Farrington (2004) Policy research and African agriculture: time for a dose of reality? *Briefing paper – ODI Natural Resource Perspectives 90,* London, Overseas Development Institute

Owens, S., J. Petts and H. Bulkeley (2006) 'Boundary work: Knowledge, policy and the urban environment', *Environment and Planning C: Government and Policy,* 24, 5, 633–43

Parsons, W. (2002) 'From muddling through to muddling up – evidence based policy making and the modernisation of British Government', *Public Policy and Administration,* 17, 3, 43–60

Porter, T. (1996) *Trust in Numbers,* Princeton, Princeton University Press

Ravetz, J. (1999) 'What is post-normal science?', *Futures,* 31, 647–53

Ravetz, J. and S. Funtowicz (1999) 'Post-normal science – an insight now maturing', *Futures,* 31, 641–46

Ritter, A. and G. Bammer (in press) 'Models of policy making and their relevance for drug research', *Drug and Alcohol Review*

Rosenstock, L. (2002) 'Attacks on science: The risks to evidence-based policy', *American Journal of Public Health,* 92, 1, 14–18

Sabatier, P. A. (1988) 'An advocacy coalition framework of policy change and the role of policy-orientated learning therein', *Policy Sciences,* 21, 2/3, 129–68

Scott, A., J. Holmes, G. Steyn, S. Wickham and J. Murlis (2005) *Science Meets Policy in Europe,* London, Defra

Shaxson, L. (2008) Who's sitting on Dali's sofa? Evidence-based policy-making. *PMPA/National School of Government practitioner exchange report,* London, Public Management and Policy Association (PMPA)

Shaxson, L. (2009) 'Structuring policy problems for plastics, the environment and human health: reflections from the UK', *Philosophical Transactions of the Royal Society: Biological*, 364, 2141–51

Snowden, D. and M. Boone (2007) 'A leader's framework for decision making', *Harvard Business Review*, November 2007

Star, S. and J. Griesemer (1989) 'Institutional ecology, "translations" and boundary objects: Amateurs and professionals in Berkeley's Museum of Vertebrate Zoology', *Social Studies of Science*, 19, 387–420

van Kammen, J., D. de Savigny and N. Sewankambo (2006) *Bulletin of the World Health Organization*, August 2006

Weingart, P. (1999) 'Scientific expertise and political accountability: paradoxes of science in politics', *Science and Public Policy*, 26, 3, 151–61

Wynne, B. (1992) 'Misunderstood misunderstanding: Social identities and public uptake of science', *Public Understanding of Science*, 1, 281–304

Young, J. and J. Court (2004) Bridging research and policy in international development: an Analytical and practical framework. *RAPID Briefing Paper*, London, Overseas Development Institute

11
Engaging Stakeholders at the Regional Level

John Ingram, Jens Andersson, Gabriele Bammer,
Molly Brown, Ken Giller, Thomas Henrichs, John Holmes,
James W. Jones, Rutger Schilpzand and John Young

Introduction

Food security in the face of global environmental change (GEC) is one of the most complex issues facing the research community at large. Although most policy-makers, scientists and funding agencies recognize the need for additional knowledge about how the various food system activities interact and how these interactions affect food security (see Chapter 2), research that is capable of adequately addressing the problem is hard to find. This is because not only are there large uncertainties in many aspects of the debate, but the debate involves a bewildering range of interested parties, or 'stakeholders'. A further complication is that food systems involve critical interactions at a number of levels on a range of scales (e.g. spatial, temporal, jurisdictional, institutional, management) (Cash et al, 2006; and see Chapters 2 and 13), each of which has its own group or groups of stakeholders. Research on the interactions between GEC and food security therefore has to recognize, and engage with, a wide range of stakeholders. This is in contrast to research on crop improvement, for instance, where the range of stakeholders is much narrower, and may remain predominantly within the research community itself. While considerable effort has been spent in improving understanding of food system–GEC interactions at the local or household level, research at the regional (sub-continental) level is far less well developed, but – as is discussed in detail in Part IV – offers important insights into food system adaptation strategies and policies. This chapter therefore addresses stakeholder engagement at regional level.

Clearly, aligning the research agenda with stakeholder needs is crucial and this requires effective dialogue (e.g. through consultancies, agenda-setting workshops and/or informal processes). Equally important, however, is the uptake of research results by the intended beneficiaries that leads on to the real value of the research. This similarly depends on continued interactions between researchers and other stakeholders. However, as stakeholder involvement complicates the research process and increases costs for all concerned, it is important to understand its importance and the value it can bring throughout a given project.

As Kristjanson et al (2009) summarize, it is important to see stakeholder engagement as an integral aspect of both the conceptualization and the life of the project. They stress the value of articulating the outcomes sought by the different stakeholders at the project outset to help bring the different actors towards a joint understanding of the overall project goals and come up with innovative strategies to achieve them (see Box 11.1). This also serves to give interested parties a tangible stake in the outcome. These aspects of research agenda alignment and uptake of outputs derive from what is a double aim of the stakeholder dialogue. The first aim concerns the formal agenda-setting. If the research is to have an impact, it is crucial that it both addresses the information needs of the intended beneficiaries and is scientifically valid. The second aim is that of a social support function that helps all stakeholders feel involved and heard. While less obvious, this aspect is no less important as effective stakeholder engagement needs to be built on trust between all concerned, which will encourage uptake of research outputs. This may be especially the case at the regional level as stakeholders may well be senior individuals with specific agendas, and in these circumstances it is crucial to be clear about whether researchers are acting as advocates or honest brokers (see Chapter 10). This social support aspect also allows researchers to obtain a better 'feel' for the context within which the research is to be conducted.

Box 11.1 *Seven propositions/principles for 'linking knowledge with action' (adapted from Kristjanson et al, 2009)*

1 Problem definition

Projects are more likely to succeed in linking knowledge with action when they employ processes and tools that enhance dialogue and cooperation between those (researchers, community members) who possess or produce knowledge and those (decision-makers) who use it, with project members together defining the problem they aim to solve.

2 Programme management

Research is more likely to inform action if it adopts a 'project' orientation and organization, with leaders accountable for meeting use-driven goals and the team managing not to let 'study of the problem' displace 'creation of solutions' as its research goal.

3 Boundary spanning

Projects are more likely to link knowledge with action when they include 'boundary organizations' or 'boundary-spanning actions' that help bridge gaps between research and research user communities. This boundary-spanning work often involves constructing informal new arenas that foster user-producer dialogues, defining products jointly, and adopting a systems approach that counters dominance by groups committed to the status quo. Defining joint 'rules of engagement' in the new arena that encourage mutual respect, co-creation and innovation improves prospects for success.

4 Systems integration
Projects are more likely to be successful in linking knowledge with action when they work in recognition that scientific research is just one 'piece of the puzzle', apply systems-oriented strategies and engage partners best positioned to help transform knowledge co-created by all project members into actions (strategies, policies, interventions, technologies) leading to better and more sustainable livelihoods.

5 Learning orientation
Research projects are more likely to be successful in linking knowledge with action when they are designed as much for learning as they are for knowing. Such projects are frankly experimental, expecting and embracing failures so as to learn from them throughout the project's life. Such learning demands that risk-taking managers are funded, rewarded and regularly evaluated by external experts.

6 Continuity with flexibility
Getting research into use requires strengthening links between organizations and individuals operating locally, building strong networks and innovation/response capacity, and co-creating communication strategies and boundary objects/products.

7 Manage asymmetries of power
Efforts linking knowledge with action are more likely to be successful when they manage to 'level the playing field' to generate hybrid, co-created knowledge and deal with the often large (and largely hidden) asymmetries of power felt by stakeholders.

Who are the stakeholders in the GEC–food security debate?

The term *stakeholder* is now commonly employed to denote 'all parties with a voluntary or involuntary legitimate interest in a project or entity' (Brklacich et al, 2007). For issues of food security, in addition to those involved in the food system activities per se (e.g. food producers, processes, packers, distributors, retailers, consumers; see Chapter 2 and Figure 2.1b), stakeholders include funding agencies, national/regional policy agencies, non-governmental organizations (NGOs), civil society groups, business (and increasingly the energy sector, vis-à-vis biofuels), individuals and communities affected by GEC, and the researchers themselves. For research projects that involve a significant natural resource management component at the local level (as is often the case in field-based, food production research), the resource managers (who are often, but not exclusively, farmers, fishers, pastoralists, etc.) themselves are usually critically important stakeholders. Indeed, methods and approaches for identifying and engaging farmers in the research process, especially in the development agriculture arena, have given rise to a wide body of literature (Chambers et al, 1989; Okali et al, 1994; Martin and Sherington, 1997; Haggar et al, 2001; Ortiz et al, 2008).

The initial problem facing researchers is to identify who the other stakeholders are (i.e. with whom researchers should aim to engage). This can be helped by being clear not only on who the intended target or beneficiary groups are (e.g. impoverished smallholder farmers; urban communities), but also on how food security research is intended to assist them. This means establishing by what route, and mediated by which institutions and structures, the research output will bring about benefit. Because of this it is key actors in *these* domains (e.g. regional policy-makers, donors) who may actually be the more important stakeholders for a given research project than the 'target' beneficiaries themselves; in other words, benefit for the 'target' beneficiaries would come about through the development of better policies at the regional level. (It is useful to note the value of the role of funding bodies in facilitating the making of these important connections as part of the funding process.)

For research at higher levels of integration on a number of different scales (e.g. spatial, political, jurisdictional; see Chapters 2 and 13) and particularly regarding food security policy (vis-à-vis food production), it is perhaps not appropriate to include individual farmers as stakeholders in the research process. However, as they (together with other members of society) are obviously among the ultimate beneficiaries of the research effort, it may well be appropriate to engage with regional organizations that represent farming groups, as this can help ensure the interests and constraints of the farming community are included in research design. Thus, in the case of GECAFS research in southern Africa, the formalization of collaboration with the Food, Agriculture and Natural Resources Policy Analysis Network (FANRPAN, which comprises national farmers' organizations; www.fanrpan.org) proved useful in this regard.

For food security research at regional level it is possible to identify four main stakeholder categories: research, government, business and civil society (see Box 11.2 and Chapter 14). As food security is a multi-factor issue, no single stakeholder has the complete answer or the power and the tools to realize the changes that will be needed. Cooperation between those involved in these stakeholder communities is required. Stakeholder dialogue necessarily plays an important role (van Tulder and van der Zwart, 2006) and can contribute to agenda-setting, the analysis of a given situation and to the creation and implementation of solutions. However, the fact that none of the stakeholders can be successful without the others presents a strong argument for further intensification of the dialogue process, going beyond consulting and informing each other (i.e. stakeholder dialogue), towards co-production of knowledge and shared responsibilities (i.e. stakeholder engagement) (Rischard, 2001; Henrichs et al, 2010). This requires multidisciplinary research teams coming together with other stakeholders to work on specific problems in the 'real

Box 11.2 *'Future of Food': A case study of multiple stakeholder dialogue*

The Future of Food initiative is an exploration of external developments that will shape the future of the global food system. This exploration is based on stakeholder dialogue between four stakeholder groups: government, science, business and NGOs. The one-day dialogue sessions are an initiative of communication consultancy Schuttelaar and Partners, and are realized in cooperation with stakeholder organizations.

A first session, in cooperation with Wageningen University and the Dutch ministries of health and agriculture, addressed the growing gap between consumer food demands and natural resources, with the role of science to bridge this gap. The second session, about the roles and responsibilities of the food retail and financial sectors in a more sustainable global food system, took place in cooperation with GECAFS, Oxford University's Environmental Change Unit, the Dutch Ministry of Agriculture and Oxfam Novib.

This dialogue is an exploratory, or horizon-scanning, exercise and is not aimed at decision-making as such. The seminar aims to explore broader views on future developments and interactions in the global food system and how to deal with them, not directly to agree a covenant or a research agenda. Relating these broader views to the long-term perspective of a sustainable, healthy and accessible food system leads to 'building blocks for a roadmap'. The participants are encouraged to make use of the resulting views and building blocks for their own innovation strategy, research agenda or policy.

A limited number of senior experts in the food sector, invited in equal numbers from the four stakeholder groups, participate. The day consists of two sets of round-table discussions, each introduced by presentations of very different viewpoints from the four sectors – an 'outside-in' approach.

The outside-in approach is chosen to stimulate the participants to look at the subject from a new angle, thereby lifting the discussion from their usual exchange of thoughts and arguments. This appeared very helpful in engendering a broad perspective.

The multi-stakeholder dialogue appears useful for the following reasons:

- All stakeholder sectors are key players in food systems, and share a sense of urgency; they feel themselves deeply involved in the three key issues of the food system: sustainability, health and hunger.
- All are committed to taking steps forward and share the vision that all players have to be involved to make genuine progress; no stakeholder can act alone or can withdraw from commitment to these key issues.
- Participants are able to collect views and concerns from the different sectors, thereby broadening their own views on the complexity of the issues in the global food system. Normally these issues are discussed separately. Bringing them together in the dialogue sessions, prompted by the expert presentations, gave the opportunity to create a more complex picture and to discuss interactions between the different issues at stake.

Box 11.3 *Engaging with stakeholders in the Competing Claims programme*

Too often, researchers blithely refer to involving 'all stakeholders'. And perhaps even worse, attempts are made to bring all stakeholders together in 'multi-stakeholder platforms'. This may be a valid approach when the issue at stake is relatively simple and has few stakeholders, but when the stakes are high and cultural differences run deep, meetings can precipitate or exacerbate conflict rather than result in useful dialogue. In work in southern Africa on 'Competing Claims on Natural Resources', focus is on food security as one important aspect of rural livelihoods that cannot be seen in isolation from other livelihood pursuits (see www.competingclaims.nl). A key concept in the approach is that local problems need to be addressed at multiple hierarchical levels to enlarge the 'solution space' within which new opportunities can be sought.

Identifying stakeholders at the higher levels in a hierarchy is simpler by definition – there are fewer players to choose from. When engaging with rural people, initial engagement must inevitably start through local officials and village leaders, though it should not be naïvely assumed that they represent the position of the majority. In particular, the poorest and most disadvantaged are the last to contribute in meetings, if they attend at all. Experience shows that it is not possible to develop a rulebook, or a standard set of methods that will work in all settings. What is critical is having an ear close to the ground, and taking time to identify marginal and excluded stakeholders and understand the positions of the different stakeholders *before* bringing them together to discuss issues at stake.

A further issue is the legitimacy of the 'outsider' researcher in local debates and problems (see Chapter 10), and this may be particularly apt in cross-cultural settings. Collaboration with local researchers, NGOs or other development agencies is necessary, but often leads the researcher to become – unwittingly – associated with such local stakeholders, compromising his/her legitimacy for another set of stakeholders. The political neutrality of the researcher is a fallacy, because already research questions tend to be posed by some parties rather than others, and inherently build on specific societal problem definitions, values and aspirations. A transparent yet rigorous approach which makes the stakes explicit is a more modest, yet realistic, approach towards becoming legitimate.

The only general rule to be drawn is that there are no quick and clean methods of identifying stakeholders. It takes time and commitment to gain useful insights, build legitimacy among stakeholders, and to contribute to development. These issues are further discussed by Giller et al (2008).

---

world' (Gibbons et al, 1994). It must be noted, however, that stakeholders play a multitude of different roles in the food system. They often have different goals and agendas that may appear to be (or really are) conflicting.

A further complication is that food systems are inherently multi-scale and multi-level, and the non-spatial scales are very relevant to food security/GEC interactions (Ericksen et al, 2009). Different stakeholders operate on different scales and levels; scale and level need to be clearly specified in research engagement activities. Identifying a discrete list of stakeholders for a given situation is therefore far from simple, and the notion held by many researchers of 'engaging with stakeholders', while well intentioned, needs to be approached with awareness of the nature and magnitude of the task, especially when working at more local levels (see Box 11.3). Indeed, the success of the project can depend very much on how this stakeholder engagement is envisioned and implemented, and who is at the table.

Finally, it is also worth noting that some key stakeholders (e.g. the business community) may sometimes be missing from the debate, and it is important to try to identify why this is the case. Is it that they cannot afford the funds or time to become engaged; or they are not allowed to be involved (perhaps for political reasons); or they are simply not interested? It is also important to try to determine the impact of their absence, and what – if anything – can be done to compensate. Certainly, given that stakeholder participation can sometimes be seen as an automatic requirement, taking on something of a 'tick-box' culture, some potentially important stakeholders may need to be persuaded to join the debate, especially if they are jaded from earlier, ineffective or disingenuous experiences. To overcome this challenge it is important to stress the benefits that engagement will bring to the stakeholder (rather than the benefits their engagement will bring to the researcher/project): how will engagement help them in their policy or business or funding planning? Ideally, reluctance should transform into a commitment to engage.

Who sets the GEC–food security research agenda, and how?

Basic research is typically disciplinary-focused, often undertaken by relatively small groups of researchers. There may be little need to engage with beneficiary groups, even if the 'end of pipe' research outputs are anticipated to be of some practical use. The more involved approach needed to address the broader issues of food security will lead to research being conducted within a more complex context. This might well be characterized by multiple biophysical and social scientific issues, a high degree of uncertainty, value loading and a plurality of legitimate perspectives of the varied stakeholders (see discussion on post-normal science in Chapter 10). Researchers trained in a given discipline which, on the face of it, addresses directly the issues they are investigating can well find themselves confronted by a range of issues in which they have no experience or training. Indeed, stakeholder engagement in the way being discussed here, and especially at the regional level, is not the norm for GEC science endeavours.

In the 'classic' GEC research project typical of the international GEC research programmes, a science plan is conceived by the scientists and published. This lays out the research need (as perceived by the research community) in terms of science output, and the relevance to the policy process and resource management is often of less importance. Where relevance to policy is indicated, it usually relates to the global level such as the United

Nations Framework Convention on Climate Change or other international conventions.

By contrast, the agriculture and food security research communities have been working with partners on the ground, farmers, policy-makers and other non-research stakeholders for many years, and lessons learnt have much to offer researchers addressing the interactions between GEC and food security.

Due to the complex and region-specific nature of both GEC and food security (see Chapter 13), GEC–food security research can be of greater value to stakeholders if set within the regional context and tailored to the needs of regional policy-makers, NGOs, businesses and resource managers. Setting a research agenda that is relevant to regional (as opposed to global and/or generic) issues needs a highly consultative and inclusive approach. Further, when conducted in regions of the developing world, the links to the development agenda, and particularly to the Millennium Development Goals, must be explicit. This necessarily means a stronger link to the development donor community, who are not traditional funders of GEC research. Again, lessons learnt by the agriculture and food security research community have much to offer. Box 11.4 shows the main steps in agenda-setting within GECAFS regional studies.

GEC–food security research agenda-setting must aspire to build a number of bridges that traditional GEC science has not well addressed. It must bridge natural, social and economic sciences; science and policy, and other stakeholders' interests; and science and development. It must also relate to the interests of each group of stakeholders. This is challenging, as it necessitates spanning disciplines, research cultures, funding modes, and even attitudes and perceptions of what constitutes science and GEC research. In fact, although often thought to be a particular problem in the developing world, the issues surrounding GEC–food security research agenda-setting are complex the world over. There are a number of reasons for this.

First, and as discussed in Chapters 1 and 8, food security involves many more issues than food production alone, and a wide range of disciplines have to be integrated to understand the full suite of issues: economics, anthropology, sociology and engineering sciences are, for example, as important as crop, animal and agronomic sciences. In the research community, biophysical scientists often assume the lead on agenda-setting for food security research, and other researchers from the social sciences of equal relevance can be left out. Alternatively, they may be invited in after the main (and often quite detailed) elements of research are decided upon; 'bolting on' social science to what is essentially a biophysical agenda generally does not work well! In contrast, the development community generally tries to take a more balanced interdisciplinary approach in agenda-setting. While outputs from disciplinary research endeavours are essential building blocks, the GEC–food security agenda must emerge from a balanced dialogue between researchers across social and natural sciences, and include other stakeholders as appropriate.

Second, the range of scales and levels pertaining to food security (see Chapter 2), and interactions between them, pose particular challenges for the GEC–food security agenda. Raising the bar from agenda-setting, which

Box 11.4 *Setting the research agenda for regional GECAFS studies (GECAFS, 2005)*

An important aspect of the GECAFS regional approach has been to ensure that the research agenda closely matches major *regional* GEC science interests (as distinct to 'international' interests), policy needs and donor priorities. The process to achieve this constituted the planning phase for each regional project (southern Africa, the Caribbean and the Indo-Gangetic Plains), and involved workshops, informal conversations and discussions with a wide range of potential stakeholders in the region. In each case it culminated in the region's GECAFS Science Plan and Implementation Strategy.

The development for each region followed a common approach. Figure 11.1 shows the main steps in the planning phase (Steps 1–3) and subsequent implementation phase (Steps 4–6).

Step 1   Working with GECAFS Scientific Advisory Committee (SAC) members, other GECAFS International Project Office (IPO) contacts and IGBP (International Geosphere-Biosphere Programme), IHDP (International Human Dimensions Programme) and WCRP (World Climate Research Programme) National Committees within the region, identify regional scientists likely to be interested in the GECAFS interdisciplinary approach and establish a GECAFS initial regional planning group. This group aimed to include members from the research, policy, NGO and private sectors.

Step 2   Working with the initial regional planning group, identify regional science, policy and potential donor interests and information needs.

Step 3   Working with the initial regional planning group, and with other stakeholders, establish GECAFS regional research questions, develop and publish the GECAFS regional Science Plan and Implementation Strategy, and establish a Regional Steering Committee.

Step 4   Working with the Regional Steering Committee and joined by Core Project/ESSP (Earth System Science Project) representatives as appropriate, establish regional research/Core Project/ESSP collaboration and jointly design and implement GECAFS analyses.

Step 5   Working with regional scientists and the policy community, and Core Project/ESSP representatives as appropriate, deliver and interpret GECAFS results in policy context.

Step 6   Integrate results across GECAFS studies in other regions to develop: (i) improved generic understanding of food systems and their vulnerability to GEC; (ii) scenarios methods; and (iii) improved decision support.

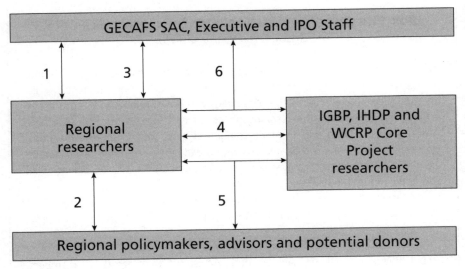

Source: GECAFS, 2005

Figure 11.1 *Key steps in design (Steps 1–3) and envisaged implementation (Steps 4–6) of GECAFS regional food systems research*

traditionally addresses agronomic issues at plot- or farm-level for a given growing season, to one addressing food security for a nation or even sub-continental region over time, is daunting. Researchers have to develop, accept and work within new conceptual models and frameworks, and relate these to policy and resource management considerations with which they are unfamiliar and often uncomfortable.

Third, and of considerable practical relevance, GEC research is usually thought of as the purview of agencies responsible for science and/or environment, whereas food security research is usually thought of in terms of agriculture or aid agencies. Bringing these two groups together, and finding a common agenda which appeals to their respective governance and donor policies, is far from easy, especially as the funders and government structures that support the respective research communities are not traditional collaborators. An encouraging development is that the international development and national security communities are now interested in becoming involved in such research, although they have limited ability to fund research without an immediate agenda for action. Within the GEC community, the 'Global Change SysTem for Analysis, Research and Training' (START) and GECAFS have both had some success on this front, while the emerging CGIAR (Consultative Group on International Agricultural Research) agenda, 'Climate Change, Agriculture and Food Security', has many aspects specifically designed to do this (CCAFS, 2009).

So who sets the GEC–food security research agenda, and how? Given the points made above, the 'who' is ideally the regional stakeholder community at large (policy advisors/makers, resource managers, researchers, NGO donors, etc.); and the 'how' is preferably by working interactively together, developing a shared vision and common understanding, and engendering trust. With such a broad stakeholder community this begs the question of how this engagement is managed, and by whom? Clearly it takes time, money and commitment, and may well result in an agenda that none of the participants anticipated. Further, the agenda-setting process needs to be flexible to inputs from science and policy developments as they emerge. This allows the agenda to encompass latest thinking, and also engenders the buy-in of a wider group (geographical and/or thematic) of stakeholders. There may also be a particular need to integrate the business sector and/or NGOs, and this particular dynamic is discussed in Box 11.5.

Some final points warrant stressing when designing and undertaking research which seeks to influence policy. It is crucial to establish the information needs of the policy process early in research planning (see below), and to develop the research programme accordingly. Further aspects of particular relevance to the science–policy debate are the requirement to: (i) establish and maintain credibility with all stakeholders; (ii) achieve practicality; (iii) demonstrate usefulness to the designated beneficiaries; (iv) provide information to end-users in a timely and accessible format; and (v) ensure acceptability by

---

Box 11.5 *Stakeholder dialogue involving the NGO and business communities*

Stakeholder dialogue can only thrive in an atmosphere of cooperation and mutual understanding, while serving the interests of the participants. There is always an aspect of power in dialogue. Certainly NGOs in stakeholder dialogue have to be able to exert such power in order to be taken seriously and to negotiate acceptable results. This power could be in the magnitude of their constituency, their high level relations, their press contacts or their cooperation with campaigning NGOs. NGOs are not, however, a monolithic entity. For example, 'watchdog' NGOs, of which Greenpeace is perhaps the most well known example, focus on agenda-setting for public opinion and openly confront companies on their deemed bad behaviour. This is in contrast to 'dialogue' NGOs (e.g. World Wide Fund for Nature, WWF), which focus on cooperation with business and other stakeholders in common analysis and finding common solutions.

While companies may not fall into as many different categories as NGOs, there are clear differences within the business sector. Who is actually representing the company or a group of companies can have a strong bearing on what they are able to contribute to the dialogue and what subsequent actions they take. In general, public affairs managers are well trained in stakeholder dialogue, but can have problems with the acceptance of dialogue results within the company, while representatives from business interest organizations have the responsibility to also take care of the less innovative of their members. Research managers feel more comfortable with scientists than with NGO campaigners. All three kinds of professionals have their own multi-stakeholder networks. It is interesting to see that these groups of networks often have very limited overlap.

end-users (Ingram et al, 2007). All these aspects will benefit from a carefully designed stakeholder engagement process which gives all stakeholders a sense of participation in, and joint-ownership of, the research process. Finally, although a number of options have been presented above, it is important to appreciate that local customs and etiquette largely dictate the best way of establishing 'who needs to know what'. Local knowledge and people are critically important.

When to engage stakeholders in research planning

Stakeholder engagement is important throughout the GEC–food security research process, not only for setting agendas. This is because the roles of non-research stakeholders include: (i) identifying the problem; (ii) helping to formulate the research agenda; (iii) being sources of information; (iv) being subjects of research; (v) being a target audience for dialogue on how to implement research results; (vi) implementing the research; and (vii) funding or co-sponsoring the research. Figure 11.2 gives a conceptual framework for organizing and understanding the complexity of stakeholder engagement organized around six interrelated science activities. These range from designing the research questions to communicating the message. All stakeholders are represented by one of the 'cards in the deck', and the three-dimensional depiction aims to capture the notion that multiple stakeholders can be involved at various points along the six research stages.

While stakeholder engagement is important for research uptake, the timing, extent and nature of engagement depend on the precise situation, as illustrated by the Inter-American Institute for Global Change Research (IAI) Collaborative Research Network Program (CRN) case studies. These ranged from studies focused on a strictly defined scientific issue, initially involving only the GEC researchers, with stakeholder engagement coming later at the communication and dissemination stages. Other studies drew on a range of stakeholders from the outset. Across the range of IAI CRN case studies no single model emerged and no single reason or motivation-driving stakeholder participation was apparent. The following important points emerged from the IAI analysis (Brklacich et al, 2007):

- Stakeholders are heterogeneous groups representing multiple interests in GEC science.
- Stakeholders choose to participate in various stages of the scientific process, seldom participating in all.
- Stakeholders' participation in GEC science needs to be founded upon a mutual understanding of their contributions to the project and the benefits they will derive.
- Stakeholders make multiple contributions to GEC science, ranging from establishing the research agenda to participating in data collection to capacity building.
- Stakeholder participation in GEC science must be in accordance with international and national law, as well as consistent with local norms for the sharing of knowledge and benefits.

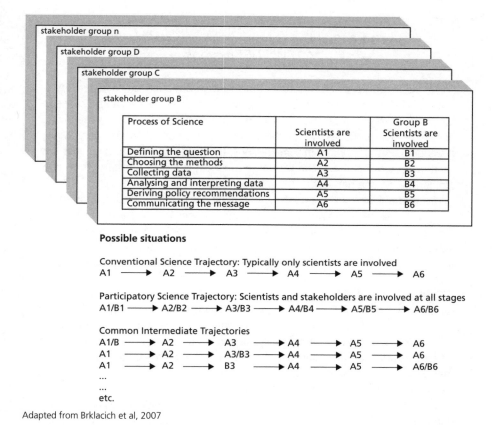

Adapted from Brklacich et al, 2007

Figure 11.2 *Organizing and understanding the complexity of stakeholder engagement*

- Stakeholder participation needs to be a planned set of activities within the GEC science process and be based upon an adaptive research design.
- GEC science must avoid overtaxing stakeholders and recognize that stakeholders have unequal capacities.
- The GEC science community has a responsibility to maintain and manage an environment that fosters long-term stakeholder participation.

As noted above, stakeholders choose to participate in various stages of the scientific process, but seldom participate in all. Brklacich et al (2007) give an example of how planning agencies both helped define a river management study in the Amazon and also then helped communicate the project's findings. Increasingly, however, there is a view that the research process cannot meet stakeholder needs unless they, in fact, participate throughout the research process. A particular example is the concept of co-production of knowledge, which is increasingly becoming a central feature of major research initiatives (e.g. UK's Living with Environmental Change programme; LWEC, 2009). An additional example is provided by the Famine Early

> ## Box 11.6 *Building consensus among stakeholders: A FEWS NET example*
>
> USAID designed FEWS NET in 1986 to provide information on food security of communities in semi-arid regions so that widespread climate-related food security crises do not occur. FEWS NET includes stakeholders who work in the public, private or non-profit sectors in the region, and whose identities vary widely depending on the location. In some regions, the meteorological communities and health care workers are at the centre; in others, it is nutrition experts from government and trade networks.
>
> The primary objective of early warning systems is to elicit an appropriate response to an identified problem. Often, this requires that all stakeholders, from the various donors of food aid, regional organizations made up of numerous national governments, to national and local governance structures in the country in question, agree that there is a problem and understand and concur on its severity.
>
> This consensus-building is very difficult and in many situations is often based on the quantitative remote sensing imagery that provides irrefutable evidence of a significant reduction in food production. Although everyone agrees that political and economic factors are usually far more important in determining food access and ultimately food security of a region, it is often the biophysical evidence that all parties can agree upon as being 'real', valid and conclusive. This puts remote sensing at centre stage in famine early-warning systems, even in an era of widespread telecommunication systems that have greatly increased information availability from remote regions. Once stakeholders agree to the nature and scope of the problem and come to the table to discuss what to do about it, then they are able to address the underlying political and economic causes of the problem through efforts to engage partners to provide income support, clean drinking water, health interventions and other responses appropriate to the situation in addition to food aid.
>
> As stakeholders continue to increase their attention and focus on food security issues outside of food availability, the pressure will grow to transform the food aid system to provide information on the wider food system's functioning during a crisis (Okali et al, 1994; Haggar et al, 2001; Brown, 2008).

Warning System (FEWS NET) project of the US Agency for International Development (USAID) (see Boxes 10.4 and 11.6). A key insight to emerge from FEWS NET concerns the importance of engaging a range of stakeholders in order to achieve consensus at the research communication stage: once they agree to the nature and scope of the problem and come to the table to discuss what to do about it, they are able to determine what actions will help relieve the most intense symptoms of the food security crisis.

How to engage stakeholders in research planning

As noted above, stakeholder engagement is of great importance in the research design phase and a combination of approaches (consultancies, agenda-setting workshops and informal approaches) can be employed to help set the agenda. Each of these is discussed below.

Consultancies. Consultancies, where researchers are hired to ascertain stakeholder views, prove particularly effective in determining information need from stakeholders who would not normally participate in, or feel comfortable at, an 'academic' brainstorming workshop. Examples of stakeholders who would be consulted for regional-level input include senior policy advisors from intergovernmental organizations or regional bodies (e.g. Southern Africa Development Community, SADC; European Commission), resource managers (e.g. operations managers of major/trans-boundary irrigation schemes), and representatives from specific target groups (e.g. farmer associations or major supermarkets). Careful selection of consultants is important: local researchers who are experienced in stakeholder dialogue (rather than international experts) usually have the best feel for the nature of the issues at regional level, and generally have the best contacts (sometimes personal, sometimes professional) and hence access to interviewees. A small team might be needed to collectively cover the main science areas. A 'down side' of this process is that interviewees may not strongly feel part of a collective agenda-setting exercise, and do not benefit from discussion with others in a workshop setting.

Agenda-setting workshops. These bring the researchers together with the various stakeholders and are commonly used in designing research projects. They have the advantage of sharing information more openly. Workshop outputs can be seen to be a product of collective discussion and consensus (vis-à-vis consultancies), and hence can have more 'standing'. This can be very important both scientifically and politically, especially if a multi-country, multi-disciplinary project is being planned. They can, however, be expensive in cash terms (especially if long-distance travel is involved), and also in time and effort in design, running and reporting. It may also be difficult to elicit attendance from senior stakeholders, such as senior government officials. There is also a risk that such workshops come up with rather long 'shopping lists' of research needs that are expressed too generally to provide the sharp focus that is needed on the key policy issues. Clear workshop objectives and skilful facilitation can overcome this potential problem.

Informal approaches. Informal approaches by researchers to other stakeholders can play a very important role in clarifying particular issues and helping to achieve 'buy-in' of key people. Important messages can often be better relayed outside the formal environment of a workshop session or interview. Workshop 'socials' and field trips are excellent opportunities for informal exchange, and a relaxed evening together can be very helpful in helping people to get to know each other better.

It was clear from GECAFS planning exercises that neither consultancies nor workshops alone delivered a clear research agenda and that some follow-up activities (such as sending drafts to technical advisors in regional agencies for their comments) were needed in all the GECAFS regional projects. Although more protracted than would normally be the case for a disciplinary science planning exercise, this process in itself had three important spin-offs: (i) it helped raise awareness of the GEC issues within the policy and other stakeholder communities; (ii) it helped raise awareness of the policy and resource management issues within the GEC science community; and (iii) it identified, and began to build, a cohort of stakeholders keen to work collaboratively.

Involving stakeholders in research planning can reveal issues that would be missed by a science-alone process. A good example emerged in GECAFS early planning discussions with senior policy-makers in the Indo-Gangetic Plains. There is a policy imperative to address the massive seasonal movement of casual labour from east to west, which brings considerable social upheaval. Hence, addressing labour issues was called for as a component of the GEC–food security agenda for the region, and research questions were developed accordingly (GECAFS, 2008). This dialogue identified a key policy problem that needed immediate solutions, along with the more general concerns about medium- to long-term GEC–food security issues. It challenged the GEC research community to incorporate issues of which they were hitherto ignorant, thereby developing an agenda of greater interest and relevance to policy imperatives. In so doing, it considerably increased the need for a larger number of disciplines to be engaged but this in turn led to greater networking.

Elements of good practice in stakeholder engagement

A number of recent studies have identified problems experienced in the management and communication of research to inform policy-making and regulation (see, for example, Holmes and Clark, 2008; Holmes and Savgard, 2008; Bielak et al, 2009; and Chapter 10). These studies have also identified elements of good practice in respect of the planning and execution of research, the communication of results and the evaluation of uptake and impact which are now discussed briefly. In addition, the Overseas Development Institution's 'Research and Policy in Development Programme' (RAPID) has published a wide range of practical frameworks and tools for researchers, policy-makers and intermediary organizations, which are targeted at developing countries (ODI, undated).

Given the importance of research influence on policy for actually making change happen, how can research-policy interactions best be enhanced? Bammer (Bammer, 2008a; Bammer et al, 2010) presents six checklists which illustrate complementary facets of this complex process:

1 Barriers to cooperation between policy-makers and researchers (Gregrich, 2003).
2 Different emphases of policy-makers and researchers (Heyman, 2000).
3 'Irrefutability' of the evidence versus the 'immutability' of policy (Gibson, 2003a).

4   Five indicators of policy-maker responsiveness to research (Gibson, 2003b).
5   Questions for researchers to think strategically about their interactions with policy-makers (Jones and Seelig, 2004).
6   Questions and suggestions for researchers on how to influence policy and practice (Court and Young, 2006; and see Table 11.1, slightly modified from the original).

As highlighted above, research is more likely to be successful in informing regional policy-making or regulation if it involves the decision-makers mandated to work at this spatial level (and/or their advisors) in the *planning* stages of research projects and programmes. Where the nature of the issue requires it (contested issues, complex systems and uncertain science as discussed above), a broader range of stakeholders should be involved. Each player may see the issue differently, reflecting what might well be their different 'world views': researchers through their disciplinary lenses; policy-makers and regulators as conditioned by the constraints and pressures they are working to; and other stakeholders influenced by their particular concerns and experiences of the issue.

If the answers generated by the research are to be meaningful to these different players, a framing of the research question needs to be arrived at through discussion that reflects their various viewpoints. Framing issues and consequent research questions based on this is inevitably selective, and hence it is important to engage with as many different kinds of problem formulation as possible (Becker, 2003; Shove, 2006).

It is important too that these interactions are sustained through the conduct of the research. If not, then as the questions faced by regional policy-makers evolve, and the research path develops according to the practicalities and consequences of discovery, the questions and the answers may drift apart. If well managed, researchers, policy-makers and other stakeholders will share ownership of the resulting knowledge and system understanding, which will improve the chances that consequent decisions are widely supported.

With regard to *communicating* the findings of research, the approach needs to use a set of communication forms and channels tailored to the audience and the circumstances. Communication through written media should be complemented with face-to-face interaction between researchers and stakeholders, allowing confidence in results to be tested and implications for decision-making to be explored.

The communication strategy needs to be well thought through and planned in advance, and a view developed on the intended impact of the communication. However, the context for communication can change quickly: it is important to anticipate changes where possible and to respond flexibly. Wherever possible, good relationships and understanding between research and user communities should be developed as a helpful precursor to research dissemination. As an example, a high-level briefing to policy-makers, funders and senior scientists from the Indo-Gangetic Plain organized to present

Table 11.1 *A framework of questions and suggestions for researchers aiming to influence policy (adapted from Court and Young, 2006)*

	What you need to know	What you need to do	How to do it
Political context	Who are the policy-makers? Is there policy-maker demand for new ideas? What are the sources/strengths of resistance? What is the policy-making process? What are the opportunities and timing for input into formal processes?	Get to know the policy-makers, their agendas and their constraints. Identify potential supporters and opponents. Keep an eye on the horizon and prepare for opportunities in regular policy processes. Look out for – and react to – unexpected policy 'windows'.	Work with the policy-makers. Seek commissions. Line up research programmes with high-profile policy events. Reserve resources to be able to move quickly to respond to policy windows. Allow sufficient time and resources.
Evidence	What is the current theory? What are the prevailing narratives? How divergent is the new evidence? What sort of evidence will convince policy-makers?	Establish credibility over the long term. Provide practical solutions to problems. Establish legitimacy. Build a convincing case and present clear policy options. Package new ideas in familiar theory or narratives. Communicate effectively.	Build up programmes of high-quality work. Action-research and pilot projects to demonstrate benefits of new approaches. Use participatory approaches to help with legitimacy and implementation. Clear strategy and resources for communication from start. Face-to-face communication.
Links	Who are the key stakeholders in the policy discourse? What links and networks exist between them? Who are the intermediaries and what influence do they have? Whose side are they on?	Get to know the other stakeholders. Establish a presence in existing networks. Build coalitions with like-minded stakeholders. Build new policy networks.	Partnerships between researchers, policy-makers and communities. Identify key networkers and salespeople. Use informal contacts.
External influences	Who are main national and international actors in the policy process? What influence do they have? What are their action priorities? What are their research priorities and mechanisms? How do they implement policy?	Get to know the main actors, their priorities and constraints. Identify potential supporters, key individuals and networks. Establish credibility. Keep an eye on policies of the main actors and look out for policy windows.	Develop extensive background on main actor policies. Orient communications to suit main actor priorities and language. Try to work with the main actors and seek commissions. Contact (regularly) key individuals.

GECAFS research findings was the more valuable as the good working relationships developed over the project life facilitated a full and frank discussion on the value of the research.

Increasingly, the benefits of, and necessity for, two-way communication with a broad range of stakeholders is recognized. In part, this reflects a shift from a top-down, directive approach to securing societal change, to one which centres on encouraging shifts in behaviour of the many individual agents who collectively can achieve the desired outcome.

A further dimension of good practice concerns the characteristics of 'robust' evidence. Clark et al (2002) consider how institutions mediate the impacts of scientific assessments on global environmental affairs (Chapter 3), and conclude that the most influential assessments are those that are perceived by a broad range of actors as having three attributes:

- *salience:* whether an actor perceives the assessment to be addressing questions relevant to their policy or behavioural choices;
- *credibility:* whether an actor perceives the assessment's arguments to meet standards of scientific plausibility and technical adequacy; and
- *legitimacy:* whether an actor perceives the assessment as unbiased and meeting standards of political fairness.

An additional point, and one that been particularly important in relation to the recent 'climate-gate' issue, concerns *transparency*: whether the research process is seen as sufficiently open.

Finally, it is worth noting that for stakeholder engagement to be effective, the personalities of all those involved are important; human nature can frustrate earnest attempts to communicate genially and find consensus. It is crucial not to 'get off on the wrong foot' so some knowledge of the proposed participants' personalities is important, especially when facilitating first-time interactions. Similarly, knowledge of any 'history' of prior interactions between stakeholders (be it good or bad) can be very helpful in designing meetings and other interactions. Setting the right atmosphere for the meeting is also important, and a range of informal and/or social activities can be a key aspect for building trust and developing friendships. These aspects not only help with the meeting itself, but can also develop a strong foundation for longer-term collaborations.

Interactions with stakeholders to enhance decision support for food security

The policy community is often the main stakeholder group of interest to researchers working at the regional level (see Chapter 13). To be of use in supporting policy formulation, research on the development and assessment of possible strategies to adapt food systems to the impacts of GEC should be elaborated in the context of the policy process. As the food security–GEC debate encompasses many complex and interactive issues, a structured dialogue is needed to assist the collaboration among scientists and policy-makers.

This can be facilitated by a variety of decision support approaches and tools, ranging from general discussions and mutual awareness-raising (including formal joint exercises such as scenario construction and analyses; see Chapter 3 and Box 11.7) to simulation modelling, geographic information systems and other tools for conducting quantitative analyses of tradeoffs of given policy options.

Application of a holistic decision support process raises awareness in the policy community of the interactions between GEC and food systems; identifies and communicates the options and constraints facing researchers and policy-makers; identifies methods and tools that best facilitate the dialogue between scientists and policy-makers related to GEC and food systems; and helps both researchers and policy-makers assess the viability of different technical and policy adaptation strategies by analysing their potential consequences (feedbacks) for food security and environmental goals.

But how can decision support best be delivered? Research on improved decision support needs to bring together a number of different approaches: 'integration and implementation sciences' to draw together and strengthen the theory and methods necessary to tackle complex societal issues (Bammer, 2005); research on how an adaptive management ethic and practice that supports the concept of sustainable development can be initiated and implemented in complex, regional or large-scale contexts (Allen, 2001); and the adaptive management approach for incorporating communications, analysis and scenarios development (Lee, 1999; Gunderson and Holling, 2002; Henrichs, 2006). Such approaches lay the foundations for delivering specific support for key policy-makers at the national and regional levels (as outlined by Lal et al,

---

Box 11.7 *Using scenario exercises to facilitate communication among stakeholders*

Scenario exercises can facilitate stakeholder involvement, thus linking research activities more closely to actual decision processes. This can be especially effective where dialogue is centred on uncertainties and complexity, and an assessment of future trends is sought – as is the case when discussing GEC interactions with medium- to long-term implications for food security. Scenario exercises have shown considerable potential to provide a mechanism for involving a range of stakeholders and for facilitating communication between them.

Generally speaking, scenario exercises can be and have been effective in supporting three main clusters in any assessment (Henrichs et al, 2010):

- The research and scientific exploration cluster (i.e. helping to better understand the dynamics of (complex) systems by exploring the interaction between key drivers).
- The education and public information cluster (i.e. providing a space for structuring, conveying and illustrating different perceptions about unfolding and future trends).

- The decision-support and strategic planning cluster (i.e. offering a platform for soliciting views about expected future developments and to analyse trade-offs between pathways).

Ideally, a scenario exercise contributes – to some degree – to all of the above by aiming to enhance *credibility* through expert knowledge (i.e. 'Is the exercise convincing?'), *salience* to stakeholders (i.e. 'Is the exercise relevant?'), and *legitimacy* in the way stakeholders are involved (i.e. 'Is the exercise inclusive and unbiased?'); also see Alcamo and Henrichs (2008). All three aspects have direct relevance to facilitating communication with and between stakeholders.

It is worthwhile noting that the discussions by which scenarios are developed (i.e. the 'process') can be as least as important as the scenarios themselves (i.e. the 'product'), because they allow uncertainties to be discussed by a range of stakeholders – see Biggs et al (2004) or Henrichs et al (2010).

Three process-related benefits particularly contribute to this (following Okali et al, 1994; Haggar et al, 2001; Henrichs et al, 2010):

1   Those who participate in a scenario development process gain better understanding of the issue via the structured dialogue between experts and stakeholders.
2   Scenario exercises offer a 'neutral space' to discuss future challenges, as uncertainty about the future has an 'equalizing effect': as no-one can predict the future, thus no-one is 'right'. This opportunity for open discussion also helps in engendering mutual respect, understanding and trust, which is crucial for building effective research teams for follow-up activities.
3   The discussion of, and reflection on, possible future trends can create the ground to reveal conflicts, common views about goals, or different perceptions about today's challenges.

A scenario exercise was conducted as part of the GECAFS research on GEC and food security in the Caribbean. The exercise's aim was twofold. First, it set out to develop a set of prototype Caribbean scenarios for research on GEC and regional food systems (i.e. develop a 'product'; as published in GECAFS, 2006). Second, the exercise aimed to initiate and facilitate an enhanced dialogue between stakeholders, including researchers and policy-makers from different countries in the region, and regional bodies, on the issue of regional food security (i.e. facilitate a 'process'; see GECAFS (2006) for more details). The enhanced communication and team-building engendered by the scenario process resulted in continued interaction between the stakeholders beyond the scenario exercise itself.

2001). These approaches rely upon a strategy that begins with identifying the key stakeholders, includes a process of reflection to develop a common understanding of the problem, and then proceeds through a joint learning process. But, and as noted above, designing such a strategy – let alone implementing it – is far from straightforward. A greater appreciation of such ideals will, however, help researchers and policy-makers work together so that the best available scientific information informs the policy process. Another way to think of decision support platforms is that they include a set of tools, methods and information that facilitates the dialogue among scientists and policy-makers in a co-learning framework. This co-learning approach is central to the engagement and decision support process. Furthermore, this approach provides a strong basis for social, biophysical and economic scientists to learn how to effectively address complex issues in a holistic, practical setting.

Innovative decision support platforms (such as the 'Questions and Decisions' (QnD) system: Kiker et al, 2005; Kiker and Linkov, 2006) will be needed for food security–GEC research, as they allow the incorporation of complex ecosystem models, and their linkage to environmental-based decision support tools, in a systematic way. Other researchers have placed particular emphasis on trade off analysis, integrating biophysical and socio-economic models in a process or dialogue with policy-makers (Antle et al, 2003; Stoorvogel et al, 2004). Antle (2003) emphasizes that the approach is a 'process', not a model per se. The form of the model and analyses are guided through stakeholder dialogue, thereby helping regional policy-makers and other stakeholders to understand and plan for impacts of GEC in the social, political and economic context in which decisions are made and policies are implemented. Decision support platforms will be used within a decision support process that combines data processing and analysis, modelling, evaluation and assessment tools, enhanced concepts (e.g. vulnerability) and policy projections (national, regional and international). The decision support process will also use a range of dissemination mechanisms (e.g. policy briefs, printed maps). No single decision support system will fit the needs of all situations, so a flexible framework will be needed. The aim of the GECAFS decision support research has been to develop approaches that will help policy-makers and other stakeholders in clear and effective ways. The scenarios exercises were very effective components in this regard, primarily by facilitating communication and mutual understanding of the range of stakeholders' world views.

Figure 11.3 shows how the various components of food systems research (i.e. on food systems and their vulnerability to GEC; adaptation options; scenario construction; and tradeoff analyses) can be brought together in a structured dialogue between scientists and stakeholders. It also shows the critical aspect of joint agenda-setting.

Assessing effectiveness of stakeholder engagement

While enhanced stakeholder engagement might be high on the researchers' agendas, and considerable efforts are made to develop links, it is important to assess its effectiveness. Ultimately, of course, the intention of research in the GEC/food security area is to bring about a change in behaviour of the intended beneficiaries, so that the outcomes of their actions become more effective in combating food insecurity – the research 'impact' (CCAFS, 2009). This can take many years to be firmly seen, and can be hard to measure, not least because of other 'confounding factors' that will influence the eventual impact of any policy. Hence, it is appropriate to examine intermediate measures of uptake and impact which can be evaluated on a shorter timeframe, are more directly determined by the research project and which can provide an early indication as to the likelihood of the eventual outcome. It is meanwhile possible to estimate the effectiveness of stakeholder engagement by undertaking a survey of stakeholders' views vis-à-vis the researchers' aspirations. Survey results can be very revealing and help set priorities for both follow-up studies and enhanced stakeholder engagement. Box 11.8 summarizes some of the questions and responses from a survey of GECAFS stakeholders.

Finally, the following questions serve as a checklist to help researchers undertake an *ex post* analysis of their interactions with the regional policy process (adapted from Bammer, 2008b):

- What was the purpose of providing research support to policy and who was intended to benefit?
- What parts of the policy system were targeted and what research was relevant?
- Who provided the research support and how did they do it?
- What contextual (i.e. broader external context) factors were important in getting the research recognized or legitimated?
- What was the outcome?

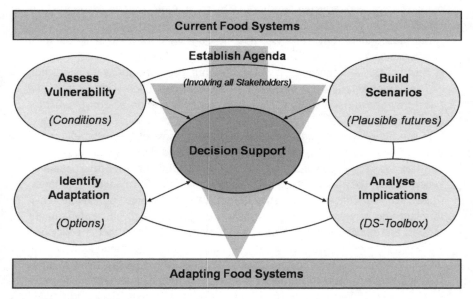

Adapted from Henrichs, 2006

Figure 11.3 *The various components of GECAFS research are brought together in a structured dialogue between scientists, policy-makers and other stakeholders*

Box 11.8 *The GECAFS stakeholder survey*

After about five years of project activities, GECAFS designed and conducted an email survey intended to assess the effectiveness of reaching out to a broad user community (i.e. stakeholders), and also to gather feedback on ways in which to better serve their needs in future.

Questions covered a range of aspects including the nature of respondents' interactions with GECAFS such as: *Does/did your engagement contribute positively to your work? Are there any specific forms of interaction that you have had with GECAFS that you found particularly useful for your work? Can you suggest any ways in which GECAFS can contribute more substantially to your work?* And more general issues such as: *Who, in your opinion, should GECAFS seek to influence with its research work and findings? Can you suggest ways in which GECAFS can strengthen its presence among national and regional stakeholders in your region? To what extent, in your opinion, does GECAFS contribute to making linkages between science and policy-making? Can you suggest any ways in which to strengthen this contribution?* Other questions cover a range of issues such as the nature and format of GECAFS workshops, the website, and the desirability of a newsletter. Feedback was received from about 30 of the ca. 100 recipients.

Almost all the participants in the survey felt that their interaction with GECAFS had contributed positively to their work in concrete ways, and many were able to cite specific examples ('*it expanded my interaction outside the notion of climate change and agricultural production to a food system as defined by GECAFS*').

Most respondents felt workshops and meetings were the most beneficial form of interaction they had with GECAFS in the past, allowing for a better understanding of GECAFS concepts, as well as presenting an opportunity for interaction with policy-makers and for networking ('*allowed interaction with some of the finest scientists in vulnerability science, environmental economist, agricultural production etc.*').

In specific relation to stakeholder engagement, most of the respondents felt that GECAFS' key target groups should be policy- and decision-making communities at the global, regional and national levels, and the science community. Policy-makers specifically identified include politicians, Permanent Secretaries, Chief Technical Officers in ministries responsible for fisheries, agriculture, environment, industry and finance, and fisher folk leaders, agriculture department officials and opinion leaders in rural communities. In addition to the policy and science communities, other target groups identified by the respondents include donors; relevant practitioners and stakeholders including representatives of agro-business, farmers, food consumers, producers and traders; international agencies such as FAO (United Nations Food and Agricultural Organization) and CGIAR; universities (researchers and students); media; NGOs; and relevant research and development institutions.

Several respondents felt, however, that GECAFS outreach to these communities was limited. Some cited possible reasons for this – including the need to have something substantial to offer before committing too much time to outreach, in terms of both findings and resources for further research; the need for funding to carry out work at the regional and national levels, as it is difficult to reach out to politicians and other decision-makers without concrete results; and the difficulties in engaging communities (such as policy-makers) whose timescales are generally more immediate than those of the research community.

Specific methods and tools suggested by the survey respondents to improve GECAFS' outreach and stakeholder interaction included:

- Identify and stick to a few themes.
- Involve stakeholders in designing research projects to create ownership of results.
- Develop smaller, more marketable projects for donor funding and greater stakeholder participation.
- Develop simulation models on food security and its socioeconomic impacts.
- Work with collaborative institutes (like CARDI) who work directly with policy-makers.
- Become a strong player in the work of the European Science Foundation.
- Develop a research programme with the CGIAR.
- Determine who the real decision-makers are in government – those who advise the policy level.
- Piggyback with relevant regional outreach activities.
- Organize regional workshops using research and papers already developed, with decision and policy-makers in attendance.

Conclusions

The ultimate aim of many GEC–food security research projects is to help people adapt to the additional problems GEC will bring to achieving food security – which is, for many, already a complex challenge. For instance, the stated goal of the GECAFS project was drafted as: 'To determine strategies to cope with the impacts of global environmental change on food systems and to assess the environmental and socioeconomic consequences of adaptive responses aimed at improving food security.' 'Determining strategies' involves more than producing science outputs – it requires very active engagement with stakeholders to discuss viability – an unviable plan (albeit scientifically robust) is not a particularly valuable strategy! So, in order to achieve ambitious goals of this nature it can be helpful to clearly differentiate between research *outputs*, research *outcomes* and research *impacts*. This helps 'break down' what might be a high-level project objective into more manageable components and clarifies the comparative roles the different stakeholders play in each.

Coping and adaptation means 'doing things differently' (i.e. changing behaviours). Real research impact will only come about if intended beneficiaries can see the benefits of making such changes, which will be most likely if they understand and trust the research process – and this will most likely be the case if stakeholder engagement is a fundamental aspect of research. This is especially important in agenda-setting and the number of different approaches discussed above (formal agenda-setting workshops, consultancies and informal approaches) can be useful. Different combinations of these approaches, or even all three, help in establishing 'buy-in' to the research process by a wider range of stakeholders than just researchers. Choosing the best approach(es), and deciding when and where implementation would be most effective, is a crucial part of research planning and stakeholder engagement at the regional level.

References

Alcamo, J. and T. Henrichs (2008) Towards guidelines for environmental scenario analysis. In Alcamo, J. (ed) *Environmental Futures: The Practice of Environmental Scenario Analysis,* Oxford, Elsevier

Allen, W. J. (2001) Working together for environmental management: the role of information sharing and collaborative learning. PhD (Development Studies), Massey University

Antle, J., J. Stoorvogel, W. Bowen, C. Crissman and D. Yanggen (2003) 'The tradeoff analysis approach: Lessons from Ecuador and Peru', *Quarterly Journal of International Agriculture,* 42, 2, 189–206

Bammer, G. (2005) 'Integration and implementation sciences: Building a new specialization', *Ecology and Society,* 10, 2

Bammer, G. (2008a) 'Checklists for assessing research-policy interactions', *Integration Insights,* 11

Bammer, G. (2008b) 'Enhancing research collaboration: Three key management challenges', *Research Policy,* 37, 875–87

Bammer, G., L. Strazdins, D. McDonald, H. Berry, A. Ritter, P. Deane and L. van Kerkhoff (2010) Bridging the research-policy gap: useful lessons from the literature. In Bammer, G., A. Michaux and A. Sanson (eds) *Bridging the 'Know-Do' Gap: Knowledge Brokering to Improve Child Well-being,* Canberra, ANU E-Press

Becker, H. (2003) *Making Sociology Relevant to Society,* Murcia, Spain, European Sociological Association

Bielak, A., J. Holmes, J. Savgård and K. Schaefer (2009) A comparison of European and North American approaches to the management and communication of environmental research, Report 5958, Stockholm, Swedish Environmental Protection Agency

Biggs, R., E. Bohensky, P. Desanker, C. Fabricius, T. Lynam, A. Misselhorn, C. Musvoto, M. Mutale, B. Reyers, R. J. Scholes, S. Shikongo and A. S. van Jaarsveld (2004) Nature Supporting People: The Southern African Millennium Ecosystem Assessment Integrated Report, Pretoria, Council for Scientific and Industrial Research

Brklacich, M., I. F. Brown, E. J. D. Campos, A. Krusche, A. Lavell, L. Kam-biu, J. J. Jiménez-Osornio, S. Reynes-Knoche and C. Wood (2007) Stakeholders and global environmental change science. In Tiessen, H., M. Brklacich, G. Breulmann and R. S. C. Menezes (eds) *Communicating Global Change Science to Society: An Assessment and Case Studies,* Washington, DC, Island Press

Brown, M. E. (2008) *Famine Early Warning Systems and Remote Sensing Data,* Berlin/Heidelberg, Springer-Verlag

Cash, D. W., W. Adger, F. Berkes, P. Garden, L. Lebel, P. Olsson, L. Pritchard and O. Young (2006) 'Scale and cross-scale dynamics: governance and information in a multilevel world', *Ecology and Society*, 11, 2

CCAFS (2009) *Climate Change, Agriculture and Food Security. A CGIAR Challenge Program*, Rome and Paris, The Alliance of the CGIAR Centers and ESSP

Chambers, R., A. Pacey and L. A. Thrupp (eds) (1989) *Farmer First: Farmer Innovation and Agricultural Research*, London, Intermediate Technology Publication

Clark, W., R. Mitchell, D. Cash and F. Alcock (2002) Information as influence: how institutions mediate the impact of scientific assessments on global environmental affairs. *Faculty Research Working Paper Series*, Cambridge, MA, John F. Kennedy School of Government, Harvard University

Court, J. and J. Young (2006) 'Bridging research and policy in international development: An analytical and practical framework', *Development in Practice*, 16, 1, 85–90

Ericksen, P. J., J. S. I. Ingram and D. M. Liverman (2009) 'Food security and global environmental change: Emerging challenges', *Environmental Science and Policy*, 12, 4, 373–77

GECAFS (2005) Science Plan and Implementation Strategy. *Earth System Science Partnership (IGBP, IHDP, WCRP, DIVERSITAS) Report No. 2*, Wallingford, GECAFS

GECAFS (2006) A Set of Prototype Caribbean Scenarios for Research on Global Environmental Change and Regional Food Systems. *GECAFS Report No. 2*, Wallingford, GECAFS

GECAFS (2008) GECAFS Indo-Gangetic Plain Science Plan and Implementation Strategy. *GECAFS Report No. 5*, Wallingford, GECFAS

Gibbons, M., C. Limoges, H. Nowotny, S. Schwartzman, P. Scott and M. Trow (1994) *The New Production of Knowledge: The Dynamics of Science and Research in Contemporary Societies*, London, Sage

Gibson, B. (2003a) *From Transfer to Transformation: Rethinking the Relationship between Research and Policy*, Canberra, Australian National University

Gibson, B. (2003b) Beyond two communities. In Lin, V. and B. Gibson (eds) *Evidence-Based Health Policy: Problems and Possibilities*, Oxford, Oxford University Press

Giller, K. E., C. Leeuwis, J. A. Andersson, W. Andriesse, A. Brouwer, P. Frost, P. Hebinck, I. Heitkönig, M. K. van Ittersum, N. Koning, R. Ruben, M. Slingerland, H. Udo, T. Veldkamp, C. van de Vijver, M. T. van Wijk and P. Windmeijer (2008) 'Competing claims on natural resources: What role for science?', *Ecology and Society*, 13, 2

Gregrich, R. J. (2003) 'A note to researchers: communicating science to policy makers and practitioners', *Journal of Substance Abuse Treatment*, 25, 3, 233–37

Gunderson, L. H. and C. S. Holling (eds) (2002) *Panarchy: Understanding Transformations in Human and Natural Systems*, Washington and London, Island Press

Haggar, J., A. Ayala, B. Díaz and C. U. Reyes (2001) 'Participatory design of agroforestry systems: Developing farmer participatory research methods in Mexico', *Development in Practice*, 11, 4, 417–24

Henrichs, T. (2006) On the Role of Scenarios in GECAFS Decision-Support. *GECAFS Working Paper 4*, Wallingford, GECAFS

Henrichs, T., M. Zurek, B. Eickhout, K. Kok, C. Raudsepp-Hearne, T. Ribeiro, D. van Vuuren and A. Volkery (2010) Scenario development and analysis for forward-looking ecosystem assessments. In Ash, N., H. Blanco, C. Brown, K. Garcia, T.

Henrichs, N. Lucas, C. Raudsepp-Hearne, R. David Simpson, R. Scholes, T. Tomich, B. Vira and M. Zurek (eds) *Ecosystems and Human Well-being – A Manual for Assessment Practitioners,* London, Island Press

Heyman, S. J. (2000) Health and social policy. In Berkman, L. F. and I. Kawachi (eds) *Social Epidemiology,* New York, Oxford University Press

Holmes, J. and R. Clark (2008) 'Enhancing the use of science in environmental policy making and regulation', *Environmental Science and Policy,* 11, 702–11

Holmes, J. and J. Savgard (2008) Dissemination and implementation of environmental research. *Report 5681,* Stockholm, Swedish Environmental Protection Agency

Ingram, J. S. I., J. Stone, U. Confalonieri, T. Garvin, P. R. Jutro, C. A. Klink, B. H. Luckman, E. Noellemeyer and P. Mann de Toledo (2007) Delivering global environmental change science to the policy process. IAI-SCOPE Synthesis. In Tiessen, H., M. Brklacich, G. Breulmann and R. S. C. Menezes (eds) *Communicating Global Change Science to Society: An Assessment and Case Studies,* London, Island Press

Jones, A. and T. Seelig (2004) *Understanding and Enhancing Research-Policy Linkages in Australian Housing: A Discussion Paper,* Queensland, Australian Housing and Urban Research Institute, Queensland Research Centre

Kiker, G. A. and I. Linkov (2006) The QnD Model/Game System: Integrating questions and decisions for multiple stressors. In Arapis, G., N. Goncharova and P. Baveye (eds) *Ecotoxicology, Ecological Risk Assessment and Multiple Stressors,* Amsterdam, Kluwer

Kiker, G. A., N. A. Rivers-Moore, M. K. Kiker and I. Linkov (2005) QnD: A modeling game system for integrating environmental processes and practical management decisions. In Morel, B. and I. Linkov (eds) *The Role of Risk Assessment in Environmental Security and Emergency Preparedness in the Mediterranean,* Amsterdam, Kluwer

Kristjanson, P., R. S. Reid, N. Dickson, W. C. Clark, D. Romney, R. Puskur, S. MacMillan and D. Grace (2009) 'Linking international agricultural research knowledge with action for sustainable development', *Proceedings of the National Academy of Sciences of the United States of America,* 106, 13, 5047–52

Lal, P., H. Lim-Applegate and M. Scoccimarro (2001) 'The adaptive decision-making process as a tool for integrated natural resource management: focus, attitudes and approach', *Conservation Ecology,* 5, 2

Lee, K. N. (1999) 'Appraising adaptive management', *Conservation Ecology,* 3, 2

LWEC, Living With Environmental Change. www.lwec.org.uk/ (accessed: 27 August 2009)

Martin, A. and J. Sherington (1997) 'Socio-economic methods in renewable natural resources research', *Agricultural Systems,* 55, 2, 195–216

ODI, Research and Policy in Development (RAPID) Toolkits. www.odi.org.uk/RAPID/Tools/Toolkits/index.html (accessed: undated)

Okali, C., J. Sumberg and J. Farington (1994) *Farmer Participatory Research: Rhetoric and Reality,* London, Intermediate Technology Publications

Ortiz, O., G. Frias, R. Ho, H. Cisneros, R. Nelson, R. Castillo, R. Orrego, W. Pradel, J. Alcazar and M. Bazán (2008) 'Organizational learning through participatory research: CIP and CARE in Peru', *Agriculture and Human Values,* 25, 3, 419–31

Rischard (2001) *High Noon: 20 Global Issues, 20 Years to Solve Them,* New York, Basic Books

Shove, E. (2006) Framing research questions: interactive agenda setting at the science-policy interface. *COST/ESF Workshop "Communicating Interests, Attitudes and Expectations at the Science/Policy Interface (CSPI): Setting Environmental Research Agendas to Support Policy,* Brussels

Stoorvogel, J. J., J. M. Antle, C. C. Crissman and W. Bowen (2004) 'The tradeoff analysis model: Integrated bio-physical and economic modeling of agricultural production systems', *Agricultural Systems*, 80, 1, 43–66

van Tulder, R. and A. van der Zwart (2006) *International Business-Society Management: Linking Corporate Responsibility and Globalization*, New York, Routledge

12
Part III: Main Messages

1 Food systems involve critical interactions at a number of levels on a range of scales (e.g. spatial, temporal, jurisdictional, institutional, management), each of which has its own group or groups of stakeholders. As the non-spatial scales are very relevant to food security/global environmental change (GEC), interactions research has to recognize, and engage with, a wide range of stakeholders; scale and level need to be clearly specified in research engagement activities.

2 Although transactions costs can be high for all involved it is important to engage as full a set of stakeholders as possible in all stages of food security and GEC studies. (In practice, as there are potentially many, this might mean identifying the *key* stakeholders.) This helps to establish the expectations and information needs of all stakeholders early in research planning and maintains their buy-in during the research process, thereby helping to establish and maintain credibility, achieve practicality, demonstrate utility, provide accessibility and ensure acceptability of research.

3 Stakeholder dialogue plays a particularly important role in agenda setting and a range of methods including consultancies, workshops and informal approaches may need to be employed. Although more protracted than would normally be the case for a disciplinary science planning exercise, this process in itself had three important spin-offs: (i) it helps raise awareness of the GEC issues within the policy and other stakeholder communities; (ii) it helps raise awareness of the policy and resource management issues within the GEC science community; and (iii) it identifies, and begins to build, a cohort of stakeholders keen to work collaboratively.

4 Knowledge brokers (or interpreters) and 'boundary organizations' have an important role to play in bridging between research and policy and in facilitating interactions between research, policy and stakeholder communities. The need for them arises from the different motivations, drivers and cultures of the research and policy communities and the time pressures faced by researchers and policy-makers.

5 Setting a GEC–food security research agenda that is relevant to regional (as opposed to global and/or generic) issues needs a highly consultative and inclusive approach. Further, when conducted in the regions of the developing world, the links to the development agenda, and particularly to the Millennium Development Goals, must be explicit.

6 As the food security–GEC debate encompasses many complex and interactive issues, a structured dialogue is needed to assist the collaboration among scientists and policy-makers. This can be facilitated by a variety of decision support approaches and tools, ranging from general discussions and mutual awareness-raising (including formal joint exercises such as scenario construction and analyses) to simulation modelling.

7 Coping and adaptation require changing behaviours. Real research impacts will occur only once intended beneficiaries see the benefits of making such changes. These benefits must therefore be deemed important, relevant and likely to happen. In addition, potential beneficiaries need to understand and trust the research process – and this will most likely be the case if stakeholder engagement is a fundamental aspect of research. But an indispensable condition for a successful stakeholder dialogue is a shared sense of urgency and ambition: all participants should feel the need to solve the problem that is at stake and to make concrete steps in that direction.

PART IV

A REGIONAL APPROACH

13
Why Regions?

Diana Liverman and John Ingram

Introduction

Global environmental change (GEC) science has traditionally been studied as separate parts of the Earth system. These include physical and biophysical aspects (e.g. the climate sub-system, the oceanic sub-system, the carbon cycle, etc.), and social, economic and/or political dimensions, which are particularly important when studying the drivers of change. An alternative approach is to study how these functional aspects interact in sub-global geographical regions. Regions are a natural level for such analysis, and especially for studies of social-ecological systems (such as food systems), as – while clearly not homogenous in all ways – they are often defined by shared cultural, political, economic and biogeographical contexts (Tyson et al, 2002).

The term 'region' is, however, ambiguous. At the coarsest level, the United Nations (UN) defines regions as continents (e.g. Latin America and the Caribbean, LAC – although this level of aggregation can produce its own problems, with many Caribbean nations often preferring not to be 'lumped in' with Latin America), but there is no standard way of dividing the world into regions. However, the term can also be used within a given continent and, in the case of Africa, official UN regions are Northern, Western, Eastern, Southern and Central. Similarly, the African Union has established seven Regional Economic Communities as the key pillars of economic cooperation within the continent. Africa has also been divided into large regions based partly on physical geography such as the Sahara and Sahel (comprising the vast western African desert and the region bordering it to the south), the Horn of Africa (with Ethiopia, Eritrea and Somalia), and the central African Congo (defined around the river basin or the forest). Even within a country the term is used to define formal administrative units (e.g. France is divided into 22 formal regions).

Geographers often point out that regions are 'socially constructed' as ways to organize the world according to perceptions, race and cultural identity, or colonial aspirations, for example (MacLeod and Jones, 2001). Such regions can change such as when, for example, the collapse of the Soviet Union resulted in several countries (e.g. Poland and Lithuania), which had traditionally been classified in the Soviet region, shifting their affiliation and regional grouping to Europe. Political scientists have also noticed how the rise of regional political and economic projects such as regional trade associations (e.g. the North American Free Trade Agreement (NAFTA) and the European Union) have created newly constructed regions (Hettne and Söderbaum, 2000).

These inconsistencies in terminology make it more difficult to address resource and governance issues and call for clarity when using the term 'region'. For global change studies – and as is the case for this book – 'regional' is usually taken to mean the sub-continental, but supra-national, level; the International Geosphere-Biosphere Programme (IGBP) synthesis *Global-Regional Linkages in the Earth System* (Tyson et al, 2002) considered regions as being southern Africa, South Asia, South-East Asia and East Asia.

Importance of the regional level

'Regional' is an important spatial level for food security, food system research and GEC considerations for several reasons.

First, regions make sense in terms of environmental change. We organize our understanding of the world around biophysical classifications that include, at a regional level, ecosystems and river basins. Examples of regional-level ecosystems include grasslands (e.g. of southern Africa and the North American Great Plains), which often convert to rangeland regions for livestock or grain production areas, or tropical forests (such as the Amazon). Agro-ecological zones often map onto these regions of common physical characteristics. Large river basins comprise regions linked by the flows of water and sediment that are often the basis for irrigated agriculture. This physical coherence of regions is the basis for environmental and food system data collection at the regional level (such as Agrhymet in West Africa). Climate and weather-related perturbations often occur at the sub-continental level with major droughts and natural disasters, for example, often spanning large areas. Pollution often affects large regions as contaminated air and water easily cross national borders.

Second, regions can have strong cultural dimensions. Proximity and common ecologies, and physical geographies, are sometimes associated with coherent cultural regions with common language, economies and social practices, including food systems that have strong regional cultural characteristics. Examples include regions of the Mediterranean (such as Tuscany) or rice cultures of Asia (such as those found in Indonesia or China). If the biophysical environment of these regions alters it can create cultural stresses, and when food systems change that can change the physical landscape as land-use changes, for example. Shared cultural traditions can be the basis for land managers with strong regional interests and identities organizing to protect landscapes and food systems.

Third, regions can be useful units for government and governance. Regional governance structures have been established in many parts of the world, such as the EU, or the Southern African Development Community (SADC). Such structures offer a 'client' for GEC/food security research and, indeed, the jurisdictional mandate of such bodies can be used to help define the geographical scope (i.e. spatial level) for a GEC/food security study. Regions have also become an appropriate spatial level for organizing peace-keeping or military security (especially in the aftermath of conflict), and/or for managing shared resources. A good example is in the Mekong River basin, where the governments of Cambodia, Lao PDR, Thailand and Vietnam formed the Mekong River Commission (MRC) to jointly manage their shared water resources and development of the economic potential of the river. The emergence of regional governance is an important reason to consider food security and environmental interactions at the regional level. In some cases, colonial powers imposed national boundaries that divided cultures and ecosystems, creating conflict or barriers, to, for example, regional mobility in response to climate variability.

Finally, intraregional trade can be significant. The friction of physical distance (e.g. transport costs, perishability) means that, where long-distant transport infrastructure is less well developed, trade is often most effective at the regional level and can enhance food security through improved intra-regional trade, strategic food reserves and transport facilities. The emergence of megacities can restructure trading systems to focus food systems across a large region on provisioning urban centres such as Mexico City or Beijing.

There are, however, challenges in taking a regional approach. While many natural science issues have been addressed at the regional level for some time, social science theories, methods and data have traditionally been better developed at the micro- or macro-levels (Rayner and Malone, 1998). This is perhaps surprising given that governance, for instance, is often central to the widespread water-related issues (e.g. the MRC example, above). Indeed, one of the affects of the rapidly increasing population and growing fears of conflict over water, has been the emergence and proliferation of 'a montage of water-related associations, programmes and organizations'; what Varady and Iles-Shih (2009) refer to as *global water initiatives*. However, they go on to say: 'because these institutions have sprung from numerous and often divergent sources, attempts to develop innovative and practical observations and recommendations have sometimes been frustrated by the sheer number of voices and diversity of approaches continually emanating from this dynamic institutional "ecosystem"'.

Nonetheless, as Wolf et al (2003) note, 'the record of acute conflict over international water resources is overwhelmed by the record of cooperation' and that 'overall, shared interests, human creativity and institutional capacity along a waterway seem to consistently ameliorate water's conflict-inducing characteristics'. So, while the example of water management at a regional level shows the potential benefits of undertaking integrated approaches (and research to support them), it is not straightforward and there is still a relative lack of studies of the social-ecological dynamic encompassed in food systems at this level.

Overcoming the mismatch between disciplinary fields at different spatial levels is, however, crucially important, as it will help fill a research gap between the many sub-national and national analyses of food production and food security (as conducted by national governments and the UN, for instance), and those at the global level (e.g. Fischer et al, 2005; Parry et al, 2005). Conducting food system research at the regional level also means that it can address both rural and urban issues, and the relationship between them. Data collection is also a challenge at regional levels if the region of interest includes parts, but not all of several nations because data, especially economic and social, is often collected at the national level and data systems may vary between countries.

GEC/food security research at different scales and levels

Moving research from local to regional

There have been a large number of experimental studies under the 'food security' banner addressing food production. Most have addressed crop or animal productivity (i.e. yield), and have reported research conducted at the experimental plot level (i.e. very local) over a growing season or perhaps a few years. However, many of the issues related to regional food production, and even more so to regional food security, operate at larger spatial and temporal levels, and warrant further research.

Aware of the need for better links between agronomic research on crop productivity at plot level and regional production, and especially over time, the last decade or so has seen agronomists begin to establish trials at landscape level (e.g. Veldkamp et al, 2001). Estimating regional production is not, however, just a matter of 'scaling up' plot-level agronomic trials, as the critically important social and institutional processes operating at higher levels need to be factored in. Put another way, studies that scale up from plot to regional level can be misleading at best and could lead to actions that impede real progress towards food security unless social and economic components are at the heart of the process. Hence, a considerable methodological challenge to be overcome at such levels is for agronomists to work more effectively with economists and social scientists, as well as with system ecologists, to capture the key economic and social processes, as well as biophysical and ecological processes at play at different spatial levels (Ingram et al, 2008). This includes not only adopting a more interdisciplinary approach but also analysing interactions among variables from one level to the other. For instance, a decrease in maize yield at the plot or field level may lead farmers to decide to shift to other crops (e.g. beans or cassava). If the shift is significant at the regional level, changes in the price of maize versus that of the alternative crops will take place. These changes in relative crop prices will trigger further changes in farmers' practices and in their adaptation of their systems to the market (Ingram et al, 2008). In contrast to agronomic studies, agricultural economic studies have often undertaken analyses at higher spatial levels, especially on economic and market implications; for example, the Institute Food Policy Research Institute's (IFPRI) IMPACT model (as discussed by Rosegrant and Cline, 2003).

Crop modellers have meanwhile been running simulations of crop yield over large areas for some time. Early approaches (e.g. Rosenzweig and Parry, 1994) used point-based estimates, scaled up using climate model output (which is only available at the higher level). More recent studies (e.g. Parry et al, 2005; Challinor et al, 2007) do model crop response at higher levels, but as they stress the influence of weather and climate, and their basis in observed relationships, large-area crop models do not currently simulate the non-climatic determinants of crop yield. These non-climatic stresses contribute to the yield gap (Challinor et al, 2009); that is, the gap between potential and actual yields. However, as such models do not encompass changes in the proportion of land under cultivation, it is not possible to estimate how regional production will actually change. Certainly, reliable information is needed on plot-level responses to environmental stresses that can be scaled up geographically, but not in isolation from the other major regional drivers of food systems. Coupling models at different spatial levels from plot to region allows the study of interactions and feedbacks among biophysical and social components at different levels. There is therefore a need to design interdisciplinary research that starts with GEC objectives at a regional level, and to build systems that facilitate better understanding of these interactions and feedbacks. The suit of 'point' (or plot-level) crop models now available (e.g. DSSAT, APSIM, SUCROS) provide a valuable foundation for such work. Regional-level studies can be greatly facilitated, and very useful information provided to social and economic models, when the point models are integrated with downscaled climate model results.

Other modelling studies at regional level address how 'mega environments' for major crops will change (e.g. for wheat; Ortiz et al, 2008), and how the biogeography of major and locally important crops, and crops' wild relatives will be affected (e.g. Jarvis et al, 2008). Cross-scale and cross-level interactions are not, however, generally included in modelling studies, other than where a spatial-scale issue has direct relevance, as is increasingly the case for multi-scale scenario studies (see Chapter 15).

In addition to considering 'up-scaling' research on food production, there is a need to also consider research at more integrated levels for other aspects of the food system. Food storage is another key determinant of food security, and is especially important during times of stress. It is, however, a complex issue, crossing a number of levels on spatial, temporal and jurisdictional (and possibly other) scales. While research has addressed the issue of strategic food reserves at village level (e.g. Mararike, 2001) and national level (e.g. Olajide and Oyelade, 2002), there is insufficient research into how best to establish long-term food reserves at a regional level. These could be a highly effective means of coping with impacts of major droughts or other stresses that manifest at the regional level, but the issues are often highly charged politically and progress can be slow. For instance, since the 1980s, SADC has considered the establishment of a strategic food reserve to deal with the growing frequency of natural disasters (see Chapter 14). Early proposals were based on considerations of enough physical maize stock for 12 months' consumption, but the SADC Council of Ministers have only recently agreed that the food reserve proposal should be revisited and should include consideration of both a physical reserve and a financial facility, supporting the notion of enhanced intraregional trade (Mano et al, 2007).

Other food system activities such as food distribution and logistics and consumption patterns also warrant further analysis at regional level. An example of an initial analysis of current knowledge and future research needs of all the major activities of the European food system is provided in the ESF/COST Forward Look on 'European Food Systems in a Changing World' (ESF, 2009).

Cross-scale and cross-level interactions for food security

The importance of scales and scaling as determining factors in many environmental and food security problems is discussed in Chapter 2. In terms of food security management, cross-scale (e.g. space, time) and cross-level (e.g. local–global; annual–decadal) interactions are crucial and have to be central to the formulation of food security policies. In general, there are three situations in which combinations of cross-scale and cross-level interactions threaten to undermine food security (Cash et al, 2006):

- *Ignorance*: the failure to recognize important scale and level interactions in food systems (e.g. distress cattle sales that reduce national price).
- *Mismatch*: the persistence of mismatches between levels and scales in food systems (e.g. food security responses planned at national level versus community level).
- *Plurality*: the failure to recognize heterogeneity in food systems in the way that scales are perceived and valued by different actors, even at the same level (e.g. local food aid programmes versus local social safety nets).

As Cash et al (2006) note, there is a long history of disappointments in policy, management and assessment arising from the failure to take into account the scale and cross-scale dynamics in social–environmental systems. For instance, the management of food systems, and the food security they underpin both over time and space, is an excellent example. Regional-level studies, especially when based on an awareness of the potential risks of ignorance, mismatch and plurality, can help identify impediments to achieving food security. This also helps to frame new research questions of direct relevance to policy formulation, for example: How would interactions among rules, laws and constitutions affect food system adaptation at different spatial levels (cross-institutional/spatial scales issues)? How would short-term changes in donor philosophy on food- or seed-aid as applied at the local level affect long-term regional self-reliance (cross-time/management scale issues)? How would implementing different short-term adaptation policies in different nations influence regional food security goals (cross-jurisdictional/management scales issues)? Box 13.1 gives a case study of scale challenges of ignorance, mismatch and plurality in relation to the distribution of emergency food aid in the 1991/92 drought in southern Africa. The situation was exacerbated by the legacy of colonial investment in transport infrastructure, which concentrated on communication lines to main ports rather than more generally within the region and between countries.

Box 13.1 *Example 'scale challenges' related to distribution of emergency food aid in the 1991/92 drought in southern Africa*

In 1991/92 southern Africa experienced one of its worst droughts, with 2.6 million square miles stricken, 86 million people affected, 20 million people at 'serious risk' and 1.5 million people displaced. In response, the international community shipped millions of tonnes of food aid to the region, with the plan to distribute to the hinterlands along six 'corridors' from the region's main ports: Dar es Salaam, Nacala, Beira and Maputo, Durban, Walvis Bay and Luanda.

While many lives were saved, the overall effort was severely frustrated by a number of scale challenges.

Ignorance
- National toll and quarantine policies vis-à-vis donor approach: *the regional response strategies to move food around the region seemed ignorant of the range of different national policies, thereby delaying moving relief food across international borders.*
- Global response vis-à-vis poor regional port management: *the massive international aid operation erroneously assumed the region's ports could unload and forward on food aid in large amounts.*

Mismatch
- Jurisdiction of the national institutions is not coterminous with supplying food to the region: *national institutions were not equipped to arrange the distribution of food at the regional level.*
- Urgency of food need poorly matched with institutional response speed: *the institutions charged with managing the crisis were unable to act at the rate needed to satisfy demand.*

Plurality
- Conflict between humanitarian requirements and commercial concerns: *the suppliers of transport and other infrastructure were usually businesses, not relief agencies.*
- Variety of objectives among donors, recipients and regional institutions: *the different objectives of many actors involved in the relief effort were not necessarily synergistic.*

Conclusions

The two-way interactions between GEC and food security manifest at the full range of spatial levels from local to global. To date, however, almost all studies have tended to focus on these two extreme levels and information for sub-global (continental or sub-continental) geographical regions is sparse. This is despite the fact that a range of options for adapting food systems to GEC and other stresses only become apparent when a regional viewpoint is adopted (e.g. regional strategic grain reserves, harmonized tariffs and taxes, or regionally managed water resources). Further, sub-global is a natural level for studies of social–ecological systems (such as food systems), as – while clearly not homogenous in all ways – they are often defined by shared cultural, political, economic and biogeographical contexts.

Research at a regional level can thus offer a range of benefits to researchers, policy-makers, natural resource managers and other stakeholders, and warrants receiving more attention in the GEC/food security debate.

References

Cash, D. W., W. Adger, F. Berkes, P. Garden, L. Lebel, P. Olsson, L. Pritchard and O. Young (2006) 'Scale and cross-scale dynamics: governance and information in a multilevel world', *Ecology and Society*, 11, 2

Challinor, A., T. Wheeler, C. Garforth, P. Craufurd and A. Kassam (2007) 'Assessing the vulnerability of food crop systems in Africa to climate change', *Climatic Change*, 83, 381–99

Challinor, A. J., F. Ewert, S. Arnold, E. Simelton and E. Fraser (2009) 'Crops and climate change: progress, trends, and challenges in simulating impacts and informing adaptation', *Journal of Experimental Botany*, 60, 10, 2775–89

ESF (2009) European Food Systems in a Changing World. *ESF-COST Forward Look Report*, Strasbourg, European Science Foundation

Fischer, G., M. Shah, F. N. Tubiello and H. van Velhuizen (2005) 'Socio-economic and climate change impacts on agriculture: an integrated assessment, 1990–2080', *Philosophical Transactions of the Royal Society Biological Sciences*, 360, 2067–83

Hettne, B. and F. Söderbaum (2000) 'Theorising the Rise of Regionness', *New Political Economy*, 5, 457–72

Ingram, J. S. I., P. J. Gregory and A.-M. Izac (2008) 'The role of agronomic research in climate change and food security policy', *Agriculture, Ecosystems and Environment*, 126, 4–12

Jarvis, A., A. Lane and R. J. Hijmans (2008) 'The effect of climate change on crop wild relatives', *Agriculture Ecosystems & Environment*, 126, 13–23

MacLeod, G. and M. Jones (2001) 'Renewing the geography of regions', *Environment and Planning D*, 19, 6, 669–95

Mano, R., J. Arntzen, S. Drimie, O. P. Dube, J. S. I. Ingram, C. Mataya, M. T. Muchero, E. Vhurumuku and G. Ziervogel (2007) Global environmental change and the dynamic challenges facing food security policy in Southern Africa. *GECAFS Working Paper 5*, Oxford, GECAFS

Mararike, C. G. (2001) 'Revival of indigenous food security strategies at the village level: the human factor implications', *Zambezia*, 28, 1, 53–66

Olajide, J. O. and O. J. Oyelade (2002) 'Performance evaluation of the strategic grain reserve storage programme (SGRSP) in Nigeria', *Technovation*, 22, 7, 463–68

Ortiz, R., K. D. Sayre, B. Govaerts, R. Gupta, G. V. Subbarao, T. Ban, D. Hodson, J. A. Dixon, J. I. Ortiz-Monasterio and M. Reynolds (2008) 'Climate change: Can wheat beat the heat?', *Agriculture Ecosystems & Environment*, 126, 46–58

Parry, M. L., C. Rosenzweig and M. Livermore (2005) 'Climate change, global food supply and risk of hunger', *Philosophical Transactions of the Royal Society Biological Sciences*, 360, 1463, 2125–38

Rayner, S. and E. L. Malone (1998) *Human Choice and Climate Change, Vol. 4: What Have we Learned?*, Washington, DC, Battelle Press

Rosegrant, M. W. and S. A. Cline (2003) 'Global food security: Challenges and policies', *Science,* 302, 5652, 1917–19

Rosenzweig, C. and M. L. Parry (1994) 'Potential impact of climate change on world food supply', *Nature,* 367, 6459, 133–38

Tyson, P., R. Fuchs, C. Fu, L. Lebel, A. P. Mitra, E. Odada, J. Perry, W. Steffen and H. Virji (eds) (2002) *Global-Regional Linkages in the Earth System,* New York, Springer-Verlag

Varady, R. G. and M. Iles-Shih (2009) Global water initiatives: What do the experts think? In Biswas, A. K. and C. Tortajada (eds) *Impacts of Megaconferences on the Water Sector,* Springer, Berlin

Veldkamp, A., K. Kok, G. H. J. De Koning, J. M. Schoorl, M. P. W. Sonneveld and P. H. Verburg (2001) 'Multi-scale system approaches in agronomic research at the landscape level', *Soil and Tillage Research,* 58, 3–4, 129–40

Wolf, A. T., S. B. Yoffe and M. Giordano (2003) 'International waters: Identifying basins at risk', *Water Policy,* 5, 29–60

14
Stakeholders' Approaches to Regional Food Security Research

John Ingram and Kamal Kapadia

Introduction

Building on the arguments for the need for research on food security and global environmental change (GEC) at the regional level (Chapter 13), this chapter focuses on how different types of stakeholders work at regional level. By considering a number of examples, it shows how three groups of stakeholders (researchers, regional policy agencies and donors) engage in 'regional science'. These groups often interact in complex ways, sometimes in support of each other, sometimes in conflict, and they can have fundamental differences in their interests in a particular region. They also all interact with other types of organizations working within and at the regional level – for instance, non-governmental organizations (see Chapter 18). In all cases, history and place are important and examples are drawn from a number of different regional projects.

Researchers

There have been numerous studies under the banner of 'food security' research, but most have addressed crop or animal productivity (i.e. yield) at the experimental plot level, or socio-economic aspects at household or local levels. Most do not claim this to be 'regional' research, although results may be 'scaled up' to derive an estimate of the issue under investigation for the region as a whole. Alternatively, researchers can take a case study approach based on the assertion that the research site/country is typical of a much bigger area. Examples can be found in some of the 'regional' projects of START's 'Assessments of Impacts and Adaptation to Climate Change' (AIACC) project:[1] a 'regional' project 'Food Security, Climate Variability and Climate Change in Sub Saharan West Africa' used Nigeria as a sample area because the country represents the climatic profile of the region. Another, 'Assessment of Impacts, Adaptation, and Vulnerability to Climate Change in North Africa: Food Production and Water Resources', used Egypt and Tunisia as the case studies. While these studies have produced excellent results at the level of investigation, they have not been studies of the region as a whole.

Other projects undertake 'regional' analyses, based on integrating in some way results from multiple research sites across a given region. These enable a better estimate to be determined of the overall value of the phenomenon under investigation (and certainly of its variation across the region) than can be derived from scaling up from sites/case studies. They do, however, pose far greater challenges in their implementation, and are much more costly (see Chapter 15). Further, they do not generally take the region as a whole as the unit of analysis.

How, then, have researchers addressed the 'regional' issue for GEC/food security research? The International Food Policy Research Institute's (IFPRI) Regional Strategic Analysis and Knowledge Support System (ReSAKSS) does focus on what policies and types of investment are needed to increase agricultural growth at the regional level (defined as Central and Eastern Africa, Western Africa, Southern Africa), but it does not specifically link into GEC. An example of a research programme that is more focused on a particular aspect of regional food security, and that takes GEC into account to some extent, is that of IFPRI's Regional Network on AIDS, Livelihoods and Food Security (RENEWAL),[2] which is bound within southern and East Africa. Although many of RENEWAL's studies have focused on particular countries, several have taken a cross-country or issue-based approach. These will be synthesized in 2010, after nine years of work, to draw out regional aspects of HIV/AIDS and hunger. The intention is to define the responses (policies and programmes) that are required to underpin food security in a context of high HIV/AIDS – entwined with other stressors including GEC.

While field research at the regional level might be relatively uncommon, recent years have seen an increase in modelling studies of impacts of climate change on agriculture (or on factors affecting agriculture; e.g. precipitation, dry days, soil moisture, etc.) for very large nations (Russia: Dronin and Kirilenko, 2008; e.g. China: Tao et al, 2009), and at continental-level (Maracchi et al, 2002; e.g. Africa: Challinor et al, 2007; Europe: Fronzek and Carter, 2007; Olesen et al, 2007). However, studies that examine food security at regional level in relation to sub-regional level are apparently missing: an analysis of 100 scientific papers over the last five years – identified by a Web of Science search on 22 May 2009 for the terms global environmental change *or* climate change, *and* food systems *or* food security *or* agriculture – revealed that only a few studies (six out of 100) are regional in scope, and even these did not study spatial level interactions. Some of the papers in other categories (e.g. continental-level analyses) also have regional components or point to the need for regional analysis; however, they do not themselves conduct regional analyses per se (i.e. the scope of the study is not regional).

Other studies (e.g. Maunder and Wiggins, 2006; Fraser, 2007; Schmidhuber and Tubiello, 2007) have addressed scientific and policy challenges and/or ways forward for studying effects of climate change on agriculture, food systems and food security more generally. There is a need to bring these types of studies together to address better the interactions between food security policies and adaptation strategies across a range of spatial levels (Ingram et al, 2008).

Regional policy agencies

As noted above, there are a range of agencies established around the world mandated to operate at the regional level. An analysis of several of these shows that, in some parts of the world, issues of food security have historically not been well coupled with environmental concerns. There is, however, a marked difference in recent years in the way agencies are approaching the GEC/food security debate, and all those analysed are now working towards adopting a closer integration of food security and GEC issues.

The Southern Africa Development Community's (SADC) key strategy document (the 15-year Regional Indicative Strategic Development Plan, RISDP) (SADC Secretariat, 2003) identifies various aspects of GEC (climate change, desertification, soil erosion and degradation, water pollution and scarcity, and depletion of forests and other natural resources caused by inappropriate agricultural practices) and urban development and population growth amongst the causes of poverty. It also reviews the region's strategies on food security and identifies 'sustainable food security' as a priority intervention area (although links between GEC and food security are not central to the strategy, as GEC is treated as an extension of disaster preparedness for food security). This does, however, give GEC a sense of urgency and disaster-related activities such as broadening and strengthening an existing early warning system, establishing a regional food reserve facility, and adopting measures to rehabilitate and bridge current and future, as well as local and regional, perspectives on environmental stress and food security. More recently, SADC's Infrastructure and Services Directorate (Water Programmes) has addressed GEC and food security indirectly through the SADC–DANIDA Regional Water Sector Programme.[3] The focus, however, is largely on climate change and water systems (e.g. through a 2008 SADC Multi-stakeholder Water Dialogue on 'Watering development in SADC: Rising above the Climate Change threat towards Security'), as opposed to food systems per se. Another more recent initiative which does explicitly make the GEC–food security link is that of the New Partnership for Africa's Development's (NEPAD) Framework for African Food Security (FAFS) (or Pillar Three of the Comprehensive African Agriculture Development Programme, CAADP).[4] This has been endorsed by a number of the Regional Economic Communities, including SADC and the Common Market for Eastern and Southern Africa (COMESA).

Africa-wide agencies also vary in their approach to GEC/food security issues: COMESA's Agricultural Marketing Promotion and Regional Integration Project (AMPRIP) has moved to a regional approach in dealing with the region's climate change/food security issues centred on trade-based strategies. CAADP, however, has no mention of GEC in its programme specification, although some CAADP workshops and other reports do, however, mention the impact of climate change on food supply and food security.

In the case of the South Asian Association for Regional Cooperation (SAARC), issues of GEC and food security were similarly addressed mostly independently of each other via its Technical Committees on the Environment, and Agriculture and Rural Development, respectively. While earlier activities, such as the commissioning of a 'Study on the Greenhouse Effect and its Impact on the Region' in 1992 did not explicitly address food security, things are now changing. In April 2007 SAARC reached an agreement for establishing a SAARC Food Bank; recognizing, amongst other imperatives, the use of a food bank 'as a means of combating the adverse effect of natural and man-made calamities' (SAARC, 2007). Further, the creation of a SAARC Climate Action Plan 2008 directly addresses food security: adaptation in agriculture and to extreme climatic events are noted explicitly (SAARC, 2008a). Most recently, the SAARC Summit (August, 2008) included a strong emphasis on climate change, especially in the context of meeting food, water and energy needs and ensuring people's livelihoods. Heads of state affirmed their resolve 'to ensure region-wide food security and make South Asia, once again, the granary of the world' (SAARC, 2008b). While the emphasis in this statement is still largely on improving agricultural production and agricultural technologies, it identified the management of climatic risks in agriculture as a focal issue. While SAARC has clearly identified the importance of the relationship between climate change and food security, it is important to recognize that SAARC's progress on many fronts has been, at best, mixed because of intraregional geopolitical tensions, largely between India and Pakistan, but also between India and its other neighbours (Rajan, 2005).

In contrast to the African and Asian examples, the Caribbean Community (CARICOM) activities pertaining to GEC and food security have been in place for some years, and are implemented through at least four regional institutions: the Caribbean Community Climate Change Centre (CCCCC, which coordinates the Caribbean region's response to climate change and is the primary repository and clearing house for regional climate change data, providing climate change-related policy advice and guidelines to the CARICOM Member States through the CARICOM Secretariat)[5]; the Caribbean Institute for Meteorology and Hydrology (CIMH); the Caribbean Environmental Health Institute; and the Caribbean Disaster Emergency Response Agency. However, most of these deal with weather and climate-related GEC, and mainly the relationship of these with agriculture and food production. Indeed, the earlier project, 'Mainstreaming Adaptation to Climate Change' (MACC), was the initial instrument for mainstreaming climate change into agriculture. Most recently, CCCCC has developed a Regional Framework for Achieving Development Resilient to Climate Change: 2009 to 2015 (CCCCC, 2009), agreed upon by Heads of Government in July 2009.

At the European level, a number of initiatives with direct relevance to the GEC/food security debate have been developed over recent years (although they focus on climate change). These have included foresight exercises (e.g. by the European Commission's Standing Committee on Agricultural Research, SCAR, established in 1974), while the EC's Framework Programme has funded a wide array of research projects from crop modelling to supply chain studies. Another study, the Forward Look on 'European Food Systems in a Changing World' (funded jointly by the Framework Programme (via COST) and the European Science Foundation), was specifically designed to identify research-for-policy needs on European food security up to about 2030 (ESF, 2009). GEC was one of the main drivers considered. Another good example from the EC is the Water Framework Directive, even though its focus is on water quantity and integrated river basin management. One of the key mechanisms being proposed is full-cost recovery by increasing water prices. While this will inevitably lead to an increase in the cost of agricultural production, and therefore a potential decrease in food affordability, it will help to improve water-use efficiency. The Common Agricultural Policy (CAP) is another EC instrument which shows how successful initiatives have been undertaken at regional (i.e. EU) level to address aspects of food security. The CAP originated from the need to increase food production in the decades after the Second World War and although its role has since changed to include the protection of small farmers and rural regions, the CAP determines to a large extent how Europe's agricultural land is used, while being a strong instrument to react to GEC.

Clearly, there is an increasing role for regional bodies to address GEC/food security issues more effectively. As noted at a recent LEAD workshop 'African Leadership on Climate Change: Challenges and Opportunities for Regional Institutions', held in Tunis in January 2009 (ENDA and LEAD, 2009), African leadership at the regional level is urgently needed. Opportunities exist for regional bodies to: (i) provide a 'regional canvas' of climate change action within which the existing disparate national initiatives could locate themselves; (ii) strengthen horizontal and vertical frameworks for regional, continental and global collaboration; (iii) ensure that the research and data basis for climate policy in Africa is considerably strengthened to enable effective action on adaptation and mitigation; and (iv) provide opportunities and funding for successful grass-roots and national initiatives to be scaled up and replicated at a meaningful level.

Donor agencies

Most donor agencies fund projects at a national or subnational level. Two bilateral agencies in particular (UK's Department for International Development, DFID; and the US Agency for International Development, USAID) have, however, placed specific emphasis on regional approaches (vis-à-vis funding a number of studies within a region at a national/subnational level).

Tackling adaptation to climate change is now a key focal area for DFID's strategy on poverty reduction.[6] This focus has begun to manifest in DFID's Regional Strategy Plans and Regional Assistance Plans. However, DFID does not have a regional plan or strategy for all the regions in which it operates (e.g. there is no regional plan for South or Southeast Asia). Further, all the regional plans do not have a fully regional focus: the Strategy Paper for Central Asia and the South Caucasus (DFID, 2000), and the Regional Assistance Plan for the Western Balkans 2004/5–2008/9 (DFID, 2004a), are still founded on country-level programmes (the regional aspect largely revolves on strategy coordination between donors working in the region). The three regional plans that are explicitly regional and mention DFID's support for programmes pertaining to GEC and food security are the Latin America Regional Assistance Plan 2004–7 (DFID, 2004b), the Southern Africa Regional Plan 2006 (DFID, 2006) and the UK Government's Regional Development Strategy for the Caribbean 2008 (DFID, 2008).

DFID's Southern Africa Regional Plan 2006 (DFID, 2006) provides a detailed rationale for a regional approach and also discusses the challenges a regional approach presents. The document lists food security and climate change as (distinct through related) areas of priority for DFID's regional programming. It recognizes the impact of climate variability on agricultural livelihoods, the need for 'significant adaptation in food production options', and the need to incorporate climate change data into food security and water policies. In 2008, DFID helped launch the LEAD-DFID Climate Change Adaptive Capacity Building Project in Africa, focused on building community resilience to climate change, developing technical best practices, strengthening community networks and influencing decision-makers.[7] In fact, the establishment and funding of OneWorld and the Regional Hunger and Vulnerability Programme (RHVP) are indicative of DFID wanting to establish regional entities to engage hunger and its underlying causes.

USAID has a strategic programme on climate change that is defined regionally (e.g. the Global Climate Change Program for Africa, and the Global Climate Change Program for Asia and the Near East),[8] although in practice the programmes tend to be implemented at a country level. USAID's chief research-oriented programmes that combine GEC and food security are called Collaborative Research Support Programs (CRSPs), wherein US land-grant universities, in collaboration with international partners, carry out the international food and agricultural research mandate of the US government. A detailed study of all current research projects funded through CRSPs reveals that many lack any GEC component, and those that incorporate GEC analysis rarely involve regional analyses.

There are, however, important exceptions within USAID's portfolio. For example, the Livestock Information Network and Knowledge System (LINKS) project in East Africa, funded through the Global Livestock CRSP, developed a regional model to aim to provide real-time information on forage availability, water resource, conflict and market-price information via the internet, radio, short message service, and other channels, to policy-makers, market traders and middlemen, and livestock producers throughout East Africa. Another project, in the Andean Highlands (funded through the Sustainable Agriculture and Natural Resource Management CRSP), aimed to improve the ability of rural communities in Andean highland (Altiplano) ecosystems to adapt to climate, market and social changes. The project analysed the interrelated changes and dynamics of climate, pests and diseases, soils, biodiversity, livelihoods and markets, and the ability to act. Such projects indicate USAID's recognition of the need to conduct GEC and food security research at the regional level to inform policy and practice.

Linking stakeholders at the regional level

The benefit to be derived from these various types of stakeholders working together on GEC/food security issues at the regional level is perhaps obvious, but finding ways in which they can interact most effectively is challenging. Each has its own mandate, objectives and operating culture, and most are more accustomed to working at a subregional level; finding common ground between the research community, the policy process and the interests of donors and other interested groups at regional level can be problematic and can hinder the stakeholder engagement process. Indeed, as Vogel et al (2007) point out, a number of problems persist, including the difficulties of developing consensus on the methodologies used by a range of stakeholders across a wide region; the slow delivery of products that could enhance resilience to change that reflects not only a lack of data and need for scientific credibility, but also the time-consuming process of coming to a negotiated understanding in science–practice interactions; and the need to clarify the role of 'external' agencies, stakeholders and scientists at the outset of the dialogue process and subsequent interactions.

It is, however, clear that endeavouring to involve all these groups, and others including non-governmental organizations, businesses and other non-state actors (see Chapter 18), in the development of new research to address regional GEC/food security issues delivers innovative agenda; as discussed in Chapters 10 and 11, the close involvement of all stakeholders in defining and designing regional research is actually key to delivering not only interesting, but policy-relevant, science.

To this end the development of the GECAFS Southern Africa Science Plan and Implementation Strategy (GECAFS, 2006) greatly benefited from close liaison with both SADC's Food, Agriculture and Natural Resources Directorate, regional researchers and regional donors. Similarly, representatives from CIMH, the University of the West Indies and the CARICOM Secretariat (with the support of CCCCC) were all closely involved in the development of the GECAFS Caribbean Science Plan and Implementation Strategy (GECAFS, 2007). Having major regional bodies (including FANRPAN, FARA, ICSU-Africa and NEPAD), and CARICOM's Council for Trade and Economic Development (COTED), then endorse GECAFS plans in southern Africa and the Caribbean regions respectively, added political weight to the regional plans. Chapter 15 discusses methods for establishing such collaborations at the regional level.

Conclusions

The number of stakeholders of various types interested in GEC and food security research at the regional level is increasing. Different combinations of these types (and sometimes all) need to be brought together to identify and debate potential adaptation options at this spatial level for improving food security in the face of GEC. A good example is the potential for establishing regional strategic grain reserves to provide food during times of crisis, as has been recently agreed in SSARC and discussed in southern Africa for many years. This is clearly a high-level policy decision, needing political will (regarding, for example, agreements on when and where to hold reserves, and how the reserve is to be managed), socio-economic and natural science (regarding, for example, food supply and analysing benefits and any drawbacks), NGO involvement (regarding, for example, food aid and development objectives) and donor involvement (regarding, for example, commitment to financing).

The regional approach can identify a wide variety of adaptation options and opportunities that only become apparent when viewed at this spatial level. To realize them, however, needs the various stakeholders who are mandated to work at this level to come together within a common framework. Providing such a framework has been a central aspect of the GECAFS regional research approach.

Notes

1   www.start.org/Program/AIACC.html (undated)
2   www.ifpri.org/renewal (undated)
3   www.sadcwater.com/index.php (undated)
4   www.nepad-caadp.net/ (undated)
5   www.caribbeanclimate.bz/news.php (undated)
6   www.dfid.gov.uk/aboutdfid/default.asp (undated)
7   www.lead.org/page/457 (undated)
8   www.usaid.gov/our_work/environment/climate/pub_outreach/ane_brochure.html (undated)

References

CCCCC (2009) *Climate Change and the Caribbean: A Regional Framework for Achieving Development Resilient to Climate Change. 2009–2015*, Greater Georgetown, Guyana, Caribbean Community Climate Change Centre

Challinor, A., T. Wheeler, C. Garforth, P. Craufurd and A. Kassam (2007) 'Assessing the vulnerability of food crop systems in Africa to climate change', *Climatic Change*, 83, 381–99

DFID (2000) Central Asia and the South Caucasus: strategy paper. *Country Strategy Papers (CSPs)*, London, Department for International Development

DFID (2004a) Regional Assistance Plan for the Western Balkans 2004/05. *Regional Assistance Plans (RPAs)*, London, Department for International Development

DFID (2004b) Latin America Regional Assistance Plan 2004–2007. *Regional Assistance Plans (RAPs)*, London, Department for International Development

DFID (2006) *Southern Africa Regional Plan*, London, Department for International Development

DFID (2008) *Ready to Grow: Helping the Caribbean to emerge as a global partner – The UK Government's Regional Development Strategy for the Caribbean*, London, Department for International Development

Dronin, N. and A. Kirilenko (2008) 'Climate change and food stress in Russia: what if the market transforms as it did during the past century?', *Climatic Change*, 86, 123–50

ENDA and LEAD (2009) *Leading the Way: A Role for Regional Institutions. African leadership on climate change in Africa*, Dakar, LEAD Africa

ESF (2009) European Food Systems in a Changing World. *ESF-COST Forward Look Report*, Strasbourg, ESF

Fraser, E. D. G. (2007) 'Travelling in antique lands: using past famines to develop an adaptability/resilience framework to identify food systems vulnerable to climate change', *Climatic Change*, 83, 495–514

Fronzek, S. and T. R. Carter (2007) 'Assessing uncertainties in climate change impacts on resource potential for Europe based on projections from RCMs and GCMs', *Climatic Change*, 81, 357–71

GECAFS (2006) GECAFS Southern Africa Science Plan and Implementation Strategy. *GECAFS Report No. 3*, Oxford, GECAFS

GECAFS (2007) GECAFS Caribbean Science Plan and Implementation Strategy. *GECAFS Report No. 4*, Oxford, GECAFS

Ingram, J. S. I., P. J. Gregory and A.-M. Izac (2008) 'The role of agronomic research in climate change and food security policy', *Agriculture, Ecosystems and Environment*, 126, 4–12

Maracchi, G., O. Sirotenko and M. Bindi (2002) 'Impacts of present and future climate variability on agriculture and forestry in the temperate regions: Europe', *International Workshop on Reduction Vulnerability of Agriculture and Forestry to Climate Variability and Climate Change*, Berlin, Springer

Maunder, N. and S. Wiggins (2006) Food Security in Southern Africa: Changing the Trend? Review of lessons learnt on recent responses to chronic and transitory hunger and vulnerability, Oxfam-GB, World Vision International, CARE, RHVP and OCHA, Discussion Draft

Olesen, J. E., T. R. Carter, C. H. Diaz-Ambrona, S. Fronzek, T. Heidmann, T. Hickler, T. Holt, M. I. Minguez, P. Morales, J. P. Palutikof, M. Quemada, M. Ruiz-Ramos, G. H. Rubaek, F. Sau, B. Smith and M. T. Sykes (2007) 'Uncertainties in projected impacts of climate change on European agriculture and terrestrial ecosystems based on scenarios from regional climate models', *Climatic Change*, 81, 123–43

Rajan, K. (2005) Renewing SAARC. *Regional Conference on 'New Life within SAARC'*, Kathmandu, Nepal

SAARC (2007) Agreement on establishing the SAARC Food Bank Periodical, Agreement on establishing the SAARC Food Bank SAARC, www.saarc-sec.org (accessed: 18 January 2010)

SAARC (2008a) SAARC Action Plan on Climate Change. www.saarc-sdmc.nic.in/pdf/publications/climate/chapter-2.pdf (accessed: 18 January 2010)

SAARC (2008b) *Colombo Statement on Food Security*, Fifteenth SAARC Summit, Colombo

SADC Secretariat, Regional Indicative Strategic Development Plan (RISDP – Draft). www.transport.gov.za/siteimgs/RISDP-16Jul2003.pdf (accessed: 20 January 2010)

Schmidhuber, J. and F. N. Tubiello (2007) 'Global food security under climate change', *Proceedings of the National Academy of Sciences of the United States of America*, 104, 19703–8

Tao, F., M. Yokozawa, J. Y. Liu and Z. Zhang (2009) 'Climate change, land use change, and China's food security in the twenty-first century: an integrated perspective', *Climatic Change*, 93, 433–45

Vogel, C., S. C. Moser, R. E. Kasperson and G. D. Dabelko (2007) 'Linking vulnerability, adaptation, and resilience science to practice: Pathways, players, and partnerships', *Global Environmental Change*, 17, 3–4, 349–64

15
Undertaking Research at the Regional Level

John Ingram and Anne-Marie Izac

Introduction

The rationale for, and benefits to be gained from, undertaking global environmental change (GEC)/food security research at the regional level are discussed in Chapter 13. However, given the varied and complex interactions between regional and national objectives, research at the regional level has to encompass considerations of multilevel dynamics. Further, many of the food security issues are based on socio-ecological interactions that need to be studied at a number of scales and levels (Chapter 2). This gives rise to three types of research questions that all need to be addressed to recognize the complex spatial and temporal dynamics within a region and to cover the varied interests of regional-level stakeholders (Chapter 14):

- *Regional-level questions*, to address issues relating to the region as a whole that cut across the range of different conditions within the region. Example: What regional-level policy instruments and strategies would reduce GEC threats to regional food security? (e.g. transboundary water agreements; intraregional trade; strategic food banks; reduction of non-tariff barriers; regional disaster management; regional licensing for agricultural inputs such as agrochemicals and genetically modified organisms (GMOs), regionally coordinated taxation and export policies).
- *Subregional-level questions,* which are researched in a set of case studies selected to represent – as best as possible – the heterogeneity of a range of parameters across the region. These case studies could be a district or even a (small) country. Example: What aspects of local governance affect the development and implementation of food system adaptation options and strategies? (e.g. vision, popular acceptance, corruption, accountability and social auditing, capacity and capability, price stability, food standards).
- *Cross-level questions*. Example: What are the key interactions between policy instruments, strategies and interventions set at different levels? (e.g. national insurance policy and regional fisheries production; land-use

regulation and local disaster management; local distribution infrastructure and intra-regional trade agreements; crop diversification and intra-regional trade; regional versus local early-warning systems).

Of the three types of questions, and as argued in Chapter 13, the subregional level and single cropping season is perhaps the best researched to date, and especially the biophysical aspects of food production. However, in terms of food security, understanding and managing the dynamic cross-level issues along all relevant scales are arguably more important. This certainly requires an interdisciplinary approach, although research on cross-level interactions along jurisdictional and institutional scales is perhaps of more interest to social scientists. This dichotomy is especially so for regional/subregional cases, where mismatch between regional and national policies can severely compromise the effectiveness of food production and other key factors determining food security. Despite the multiscale, multilevel nature of food systems, different scales are usually singled out as important at different levels, and which might act as bottlenecks to cross-level interactions.

Matching research to regional information needs: Who is the 'client'?

Several arguments in support of undertaking GEC/food security research at the regional level have been presented in Chapter 13. While Chapter 14 identifies a varied set of researchers, policy agencies and donors operating at the regional level, the notion of regional research does raise two important questions: 'Who in particular wants to use research results at this spatial resolution?'; and 'What do they want to know?'. While there might well be academic interest in such studies, their utility in assisting in policy formulation at the regional level and their relevance to regional resource management would be slight if there is no obvious 'client'. Further, and as discussed in Chapter 11, even if identified, these clients need to be engaged in helping to set the appropriate research agenda; it is important that research outputs match the information needs in relation to managing common agro-ecological zones, shared river basins, common problems, etc. Further, identifying groups and agencies that need information at these spatial levels helps to determine the 'client'.

Identifying the 'client' at regional level can be relatively easy if, by chance, there is a formal entity whose geographical mandate approximately matches the geographical area of interest from a GEC viewpoint (Chapter 14). Often such entities are economic and/or political groupings, such as the Caribbean Community (CARICOM, which includes many of the Caribbean nations), and which nests well with regional studies of GEC-induced changes in hurricane track and intensity; or the South Asian Association for Regional Cooperation (SAARC), with reference to Asian 'brown cloud' studies. Another good example can be found in Southern Africa, where subcontinental studies on, for example, the regional transport of air, human vulnerability and biodiversity loss match much of the Southern Africa Development Community (SADC) region. While many such bodies have a clear food security mandate, and there may be a clearly expressed need for GEC/food security research for a given region and/or by a given body, translating this into a practical research package is daunting.

One of the first challenges is to establish with which individual(s) in such organizations to engage. Often there is a unit related to food security (e.g. in SADC there is the Directorate for Food, Agriculture and Natural Resources, FANR). But given the major dependence on agriculture in many parts of the world, coupled with what might be chronic (or worse, acute) food insecurity, staff in such units are often overstretched and under-resourced. Despite the agreed need (and usually sincere desire) to engage in planning discussions and follow-up activities (Chapter 11), there is simply insufficient human capacity to do so; staff often need to respond to immediate crises (e.g. a deepening drought or an imminent hurricane) rather than concentrate on more strategic planning issues. Holding planning meetings in locations where representatives from such bodies would most likely be able to participate (e.g. for southern Africa in Gaborone, where SADC-FANR is based) can help gain input, and hence an insight, into policy interests. An innovation developed by the International Food Policy Research Institute's (IFPRI) RENEWAL project has been the establishment of Advisory Panels to ensure 'in-reach'. This model involves explicitly including policy-makers and other stakeholders in the research from inception, ensuring that the project 'asked the right questions' that were relevant and important (Chapter 11). These panels were established in the main countries in which RENEWAL operates. Although an attempt was made to set up a similar structure at the regional level, it became evident that staff of relevant directorates (especially FANR) are heavily overstretched and the effectiveness of this was questionable.

Despite the number of examples discussed above, it is questionable whether these tactical approaches really add up to a strategic decision at a political level to engage in systematic analysis and preparedness for GEC/food security issues. An education/capacity-building effort at a political and decision-making level that involves the food industry and other stakeholders appears to also be needed (this, and other issues related to stakeholder dialogue in general, are discussed in Chapter 11). It is important to realize, however, that regional bodies such as EU or CARICOM comprise member states, each of which has its own national concerns and goals. What might appear a logical way forward for the region as a whole might be thwarted by political and economic concerns of individual members, or by conflicts between them, such as between India and Pakistan. For instance, since the 1980s, SADC has been considering the establishment of a strategic food reserve to deal with the growing frequency of natural disasters. Early proposals were based on considerations of enough physical maize stock for 12 months' consumption. Despite this, most government reserves were at record low stocks at the 2002/3 marketing year (Mano et al, 2003, 2010; Kurukulasuriya et al, 2006). This regional/national dynamic adds a further complication to identifying the regional 'client'.

Methods to engender research at the regional level

Within a given region there are often many research projects working at the national or sub-national level addressing aspects of GEC and food system research (e.g. social, agronomic, fisheries, policy, economics, ecological and climate sciences). If integrated, these individual projects could be very relevant both to the broader, interdisciplinary GEC/food security agenda and to higher-level analyses of value to policy development at the regional level. Such integration depends on effective networking.

Analysis of a number of international projects has identified good examples of how to engender such networks, the importance of team-building and standardized methods, the value of using integrated scenario approaches for facilitating regional-level analyses, possible ways to overcome some of the many methodological challenges for research at regional level, identifying case study sites and the value of linking regional research within the broader international context. These are discussed below.

Encouraging regional research networks

The Assessments of Impacts and Adaptations to Climate Change project (AIACC) focused on training and mentoring developing country scientists to undertake multisector, multicountry research. This addressed a range of questions about vulnerabilities to climate change and multiple other stresses, their implications for human development and policy options for responding and the information, knowledge, tools and skills produced by AIACC research enhanced the ability of developing countries to assess their vulnerabilities and adaptation options. A key aspect of AIACC was the development of regional research networks and to this end AIACC was structured in such a way as to encourage interactions across research disciplines, institutions and political boundaries, and enable more effective south–south exchange of information, knowledge and capacity, and through that process engender network building. This approach, replicated across such a large number of assessments and in contrasting research environments, generated a number of key insights that can inform 'good practice' recommendations for encouraging regional research networks. These include the need to consider broad criteria in selecting research and assessment teams; the value of coordinating multiple climate change assessments under the umbrella of a larger project; the value of providing flexible, bottom-up management; and the need to promote multiple, reinforcing activities for capacity-building (Box 15.1).

Box 15.1 *Recommendations from the AIACC project for encouraging regional research networks*

The Assessments of Impacts and Adaptations to Climate Change project (AIACC) was a global initiative developed in collaboration with the United Nations Environment Programme (UNEP)/World Meteorological Organization (WMO)

Intergovernmental Panel on Climate Change (IPCC), and funded by the Global Environment Facility to advance scientific understanding of climate change vulnerabilities and adaptation options in developing countries. It was completed in 2007. Key lessons from the AIACC project on simultaneously achieving regional network building and capacity development included the following:

- *Consider broad criteria in selecting research and assessment teams.* The peer-review process of selecting proposals for the AIACC project considered the need for representation of countries with low capacity as a co-criterion to scientific merit. This inclusive selection approach helped to broaden the reach to least developed countries, where there are substantial knowledge and capacity gaps. The presence of a strong technical support team within the project and the project's emphasis on capacity-building helped to support the needs of teams from low-capacity countries.
- *Coordinate assessments.* Execution of multiple climate change assessments under the umbrella of a larger project produced synergistic benefits. The AIACC project provided numerous opportunities for the different teams to interact with each other through regional workshops, synthesis activities, joint training activities, peer-review of each other's work and electronic communication. Moreover, executing a group of assessments together also made it possible for investigators from multiple projects of similar design to compare results from across the projects and to identify and synthesize common lessons.
- *Provide for flexible, bottom-up management.* The teams were given wide latitude to set their specific objectives, focus on sectors and issues of their choosing and select the methods and tools to be applied. This allowed for a high degree of innovation and matching of the focus and design of each assessment to the priorities, capabilities and interests of the teams, and it allowed for flexibility in adapting to shifting priorities within the assessment. The flexible and 'bottom-up' approach to project management created good working relationships and respect among the participating institutions and was a key factor in the overall performance of the project.
- *Promote multiple, reinforcing activities for capacity-building.* A comprehensive programme of learning-by-doing, technical assistance, group training, self-designed training and networking was demonstrated to be effective at building capacity. Efforts were made to utilize the expertise of developing country participants to assist with training and capacity transfers to their colleagues. This worked well and even led to a number of training workshops organized by some of the teams for colleagues in other projects. A substantial portion of the capacity-building resulted from the cross-project learning and sharing of methods, expertise, data and experiences. The central role assumed by regionally-based capacity-building and regional research networks helped to ensure greater sustainability and achieve a wider impact than is generally the case with north–south transfers of expertise and capacity development.

The importance of team-building and standardized methods

The breadth and complexity of GEC/food security research at the regional level necessitates bringing together a group of researchers (with varied skills) and other stakeholders. Such groups can develop into strong research networks but this depends on careful team-building. This was particularly important in the Alternatives to Slash-and-Burn Program (ASB, see Box 15.2).

Box 15.2 *Alternatives to Slash-and-Burn (ASB) Program*

A multi-institutional and global research programme launched in 1993, the *ASB Program* provides a successful example of a global programme which has regional sites in three continents. ASB has been focusing on tropical forest margins to develop more environment-friendly farming techniques that would result in local and regional food security and on slowing deforestation at forest margins. It has now grown to a global partnership of over 80 institutions, conducting research in 12 tropical forest biomes (or biologically diverse areas) in the Amazon, Congo basin, northern Thailand, and the islands of Mindanao in the Philippines and Sumatra in Indonesia. Its efforts are directed towards curbing deforestation, while ensuring that those living in poverty benefit from nature's environmental services and achieve food security.

A partnership of institutions around the world, including research institutes, non-governmental organizations, universities, community organizations and farmers' groups, ASB operates as a multidisciplinary consortium for research, development and capacity-building. ASB applies an integrated natural resource management (iNRM) approach to analysis and action (Izac and Sanchez, 2001) through long-term engagement with local communities and policy-makers.

ASB undertakes participatory research and development of technological, institutional and policy innovations to raise productivity and income of poor rural households in the humid tropics, whilst slowing down deforestation and enhancing essential environmental services. Poverty reduction in the humid tropics depends on finding ways to raise productivity of labour and land, often through intensification of smallholder production activities. Although there are some opportunities to reduce poverty while conserving tropical forests, tropical deforestation typically involves tradeoffs among the concerns of poor households, national development objectives and the environment (De Fries et al, 2004; Palm et al, 2005). ASB partners work with households to understand their problems and opportunities. Similarly, consultations with local and national policy-makers bring in their distinctive insights. In this way, participatory research and policy consultations have been guiding the iterative process necessary to identify, develop and implement combinations of policy, institutional and technological options that are workable and relevant.

The participatory on-farm work is undertaken at ASB 'benchmark' sites, established in each of the regions. These are areas (roughly 100km²) of long-term (i.e. more than ten years) study and engagement by partners with households, communities and policy-makers at various levels. All benchmark sites are in the humid tropical and subtropical broadleaf forest biome (as mapped by WWF, the World Wide Fund for Nature). The most biologically diverse terrestrial biome by far, conversion of these forests leads to the greatest species loss per unit area of any land-cover change.

ASB focuses on the landscape mosaics (comprising both forests and agriculture) where global environmental problems and poverty coincide. ASB's multisite network helps to ensure that analyses of local and national perspectives, and the search for alternatives, are grounded in reality.

An analysis of the lessons learned in ASB shows that the importance of team- and network-building at each one of the 'benchmark sites' cannot be overstressed. Much care was taken in the selection of research sites representative of the major regions, globally, where slash-and-burn agriculture is important. Each site thus encompasses a broad range of biophysical and socio-economic conditions and is representative of conditions prevailing throughout different regions. Standardized methods for site characterization and for undertaking both biophysical and socio-economic research at each site, were designed and discussed with the multi-institutional and interdisciplinary teams at each benchmark site. The core team of scientists who conceived the research agenda of ASB, which includes the need to work in parallel in different sites representative of a range of regional conditions, did not, however, foresee how difficult the initial team- and network-building was going to be. Even though all the institutions involved subscribed to the ASB research agenda, and had been attracted to the ASB Consortium because they found this research agenda compelling, teams at each site took an average of two years to coalesce. The core team did not foresee that it would take this long for scientists from different institutions and different disciplines to successfully work together in a regional mode, but on global questions. In hindsight this seems naïve, but at the time all the scientists were focusing on the contents of the research and just did not think about what it takes to build a team in one location (let alone in multiple locations in parallel). The transaction costs of learning to work together at each benchmark site were compounded by those associated with the need for overall coordination and communication. The different teams at each site consisted of scientists from research institutions (national and international), from NGOs and from universities. Some had previous experience with participatory research methods; others had none. It quickly became obvious that it was essential to reconcile the objectives of the programme with the expectations of each team member at a given site. The differing roles of these participants required almost constant renegotiations on the part of the overall coordinator.

One dimension of the ASB approach, the use of 'standardized' methods at each site to facilitate cross or interregional analyses of results, proved difficult to implement at first. Each benchmark team considered that the set of methods proposed by the programme needed to be significantly amended to account for the particularities of their own site. The analysis of data across the benchmark areas and the global results were indeed not a very strong motivation for some of these teams, until it was agreed that the global or cross-site analysis would be undertaken by all interested scientists, no matter whether they were located at a given site or were part of the initial core team. A geo-referenced database was developed to facilitate the synthesis of results and the sharing of information across the regional teams. After a few years of data collection at each benchmark area, and once some of the regional results started being analysed from a global perspective, the regional teams became almost more interested in the global analysis than in the production of a full analysis of their own data.

Using integrated scenario analyses for facilitating regional-level analyses

Scenario development and analysis has already been successfully used at a global level to help reveal and address knowledge gaps about the plausible future interactions between GEC and a number of ecosystem goods and services (e.g. food production or water availability or climate regulation). Such studies are often called 'integrated' as they: (i) include social, economic and environmental processes and scientific disciplines; (ii) cover multiple levels on multiple scales; and (iii) strongly involve stakeholders (see Chapter 3 and Box 11.7). Such scenarios can be either qualitative (stories) or quantitative (models), or both.

Scenario analyses conducted at the regional level help to systematically explore policy and technical options at the appropriate level by providing a suitable framework for: (i) raising awareness of key environmental and policy concerns; (ii) discussing viable adaptation options; and (iii) analysing the possible consequences of different adoption options for food security and environmental goals. These can be based on scenarios developed at the global level (e.g. the Intergovernmental Panel on Climate Change (IPCC) Special Report on Emissions Scenarios (IPCC, 2000); the Millennium Ecosystems Assessment (MA, 2005); and the UNEP's Global Environment Outlooks (GEO-3 and GEO-4: UNEP, 2002, 2007)), but such analyses do not necessarily feature issues that are of particular relevance at the given regional level (Zurek and Henrichs, 2007).

Downscaling global scenarios to national (or even local) level has been considered by a number of authors (e.g. Lebel et al, 2006; Biggs et al, 2007; Kok et al, 2007), but, while a commonly agreed approach is still lacking, downscaling methods and theories are becoming more common (e.g. Zurek and Henrichs, 2007). Upscaling has, however, proven to be more difficult and remains one of the largest challenges. Some argue that global downscaling limits the creativity and diversity of regional scenarios and call for more upscaling efforts. Others argue that upscaling will fail because of the lack of a common framework in terms of drivers, time horizon, definitions, etc. (see Alcamo et al, 2008). So, while there are a large number of detailed global scenarios available, their potential has been undervalued for developing scenarios at a regional level.

The current state-of-the-art is to embark upon a cross-level methodology in which global scenarios are first downscaled and used to produce regional or local scenarios without being prescriptive, after which local scenarios are used to enrich the existing global storylines in an iterative procedure, often using qualitative storylines as well as quantitative models. Creating regional scenarios is not just a matter of 'downscaling' the information available in global scenarios (e.g. climate change projections) for regional use; some information (such as trends in trade) will have been built up from lower levels. Other information will be new and will need to come directly from the region in question (Zurek and Henrichs, 2007). Regional 'storylines' of plausible futures can share some of the key assumptions with global-level storylines, that is, be coherent with global assumptions, yet regionally 'enriched', as was the case for the GECAFS Caribbean exercise (Box 15.3). Similarly, the Southern African Sub-Global Assessment (SAfMA) (Biggs et al, 2004) adapt existing scenarios, stressing governance as a major driver and developing two regional storylines: African Patchwork and African Partnership. These can be mapped to the MA Global Scenarios.

Box 15.3 *The GECAFS Caribbean Scenarios Exercise*

In 2005, with funding from the International Council for Science (ICSU) and the United Nations Education, Science and Cultural Organization (UNESCO), GECAFS, in collaboration with the UN Food and Agriculture Organization (FAO), the MA, the European Environment Agency (EEA) and UNEP, developed the conceptual frameworks and methods necessary to formulate a set of prototype scenarios for researching the interactions between food security and environmental change at the Caribbean regional level. These scenarios were specifically designed to assist analyses of possible regional policy and technical interventions for adapting food systems to environmental change so as to explore the medium- and long-term prospects for given adaptation options for food security. The innovative operational framework was based on theoretical advances in the notion of food systems and their vulnerability to GEC, and downscaling global scenarios to regional level.

The Caribbean scenarios exercise involved about 30 people, including social and natural scientists from regional research institutions (e.g. the University of the West Indies (UWI) and the Caribbean Institute for Meteorology and Hydrology (CIMH)); social and natural scientists from national research institutions (e.g. universities and national laboratories); policy-makers from regional agencies (e.g. the Caribbean Community Secretariat (CARICOM), Inter-American Institute for Cooperation on Agriculture (IICA)); policy-makers from national agencies (e.g. Ministries of Agriculture); international agencies (e.g. FAO, UNEP); and was facilitated by the GECAFS scenarios group. A number of key steps were involved:

1 Identifying key regional GEC and policy issues, based on an initial stakeholder consultation workshop involving regional scientists and policy-makers.
2 Drafting a set of four prototype regional scenarios (Global Caribbean, Caribbean Order from Strength, Caribbean TechnoGarden and Caribbean Adapting Mosaic) in a first regional workshop, which were then elaborated upon in a follow-up writing exercise by regional authors. These were based

on the broad rationale, assumptions and outcomes of the MA scenarios exercise, but allowing for regional deviation where needed.

3 Describing developments per scenario for key aspects of the food system, the focus of a follow-up regional workshop involving most of the first regional workshop participants.

4 Systematically assessing food system developments per scenario, and presenting outputs graphically as part of a second regional workshop. This involved describing the main developments per scenario for each food security element, systematically assessing each development per scenario for each food security element, and finally plotting each assessment (see Figure 15.1).

The scenario exercise delivered a number of related outputs: it integrated (i) improved holistic understanding of food systems (axes on graphs) with (ii) vulnerability (change of position along axes); with (iii) policy interpretation of future conditions (comparing four graphs); and with (iv) adaptation insights at the regional level for improving overall food security (where to concentrate effort on enlarging the polygon areas of each graph).

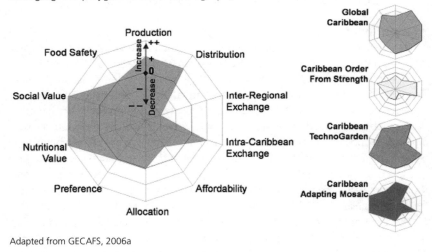

Adapted from GECAFS, 2006a

Figure 15.1 *Indicative food security diagrams for four Caribbean scenarios*

Methodological challenges for research at regional level

Research undertaken at the regional level embodies a number of methodological challenges that need to be addressed at the outset; otherwise they become bottlenecks. In the biophysical sciences, data collection takes place at specific geographical and physical locations, and therefore at specific points in time. Given the spatial heterogeneity, or the spatial patterns, in most biophysical parameters of relevance to food security and to the interactions between GEC and food security within a region, data needs to be collected using statistical methods that allow extrapolation from a relatively limited number of data points (relative to a whole region). For instance, measures of soil fertility – however a given project defines soil fertility and its measures – need to be collected in such a way that the heterogeneity of soil fertility at the subregional/watershed level is captured, and extrapolation of the measures to a regional picture of soil fertility is feasible and meaningful. This is, in itself, not a straightforward exercise. Methods have been developed to conduct such data collection and analyses of spatial patterns across regions; for instance, using geostatistics (Coe et al, 2003).

In food security and GEC studies, this is, however, complicated by the fact that other system parameters within a region (e.g. farmers' access to roads and markets) also vary, but at a different spatial level. As a rule, spatial patterns in biophysical and socio-economic parameters within a region occur over different levels on both spatial and temporal scales. Since the data collected will have to be analysed in an integrated manner to arrive at a meaningful and useful picture of GEC/food security interactions within a region, a 'silo' type of analysis, in which biophysical data are analysed separately from socio-economic data, is not a feasible option.

To enable researchers to conduct a scientifically robust integrated analysis of all the data collected, as is essential in GEC/food security studies, a geographical unit of analysis, which is meaningful from both a biophysical and a socio-economic perspective, needs to be identified by the scientists involved. In the ASB programme, for instance, the scientists finally agreed upon a set of 'land use categories' that were used throughout all the ASB benchmark sites, and that were essential in the extrapolation of the data collected to regions. In the GECAFS work in the Indo-Gangetic Plain, the agreed unit of analysis for each case study was an administrative district.

In addition to these requirements for data collection and integrated analysis of data at the regional level, methods capable of investigating interrelationships among different types of analysis and capable of synthesizing and analysing the key economic, environmental, agronomic and biophysical issues at stake, and at the correct resolutions, are also needed. A range of models and mathematical methods exist that provide relevant tools for this, but all have limitations of course (Coe et al, 2003; van Ittersum and Wery, 2007). Scientists conducting GEC/food security studies thus need to carefully select the most appropriate tools, given the specific objective of their research.

Identifying case study sites

Research designed to address the regional-level, subregional-level and cross-level questions identified above will give insights into how food systems operate across a region, how this diversity affects food security across a region, and what possible adaptation strategies can be considered both locally and for the region as a whole. However, selecting case-study sites across a diverse region necessitates a compromise between optimal scientific and practical considerations. One aim can be to build on ongoing research infrastructure and research sites, and existing data, rather than establishing research sites *de novo*. An initial useful step can be to survey existing work in the region to see what to build on, and also to show where new studies can add value to others by 'mapping' them onto the research structure of the new study. This type of information can be used to identify research projects for which suitable socio-economic and environmental data are already available, and would help to build up regional research networks (as have been very successfully developed by the Inter-American Institute for Global Change Research, IAI; see Chapter 11). It would also enable an analysis of existing work that can be integrated to help address the research agenda and to identify where the major gaps appear to lie. Box 15.4 illustrates the process of case-study site selection in the GECAFS work in the Indo-Gangetic Plain.

Box 15.4 *Identifying research sites for multilevel research*

Choosing case-study sites is a critical part of multilevel research, and should ideally be based on discussion among all stakeholders. For GEC/food security research, some selection criteria could include:

- lie-along gradients of, for example, anticipated temperature and precipitation change, or current and anticipated grazing pressure;
- sufficient representation of different governance (e.g. land tenure) arrangements;
- sufficient representation of the region's principal farming systems (Dixon et al, 2001);
- sufficient representation of key drivers in regional scenarios (see Box 15.3);
- building on work where interventions have been shown to be effective.

The aim is to identify a set of sites that are individually representative of the specific subregion/selection criteria, and which, when considered as a whole set, give a good representation of the region overall and the heterogeneity within it. This might need an integrative approach, as proved to be the case in selecting GECAFS research sites in the Indo-Gangetic Plain (IGP).

A workshop was convened to identify five case-study sites reflecting the socio-economic and environmental diversity of the IGP. Administrative districts were chosen as this was the unit for which socio-economic data was generally collected. Initial suggestions were presented for a site in each of the five rice-wheat subregions across the IGP (as identified by the Consultative Group on International Agricultural Research (CGIAR)-supported Rice-Wheat Consortium). Criteria included identifying where the effects of various GEC drivers will be most extreme and where those living in poverty will be worst hit. The likelihood of data availability was also an important consideration. When all five sites had been presented, they were considered in terms of balance of GEC drivers and food system variables (so as to assist with synthesizing across the sites), and one different site (in Bangladesh) was chosen. The final set (see Figure 15.2) captures the heterogeneity across the IGP, while also being of relevance to district-level planners.

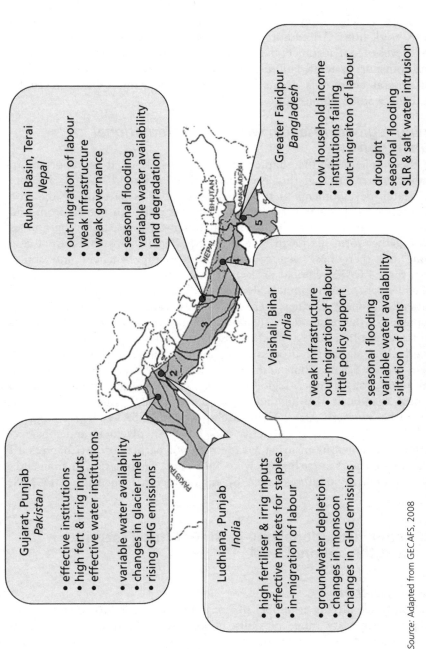

Source: Adapted from GECAFS, 2008

Figure 15.2 *Location of GECAFS-IGP case studies identified in relation to the main rice-wheat growing area, and showing major socio-economic characteristics and GEC concerns*

Regional synthesis and integration workshops can draw together case-study research output to address regional synthesis questions. The survey of recently completed, ongoing and/or imminent region-wide research activities can indicate which outputs from other projects can be best included in synthesis workshops. An additional approach is to undertake analyses at the regional level of, for instance, market and physical infrastructure, food storage and transport systems, land-use conversion, etc. These give a good insight of the general conditions across the region which cannot be achieved by synthesizing a small number of case studies.

Defining regional research within the international research context

While the emphasis of the chapter is the regional level, it is important to remember that research at this level can (and often should) usefully interact with research developments at the global level. This is particularly the case where international research structures exist that can both benefit from and contribute to regional research. The international GEC research programmes (which together form the Earth System Science Partnership, ESSP) are a case in point and there has been considerable mutually beneficial interaction between the GECAFS regional projects (being sponsored by ESSP) and these global endeavours. Such contacts also help with international networking and capacity development. Further, there can be great value gained by the iterative development of the regional (i.e. place-based) and conceptual (i.e. non-place-based) research often more typical of some international endeavours.

Figure 15.3 shows the process of research development for the GECAFS southern Africa Science Plan and Implementation Strategy, which involved several interactions between regional stakeholders with developments on the conceptual agenda. It also highlights interaction with another GEC research initiative (the Southern Africa Vulnerability Initiative, SAVI). These steps all depended heavily on strong and active stakeholder involvement and considerable effort was taken in ensuring a wide range of participation (Chapters 10 and 11). Key outputs at each stage were built upon, culminating in an agenda seen as of high regional relevance by major regional agencies (GECAFS, 2006b).

Establishing institutional buy-in for GEC/food security research at the regional level

Research partners

In addition to identifying a number of researchers who need to come together to bring the necessary range of skills, effective GEC/food security research needs the close involvement of other stakeholders in planning and delivery. While the need for such involvement may be fully accepted (Chapters 10 and 11), it can be hard to identify the right partners and even harder to engage them meaningfully. One way to help is to try to identify existing institutions which individually can fulfil different key roles in the research. These could,

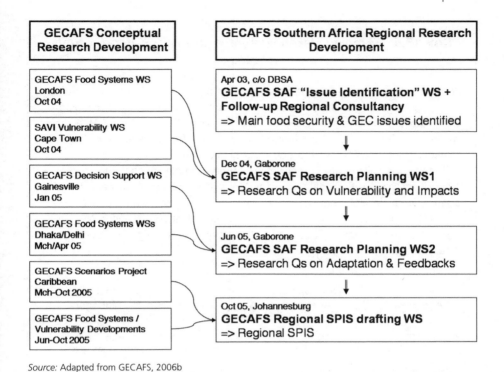

Source: Adapted from GECAFS, 2006b

Figure 15.3 *Development of the GECAFS-Southern Africa Science Plan and Implementation Strategy*

for instance, help to provide scientific visibility and credibility within the region, or act as 'boundary organizations' (Chapter 10) between the main research endeavour and other, more distant, stakeholder communities (e.g. individual farmers). Establishing institutional buy-in from a range of stakeholders can add a powerful dimension to the research itself, and can also be very useful in outreach as such partners are often well connected to a wide range of beneficiaries. Posing a clear question, and drafted in terms to which potential partners can easily relate (e.g. 'how can Southern African food systems be adapted to reduce their vulnerability to GEC?), helps both attract interest and identify the role each can play. Box 15.5 shows how a range of stakeholders thus 'bought-in' to a GEC/food security research question in the region. The signing of Memoranda of Understanding (MoUs) and/or the formal endorsement of research plans can be a powerful way of demonstrating this buy-in to other potential stakeholders, including donors.

Donors

Several major organizations are now embracing the food systems concept (Chapter 2) for advancing food security research at regional level. Notable examples include the FAO (2008), UK government (Defra, 2008), European Science Foundation (ESF, 2009), Dutch government (NWO, undated) and the

Box 15.5 *Mapping stakeholder interests in a GEC/food security research*

A major emphasis of GECAFS regional research planning has been the identification of a range of stakeholder groups in each region and the clear mapping of their respective interests in relation to the overall endeavour. Figure 15.4 maps the key interests of a number of different southern African stakeholder groups in relation to a fundamental GEC/food security question.

The two key collaborative organizations in this example are: (i) the International Council for Science Regional Office for Africa (ICSU-ROA), which helps link the science and development agencies with policy-makers; and (ii) the Food, Agriculture and Natural Resources Policy Analysis Network (FANRPAN), which helps link the development agencies, policy-makers and resource managers (principally farmers).

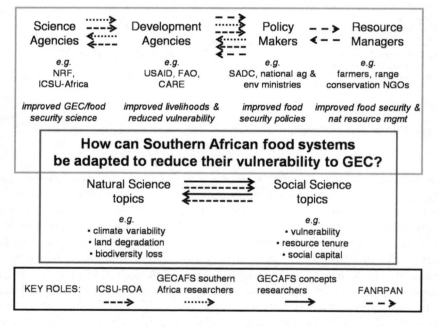

Figure 15.4 *Key interests of a number of different southern African stakeholder groups in relation to a fundamental GEC/food security question*

CGIAR's new initiative 'Climate Change, Agriculture and Food Security' (CCAFS, 2009). Despite this, raising funds for GEC/food security research at a regional level is not easy. This can be due to a number of factors.

First, donors are most often mandated to operate on a bilateral basis at a national level and efforts to 'regionalize' projects generally has to involve a synthesis across multiple projects as they come to an end. Also, regional-level projects, and especially those that try to link across spatial and/or temporal levels, can be deemed too unfocused.

Second, research on food security, let alone when coupled with GEC issues, is highly complex and full of uncertainties. Research designed to 'grasp the nettle' is therefore highly complex and involves a large number of parameters and collaborators. While it might address the stated aims of development agencies better than research on, say, food production, the inherent complexity means it is hard to fit within funding portfolios (which might be structured in terms of agriculture development, policy, governance, science, etc.), and is also deemed 'high risk'.

Third, given national agendas and institutional mandates, it can be hard to identify a champion for regional studies who is able to devote the necessary time to lobbying donors and strategic partners. The importance of this aspect cannot be underestimated, as the need to 'persuade' donors not only of the value in what they might see as 'high risk' studies, but also its new research concepts (i.e. the food system, vis-à-vis agriculture) as underpinning food security, takes a great deal of time and energy.

Fourth, research results will only be realized after a few years. This is because most researchers in the food security domain are trained in agricultural issues, and there is a need for a new cadre of researchers to become conversant with food security concepts and food system analysis. Next, the food system(s) need describing and how, where and when they are vulnerable to GEC made explicit. Only then can research on food system adaptation begin in earnest.

Fifth, the complexity of the research coordination and administration necessitates a well-organized organization to host and administer large grant(s). This organization needs a regionally recognized regional mandate, in which donors will need to have full confidence; such organizations might be hard to identify.

Finally, holistic, on-the-ground research, in a number of case studies across a region, over a number of years, is expensive. Experience in the GECAFS regional projects shows that, while donors strongly supported the planning exercises around the world, raising the sums needed for research implementation (1–2 orders of magnitude higher than for planning) has not thus far proved possible.

As political and science pressure grows for action on the GEC/food security agenda, the hope is that donors will begin to support to a greater extent this more complex type of research.

Conclusions and recommendations

There are numerous challenges related to research at the regional level: cross-scale and cross-level issues; identifying and integrating results from case-study sites; building research networks; establishing institutional buy-in; and raising funds. However, these challenges must be overcome as GEC/food security research at the regional level will deliver considerable benefits to a range of stakeholders and ultimate beneficiaries, which would not be apparent by restricting research to local or global levels.

Research conducted within AIACC, GECAFS, ASB, RENEWAL and other projects has identified a number of key factors that can help in terms of research framing:

- Three types of research questions (regional-level, subregional-level and cross-level) all need to be addressed to understand the complex spatial and temporal dynamics within a region and to cover the varied policy interests of regional stakeholders.
- A useful initial step is to survey existing work in the region to see what can be built on, and also to show where new studies can add value to others by 'mapping' them onto the research structure of the new study. This also helps identify potential members of a research team.
- A geographical unit of analysis, which is meaningful from both a bio-physical and a socio-economic perspective, needs to be identified and agreed upon.

It is also clear that for the research to have a good chance of having significant policy-relevant outcomes (i.e. not just science outputs), it is crucial to establish buy-in from regional policy agencies. A number of lessons in this area have also emerged:

- holding planning meetings in locations where representatives from such bodies would most likely be able to participate helps gain input, and hence an insight, into policy interests;
- the establishment of advisory panels including representatives from such agencies helps to ensure 'in-reach';
- scenario analyses conducted at the regional level help to systematically explore policy and technical options, and provide a valuable means of integrating the policy dimension;
- having a clear question posed and drafted in terms that potential stakeholders can easily relate to, helps both attract interest and identifies the role each can play;
- the signing of MoUs and/or the formal endorsement of research plans can be a powerful way of demonstrating this buy-in to other potential stakeholders, including donors.

Ultimately, of course, success will depend on establishing good working relationships with a range of stakeholders, so as to set and undertake an agenda that is both scientifically exciting and relevant to improved regional food security policy and resource management.

References

Alcamo, J., K. Kok, G. Busch and J. Priess (2008) Searching for the future of land: scenarios from the local to global scale. In Alcamo, J. (ed) *Environmental Futures: The*

Practice of Environmental Scenario Analysis. Developments in Integrated Environmental Assessment – Volume 2, Amsterdam, Elsevier

Biggs, R., E. Bohensky, P. Desanker, C. Fabricius, T. Lynam, A. Misselhorn, C. Musvoto, M. Mutale, B. Reyers, R. J. Scholes, S. Shikongo and A. S. van Jaarsveld (2004) Nature Supporting People: The Southern African Millennium Ecosystem Assessment Integrated Report, Pretoria, Council for Scientific and Industrial Research

Biggs, R., C. Raudsepp-Hearne, C. Atkinson-Palombo, E. Bohensky, E. Boyd, G. Cundill, H. Fox, S. Ingram, K. Kok, S. Spehar, M. Tengö, D. Timmer and M. Zurek (2007) 'Linking futures across scales: a dialog on multiscale scenarios', *Ecology and Society*, 12, 1

CCAFS (2009) *Climate Change, Agriculture and Food Security. A CGIAR Challenge Program*, Rome and Paris, The Alliance of the CGIAR Centers and ESSP

Coe, R., B. Huwe and G. Schroth (2003) Designing experiments and analysing data. In Schroth, G. and F. L. Sinclair (eds) *Trees, Crops and Soil Fertility – Concepts and Research Methods*, Wallingford, CAB International

Defra (2008) Ensuring the UK's food security in a changing world, a Defra Discussion Paper, London, Department for Environment, Food and Rural Affairs

De Fries, R. S., J. A. Foley and G. P. Asner (2004) 'Land-use choices: balancing human needs and ecosystem function', *Frontiers in Ecology and the Environment*, 2, 5, 249–57

Dixon, J., A. Gulliver and G. Gibbon (2001) *Farming Systems and Poverty: Improving Farmers' Livelihoods in a Changing World*, Rome and Washington, DC, FAO and World Bank

ESF (2009) European Food Systems in a Changing World. ESF-COST Forward Look Report, Strasbourg, ESF

FAO (2008) Climate change and food security: a framework document, Rome, Interdepartmental Working Group on Climate Change of the FAO

GECAFS (2006a) A Set of Prototype Caribbean Scenarios for Research on Global Environmental Change and Regional Food Systems. *GECAFS Report No. 2*, Wallingford, GECAFS

GECAFS (2006b) GECAFS Southern Africa Science Plan and Implementation Strategy. *GECAFS Report No. 3*, Oxford, GECAFS

GECAFS (2008) GECAFS Indo-Gangetic Plain Science Plan and Implementation Strategy. *GECAFS Report No. 5*, Oxford, GECAFS

IPCC (2000) Special Report on Emissions Scenarios: a special report of Working Group III of the Intergovernmental Panel on Climate Change, New York, IPCC

Izac, A.-M. N. and P. A. Sanchez (2001) 'Towards a natural resources management paradigm for international agriculture: the example of agroforestry research', *Agricultural Systems*, 69, 1–2, 5–25

Kok, K., R. Biggs and M. Zurek (2007) 'Methods for developing multiscale participatory scenarios: insights from southern Africa and Europe', *Ecology and Society*, 13, 1

Kurukulasuriya, P. R., R. Mendelsohn, R. Hassan, J. Benhin, T. Deressa, M. Diop, H. Eid, K. Yerfi-Fosu, G. Gbetibouo, S. Jain, A. Mahamadou, R. Mano, S. Kabubo-Mariara, S. El-Marsafawy, E. Molua, S. Ouda, M. Ouedraogo, I. Séne, D. Maddison, S. Niggol-Seo and A. Dinar (2006) 'Will Africa survive climate change?', *World Bank Economic Review*, August 2006

Lebel, L., J. M. Anderies, B. Campbell, C. Folke, S. Hatfield-Dodds, T. P. Hughes and J. Wilson (2006) 'Governance and the capacity to manage resilience in regional social-ecological systems', *Ecology and Society*, 11, 1

MA (2005) *Millennium Ecosystem Assessment,* Washington, DC, Island Press

Mano, R., B. Isaacson and P. Dardel (2003) Identifying Policy Determinants of Food Security Response and Recovery in the SADC Region: The case of the 2002 Food Emergency. FANRPAN Policy Paper, Pretoria, FANRPAN

Mano, R., J. Arntzen, S. Drimie, O. P. Dube, J. S. I. Ingram, C. Mataya, M. T. Muchero, E. Vhurumuku and G. Ziervogel (2010) 'Global environmental change and food systems in southern Africa: the dynamic challenges facing regional policy', submitted to *Environmental Science and Policy*

NWO (undated) Global Food Systems: A challenging approach to food security in developing countries (a WOTRO research programme), the Hague, Netherlands Organisation for Scientific Research

Palm, C. A., S. A. Vosti, P. A. Sanchez and P. J. Ericksen (2005) *Slash-and-Burn Agriculture: The Search for Alternatives,* New York, Columbia University Press

UNEP (2002) *Global Environment Outlook 3: Past, Present and Future Perspectives,* Nairobi, United Nations Environment Programme

UNEP (2007) *Global Environment Outlook 4: Environment for Development,* Nairobi, UNEP

van Ittersum, M. K. and J. Wery (2007) Integrated assessment of agricultural systems at multiple scales. In Spiertz, J. H. J., P. C. Struik and H. H. van Laar (eds) *Scale and Complexity in Plant Systems Research: Gene-Plant-Crop Relations,* Berlin, Springer

Zurek, M. and T. Henrichs (2007) 'Linking scenarios across geographical scales in international environmental assessments', *Technological Forecasting and Social Change,* 74, 8, 1282–95

16
Part IV: Main Messages

1 Sub-continental regions are a natural spatial level for studies of social–ecological systems (such as food systems), as they are often defined by shared cultural, political, economic and biogeographical contexts.

2 The interactions between global environmental change (GEC) and food security manifest at the full range of spatial levels from local to global. Studies have, however, tended to focus on these two extreme levels and information for sub-global (continental or sub-continental) geographical regions is sparse despite the fact that many food system adaptation options emerge when a regional viewpoint is adopted. Further, sub-global is a natural level for studies of social-ecological systems (such as food systems), as – while clearly not homogenous in all ways – they are often defined by shared cultural, political, economic and biogeographical contexts.

3 At least three main groups of stakeholders are interested in research on the interactions between GEC and food security at regional level: researchers, the regional policy agencies and donors. However, they are not well integrated despite the fact that many of the potential adaptation options for enhancing food security will require close collaboration between them.

4 It is important to establish the legitimacy of undertaking research at regional level. This can be in terms of political mandate, science interest, and/or resource managers' information need, as relating to issues that emerge when one takes a regional perspective.

5 The importance of spatial, temporal, jurisdictional and other scales and scaling as determining factors in many environmental and food security problems is now well recognized, and both scientists and policy-makers are increasingly aware that finding solutions requires consideration of various scales. Interactions across scales, and across levels within each, are not, however, generally included in GEC/food security studies, other than where a spatial level issue has direct relevance, as is increasingly the case for multilevel scenario studies.

6 While many regional policy agencies have a clear food security mandate, and while there may be a clearly expressed need for GEC/food security research for a given region and/or by a given body, it can be hard to identify with whom in given organizations researchers should aim to be engaging. Further, once identified, the key individuals may be already overcommitted and/or responding to immediate crises.

7 An education/capacity-building effort at a political and decision-making level that involves the food industry and other stakeholders appears to be needed. It is important to realize, however, that regional bodies such as the EU or the Caribbean Community (CARICOM) comprise member states, each of which has its own national concerns and goals. What might appear a logical way forward for the region as a whole from a scientific viewpoint might be thwarted by political and economic concerns of individual members, or by conflicts between them.

8 Many of the food security issues are based on socio-ecological interactions that are too complex to study at a high level because they are dependent on subregional conditions. This gives rise to three types of research questions (regional-level, subregional-level and cross-level), which all need to be addressed to cover the varied policy interests of regional stakeholders, and recognize the complex spatial and temporal dynamics within a region. Of the three types of questions, the subregional level is the best researched to date, while cross-level (and cross-scale) issues are the most important issue as understanding of the dynamic cross-level interactions is weak – yet these are often paramount in food security issues.

9 Scenario development offers one of the important methodologies to study GEC–food security interactions. With the further expansion of scenario studies to address food security, the non-spatial issues related to 'scale' will be on the forefront of the discussions. In addition to helping to communicate the food security outcomes of plausible near- to medium-term futures, scenario analyses help to: (i) integrate social and natural science in scenario design (climate, natural resource management, regional governance and policy dimensions); (ii) down-scale global scenarios to regional level, and enrich them with regionally relevant features related to both policy and science; and (iii) establish effective dialogue between GEC researchers and the policy process. This raises mutual awareness in science and policy communities of each other's interests, possibilities and limitations.

10 Integrated, on-the-ground research, in a number of case studies across a region, over a number of years, is expensive. Experience shows that, while donors strongly support planning exercises for GEC/food security research, raising the sums needed for research implementation (which can be one to two orders of magnitude higher than for planning) is challenging. As political and science pressure grows for action on the GEC–food security agenda, the hope is that donors will begin to support to a greater extent this more complex type of research.

PART V

FOOD SYSTEMS IN A
CHANGING WORLD

17

Food, Violence and Human Rights

Hallie Eakin, Hans-Georg Bohle, Anne-Marie Izac,
Anette Reenberg, Peter Gregory and Laura Pereira

My main message to you today is simple; in order to cope with
soaring food prices, supply must adjust to demand. For this to
happen, trade will help. Easier, more open trade can strengthen
the production capacity of developing countries, rendering them
less vulnerable.

Pascal Lamy, World Trade Organization (WTO) Director-General,
High-Level Conference on World Food Security, Rome, 3 June 2008

At a time when chronic hunger, dispossession of food providers
and workers, commodity and land speculation, and global
warming are on the rise, governments, multilateral agencies and
financial institutions are offering proposals that will only deepen
these crises ... Actions by some governments and top UN
leadership ... constitute an assault on small-scale food providers
(among whom women are in the forefront) and the natural
commons.

Statement by the International NGO/CSO Planning Committee for
Food Sovereignty, Platform for Collective Action, Forum Terra Preta,
Rome, 4 June 2008

Rising risks of violence associated with food and global environmental change

The current debate on global environmental change (GEC)-induced security
risks suggests that climate change will exacerbate existing resource conflicts
that might lead to violence, particularly around food, land and water (WBGU,
2008). An extreme position is that 'climate wars' may break out, driven by the
inequitable distribution of loss and harm associated with climate change

(Welzer, 2008). Catastrophic scenarios have been put forth, entailing dwindling resources, natural disasters, spreading epidemics and plummeting agricultural yields that trigger economic collapse, political turmoil and the destabilization of entire regions (Dyer, 2008). One recent media story claimed that 'More than hundred countries with 2.7 billion people are at high risk of political chaos and climate change-induced violence' (McKie, 2007). Conflicts around food often play centre stage in these violent scenarios, following the hypotheses of supply-induced scarcity as a driver of violence (Homer-Dixon, 1999). One recent study, for example, found a statistically significant linkage between temperature and civil conflict in Africa, and explains this linkage by the implications of temperature stress on agricultural production and livelihood security (Burke et al, 2009).

The uncertainties in these scenarios are large and scholars are not in agreement as to the extent to which GEC is a primary or direct source of violent conflict (Hauge and Ellingsen, 1998; Le Billon, 2001; Theisen, 2008). Nevertheless, current trends related to GEC and food do give rise to concern. The global market for food is intimately related to markets for increasingly scarce global resources of land, water and energy. The close interaction of these markets implies an accelerated transfer of risk across geographic space and among populations, with far-reaching impacts on all elements and activities in food systems at all levels on all scales of analysis. Climate uncertainties, coupled with economic and political surprises, are likely to exacerbate circumstances where violence is already a concern in the food system. Violence can exist in terms of direct (physical) forms (e.g. armed conflict), but also includes what Johan Galtung (1969) has termed 'structural violence', meaning chronic economic marginalization, social exclusion, disempowerment and other forms of indirect violence to which vulnerable people are exposed. A recent examination of the global food crisis in 2007–8 has, for example, shown how the hike in global food prices has pushed many millions of people into hunger, deprivation and poverty, and how it has sparked riots and protests in several countries including Haiti, Ivory Cost and Indonesia (Bello, 2009). Recent reports of global 'land-grabbing' by food-insecure countries mirror the expectation of future accelerated competition for agricultural land that might lead to violent food crises.

While the association of violence with food is not new – food has been used as a weapon and political tool throughout history (Davis, 2001) – the combined effects of globalization and GEC (Liechenko and O'Brien, 2008) within relatively inflexible or inadequate institutional frameworks now highlight food-related violence as a potential rising concern of this century. This chapter explores emerging evidence for a change in food-related conflict and violence as a result of accelerating trends in environmental change and globalization; and argues that underlying much of the rising violence is a fundamental incompatibility associated with the meaning of food, and how food is codified in both informal and formal institutional arrangements governing food and resource transactions, and the participation of different social groups in such institutions. It examines the contrasting dominant discourses embodying food systems, and the new institutional arrangements in global food governance that will be required to reduce conflict and violence associated directly and indirectly with food security.

The emergence of food violence as a global concern

According to Galtung's initial conceptualization, violence is 'the cause of the difference between the potential and the actual' (Galtung, 1969) and what impedes this difference from being reduced. Violence often occurs when an identifiable actor induces harm on another, but also may emerge where no specific actor is identifiable. In this latter case, violence is embedded in systemic and structural conditions of human–environmental relationships: inequities and gross distortions in the dynamics of resource access, control and distribution resulting in chronic and systemic physical (e.g. hunger), but also social and cultural harm (Galtung, 1969).

Environmental security theorists posit that armed conflict will emerge from conditions of environmental scarcity, induced by environmental degradation or shocks, which in turn can enhance opportunities for the capture of valuable resources by elite or powerful classes and/or the displacement of populations into ecologically marginal zones where competition over resource access can be acute (Homer-Dixon, 1999; see also discussion in Theisen, 2008). Others have argued that armed violence in relation to resource concerns is almost always caused by poor governance, ingrained social inequities and the actions of the political elite rather than directly by specific environmental triggers (Peluso and Watts, 2001). While scholars will continue to debate the precise causal relationships in the environment–violence–governance relationship, it is increasingly clear that relationships among these factors do exist and are quite complex. Today, the specific risk of food-related violence is increasing as a result of the coupling of accelerated GEC with inadequate institutional arrangements for addressing the complexity of food and its multiple meaning in today's globalized world.

This chapter focuses on two primary mechanisms by which environmental change might increase food system violence: first, by directly or indirectly exacerbating conditions of scarcity in food supplies among vulnerable populations; and second, by contributing to social processes that are already leading to increased marginalization, impoverishment or injustices of population in relation to their access to food. In the first case, of most obvious concern in relation to incidences of direct violence are environmental shocks, particularly those that cause volatility in local and global food supplies such as droughts and floods, but also those associated with disruptions in transport and food distribution. These shocks can exacerbate and deepen conditions of structural violence, as well as precipitate events of food-related direct conflict. The risk of direct violence from environmental shocks rises when such shocks occur in contexts where structural violence is already prominent – for example, conditions of political instability, high social inequality and chronic problems of resource access (Homer-Dixon, 1999). Recent examples of such problems can be found in relation to the failure of food distribution and supply networks in Myanmar in the aftermath of Cyclone Nargis in 2008 (WFP,

2008), and in New Orleans following Hurricane Katrina; the rise of armed conflict between pastoralists competing for dwindling water resources in the 2009 drought in North-eastern Kenya (Bello, 2009); and violence over access to water in drought-affected Aden, Yemen in August 2009 (SANNA/Reuters, 2009). All of these cases demonstrate how environmental shocks can inflame conditions in which political, social and economic inequities are chronic.

In addition to the environmental shocks described above, food system violence can be provoked by environmental scarcity associated with changes in ecosystem services. One issue that is of increasing concern to analysts and to society at large is the availability of arable land of sufficient quality to meet future food demands. Since 1950, land under agricultural production has decreased on a per capita basis from 0.8ha/person to 0.1ha/person today, and continued population increase as well as changes in food habits will increase the pressure on land further in spite of increased yield levels (Seto et al, in press). Currently 11 per cent of the world's land surface is used for crop production (Turner II et al, 2007) and approximately double that amount is used for pastures. The United Nations Food and Agriculture Organization (FAO) (2003) estimates that 25 per cent of available land could be used for production, but such figures are highly uncertain and subject to discussion. For example, as climate conditions in some regions become increasingly arid, the use of freshwater sources for irrigation will also increase. With 85 per cent of extracted fresh water currently used in agricultural production, the expected future scarcity of global water resources will pose challenges to food production, particularly in relation to any assumptions associated with the viability of irrigated farming (Rosegrant et al, 2002).

Whether or not any additional land brought into production will assist in meeting increasing food demand depends in part on the implications of agricultural extensification on associated ecosystem services. Significant losses in biodiversity, irreversible levels of soil erosion, dramatic deforestation and water-quality deterioration are only a few examples of natural resource degradation that have been associated with the rate of agricultural land-use expansion since 1950 (MA, 2005a). These environmental consequences indicate that the future of food availability is not just a factor of land but also the ability of society to maintain the productivity of the natural resource base through appropriate technologies and production practices.

Given the numerous social and environmental tradeoffs associated with agricultural expansion, some scholars expect that as little as 20 per cent of future increases in crop production will come from the cultivation of new land; the majority must come from increased yields (67 per cent) and higher cropping intensity (12 per cent) (Bruinsma, 2003). The main means of intensifying crop production (producing more of the desired products per unit area of land already used for agriculture or forestry) will be through increased yields (increasing crop productivity), together with a smaller contribution from an increased number of crops grown in a seasonal cycle (increasing cropping intensity) (Gregory et al, 2002).

Yet increasing agriculture production intensity is also problematic. As discussed in Chapter 1, agricultural intensification has also been known to degrade the environmental services on which further production depends (Gregory et al, 2002), and in some cases has also been a direct trigger of violence where social inequities are already prevalent. In the past, concerted efforts to increase yields through intensification have indirectly and in some cases directly resulted in physical conflict over access to critical resources such as land, water and technology, as well as structural violence in terms of the exclusion of some populations from the means of production. In the 1970s and 1980s, for example, the mechanization of farming in the Sahel pushed nomadic pastoralists from their traditional lands. The resulting competition for increasingly scarce pasture was one factor that contributed to an increase in violence among various pastoral groups, and, eventually, among pastoralists and sedentary farmers over land (Suhrke and Hazarika, 1993). Violence of both physical and structural natures has also been associated with the dissemination of green revolution technologies. Shiva claims that Indian smallholders were priced out of production by the introduction of green revolution technologies, and that the harnessing of water and land for agricultural intensification produced violent conflict among dispossessed farmers and water infrastructure providers in the Punjab (Shiva, 1991). Today, fear of similar negative outcomes surrounds the current controversy over the introduction of genetically modified seeds in developing nations, which some populations perceive as a threat to their health and to the ecological resources that they value, while other populations raise concerns about rights, access and control of the technology (Paarlberg, 2000).

While there is no necessary linkage between agricultural intensification or extensification per se and violence (in fact, depending on the specific mode of intensification and resources available, intensification might alleviate conflict in areas where land scarcity is the foremost concern), any path chosen to meet the needs of future food consumers will likely have significant environmental, social and cultural consequences. Whether these consequences induce violence will have a lot to do with existing circumstances of livelihood and food security, and with the adequacy of existing institutions to mediate conflicts as they emerge and to address concerns over inequitable distribution of the costs and benefits of changes in the food system.

The possibility of involuntary migration associated with GEC, such as increased drought or sea-level rise is also increasing the risk of food-related violence. In the past, localized food shortages have sometimes been accommodated by migration to other adjacent areas or, in instances such as the Irish potato famine of the mid-1800s, to other countries and continents (Spitz, 1978). Today, resource scarcity, border controls and restrictions on population movements are increasingly making voluntary migration difficult. Nevertheless, involuntary population resettlement as a result of climate change may force the redistribution of productive resources to accommodate migrants. The threat of violence is already associated with such migrations. The mechanization of Sahelian agriculture, for example, not only spawned violence among pastoralists and farmers, but also induced a wave of migration to urban centres such as Khartoum. Not only did this migration contribute to unregulated urbanization but also exacerbated existing ethnic tensions between various groups now living in high-density, urban centres. In some cases food riots resulted (Suhrke and Hazarika, 1993). Similar situations have been seen in migration from Bangladesh to Bombay (Homer-Dixon, 1999) and in Thailand, where deforestation in the north-east area of the country drove migration into Bangkok (Suhrke and Hazarika, 1993).

Violence in food systems has come to the fore in the context of increasingly complex and coupled markets for food-related resources. Food system globalization has numerous benefits, including increased efficiencies in moving commodities across geographic regions to satisfy emerging demands and to smooth price volatility and supply shocks (Pingali, 2007). Globalization and the commoditization of food may well serve to quell violence, particularly violence associated with surprise and shocks (see Chapter 20). Nevertheless, there is also validity to concerns that processes of globalization may contribute to both increases in physical violence as well as to reinforcing conditions of structural violence. Many scholars have written on the implications of globalization for global inequality, and concluded that market integration has not uniformly benefited the world's poor or improved food security (Gledhill, 1995; Cornia, 1999; Davis et al, 2001; Milonavic, 2003; Wade, 2004). The World Bank's data indicate a widening income gap between wealthy and poor countries since 1960 (World Bank, 2001), although international income inequality may be diminishing in the most recent period of 2001–6. This inequality is also reflected in the concentration of wealth within nations, and between urban and rural populations (Wade, 2004).

In a reflection of what Homer-Dixon has termed 'resource capture' (Homer-Dixon, 1999), some countries have sought to ensure their own security of food supplied through the direct purchase (via government-owned companies) of millions of hectares of land in other countries (a compilation of media accounts of such transactions can be found at www.farmlandgrab.org). Such land purchases can be interpreted as both a response to, and a provocation of, resource scarcity and food security at the local to national level. The violence associated with such moves is typically structural: it accentuates existing inequities in power over resource access and control, and may have the effect of marginalizing within-country populations from production opportunities. India, for example, recently invested US$4 billion for land in Ethiopia to be used to grow maize, sugar cane, rice and other crops to supply its domestic food markets (Nelson, 2009). In other cases, private entities within countries and between countries are driving the observed increase in land transactions (von Braun and Meinzen-Dick, 2009). In some cases these transactions have a relatively long history. China, for example, has been leasing land from Cuba and Mexico for over ten years (von Braun and Meinzen-Dick, 2009).

International transactions in land do not necessarily have to result in exploitation or adverse social outcomes. Nevertheless, in the current context of rising food prices and increased resource scarcity, these land purchases have raised contentious questions over how these transactions are regulated, whose interests are represented, and the implications of these land transactions for food insecure populations. In some cases, these transactions have resulted in direct violence and conflict. For example, the competition between land for food and land for biofuels has already produced instances of local violence, as in the case of indigenous communities in Indonesia protesting against deforestation for oil palm production (Casson, 1999). In Madagascar, land purchases by South Korea for maize and oil palm production were cancelled after the international media highlighted concerns about the displacement of poor farmers and potential impacts on biodiversity (Ryal and Pflanz, 2009; Vidal, 2009). Saudi Arabia has reportedly purchased land in Sudan to secure food supplies in the face of declining domestic water resources for production (GRAIN, 2008). While these land deals are officially sanctioned by Sudan, there are concerns that the transactions disenfranchise vulnerable populations in Sudan, for whom access to limited land and water resources is a matter of survival (see Box 17.1).

Box 17.1 *Land for fuel or food?*
Conflict and food security implications

The International Food Policy Research Institute (IFPRI) estimates that, since the spike in world grain prices over the 2007–8 period, a wave of new land-for-biofuel transactions has occurred involving 15–20 million hectares and deals between private commercial interests, governments of industrialized nations and developing countries (von Braun and Meinzen-Dick, 2009). Meeting the demand for biofuels requires significant land area, and has put fuel production in direct competition with other agricultural products (e.g. when maize and soybean are produced for biofuel) and with natural habitats (e.g. to establish palm oil plantations in tropical forest areas). Among the largest known transactions are purchases by the Chinese government, via ZTE International, of 2 million ha in Zambia for jatropha production and 2.8 million ha in the Democratic Republic of Congo to create the world's largest oil-palm plantation. In the domain of private sector–government transactions for biofuel farming, SunBiofuels of the UK has purchased land in Ethiopia and Tanzania, and a Swedish company, Skebab, has secured land in Mozambique.

The outcomes of these transactions are not yet apparent. Nevertheless, given the fragile state of democracy and informal nature of land tenure institutions in many of the countries where such transactions are being implemented, there are concerns that the interests of domestic populations – both producers and consumers – may be poorly represented, creating conditions for violence and conflict. In Ghana, for example, Norway's purchase of land to build the world's largest jatropha plantation has resulted in extensive deforestation and loss of valuable shea nuts (von Braun and Meinzen-Dick, 2009). In this case it appears that the most negatively affected population are the women who traditionally harvest the shea nuts and depend on these harvests for their livelihoods (FAO, 2009a).

Where institutions are in place, conflict emerging from the land transactions can be addressed through formal legal means. In a recent case, a farm association in Pakistan (the Pakistan Kissan Board) challenged in the Lahore High Court a government proposal to sell or lease millions of hectares of agricultural land to foreign interests (Staff, 2009). The association charged that the transactions would amount to depriving the people of rich and fertile soil and would threaten the sovereignty of the country. Where legal recourse is absent or unreliable, violence is far more likely. In Madagascar, for example, the announcement of a 99-year contract to lease 1.3 million ha to South Korea's Daewoo Corporation to grow corn and palm oil was apparently an important factor in causing the recent overthrow of the national government.

In recognition of the high risks involved in these transactions and potential significant inequities that could result, some countries have taken steps to protect their resources and populations from unscrupulous private transactions. Thailand, for instance, has enacted legislation to prevent 'land grabs' over protests regarding local stakeholders' (such as farmers) customary land rights being violated by such transactions (Nelson, 2009). The initial reports emerging from these transactions suggests not only that these land deals are a potential source of conflict and violence, but also that existing institutions are poorly equipped to regulate them, raising important concerns of social justice and food security.

While the international land markets may be poorly regulated, food commodity trade is subject to greater scrutiny and negotiation through international trade institutions. Nevertheless, in industrialized countries, market and trade policies are often used to ensure the food supplies and livelihood security of domestic consumers and producers, at the expense of market access by developing countries (Valdes, 1980). As is argued below, the pervasive treatment of food, first as a tradable commodity, and second as a subject of cultural, social and political interest, has given rise to institutions that favour commercial interests and that shift power to commercial actors (Goodman and Watts, 1997). The concentration of control of production technology and distribution systems in the hands of relatively few companies has led to circumstances in which scarce resources are allocated on the basis of where profits and commercial opportunities are the greatest, rather than where needs and rights related to food are acute.

Some scholars have attributed the loss of competitiveness for some domestic products in relation to imported goods in many developing countries to the high level of agricultural subsidies in Europe and the USA (Anderson and Martin, 2005). While the availability of less expensive imported food can reduce food insecurity and mediate risk of violence, in other circumstances perceptions of market inequities are an increasing source of conflict, as has been repeatedly demonstrated with the sometimes violent protests surrounding the Doha Round negotiations of the WTO, as well as with the social movements provoked by bilateral trade agreements such as the North American Free Trade Agreement. In one case, an estimated 200,000 farmers took to the streets in January 2008 in Mexico City to protest against a government decision to remove trade barriers against US maize, beans and sugar (Tobar, 2008). In Europe, thousands of dairy farmers staged violent demonstrations in June 2009 in the face of falling global milk prices and the EU's decision to lift milk production quotas, exposing producers to free-market forces.

In some countries where private sector interests are particularly well organized, agriculture and food policies may also favour protecting commercial interests at the expense of domestic food security and public health. This has been illustrated in the debates surrounding food policy in the USA and UK, in which some have argued that policy tends to support the production of foods that do little to enhance the dietary health of the population, or which create 'food deserts' where poor populations have no easy access to healthy produce and food products (Wrigley, 2002; Wrigley et al, 2003; Hawkes, 2006; Blanchard and Blanchard, 2007).

In summary, while climate shocks are not always the foremost or principal driver of violence in food systems, the existing trends in extreme events, land use and land markets, agricultural expansion and intensification suggest that GEC can exacerbate existing chronic conditions of structural violence, and, in some cases, directly trigger physical violence and unrest in relation to resource access. The twin processes of GEC and globalization independently can lead to violent outcomes for food insecure populations. The potential synergistic interactions of both processes are particularly worrying.

Underlying causes of violence: Contradictory views on food

Underlying the confluence of violence around food and food security are distinct and not necessarily congruent perspectives on the meanings that different populations associate with food, the control of such meaning in popular discourse, and the relation of food to society and the environment. While the disparate and changing meanings associated with food, and the power relations that circumscribe food meaning, have been discussed at length (McMichael, 1994; Goodman and Watts, 1997), the incompatibility of diverse meanings is now highlighted in the confluence of three circumstances: (1) lengthy, globalized food chains; (2) increasing risk from GEC; and (3) the persistent and increasing social inequality. Each of these three circumstances supports a distinct discourse related to the meaning of food, and it is the clash between these contradictory meanings that, in the view of the authors, is contributing to the rising risk of violence associated with food systems and food security.

In the discourse on globalization and global food chains, food is viewed as a global *commodity* (see Figure 17.1), for which there is a demand without differentiation across a market (i.e. 'wheat is wheat'). This perspective is what McMichael refers to as the focus on the 'exchange-value' of food (McMichael, 1994), supplanting or challenging value systems in which culture, place and the use of food predominate. Food consists of a diversity of products that are bought and sold competitively, and the market system rewards efficiencies of volume and cost, and favours product differentiation and, in some cases, quality (Watts and Goodman, 1997). The focus on entitlement failures as primary drivers of food insecurity has particular explanatory power in this perspective, where the economic value of food is paramount. In its latest report *The State of Agricultural Commodity Markets* (FAO, 2009b), the FAO has addressed exactly these tensions between the winners and losers in global agricultural

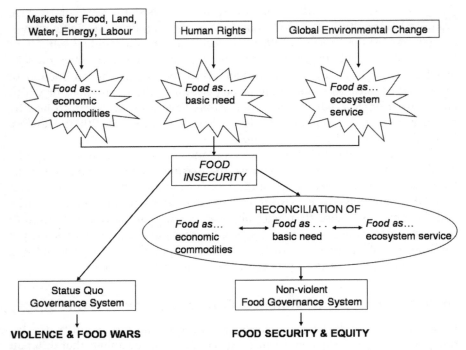

Figure 17.1 *Drivers of violence in food systems and avenues for non-violence food governance*

commodity markets, by searching for ways of making food commodity markets work for developing country and industrialized participants alike.

In the discourse on GEC, food is quite differently viewed as an *environmental service* (see Figure 17.1). The concept of environmental services, which was formalized by the 2005 Millennium Ecosystem Assessment (MA), views natural ecosystems as performing fundamental life-support services upon which human life depends, including provisioning of food, water and energy, supporting of the nutrient cycle and regulating of pests and diseases (MA, 2005b). Under the impact of GEC, the provisioning, supporting and regulating services of ecosystems are altered, and in some cases put under stress with adverse effects on the quality and quantity of food, water, land and health. Food provisioning is of course not simply a victim of the adverse effects of GEC, but, as discussed above, is also one of the most important drivers of GEC and ecosystem service modification (Gordon et al, 2008).

The two discourses above are at odds: the commoditization and globalization of food aims to standardize and secure production by controlling as much as possible the environmental conditions in which food is produced through advanced technology (McMichael, 1994; Buttel, 1997). With economic yields foremost in mind, industrial agro-ecosystems have historically been managed to enhance productivity by using introduced and synthetic inputs to overcome biochemical or physical limitations posed by the natural environment (Atkins and Bowler, 2000). In contrast, approaching food provisioning as an ecosystem service opens the way for ecological perspectives on resource management. Gordon et al (2008), for example, argue for maximizing the functional and response diversity of species and enhancing the capacity of agricultural systems to persist in conditions of high disturbance and environmental volatility. Similarly, Tilman et al (2002) argue for viewing agriculture production not on the level of the field but rather as part of a complex landscape in which agricultural activities both procure and provide ecosystem services.

Together, the commoditization of food in a globalized food market (which creates winners and losers), and the increasing pressure of GEC on ecosystem services (also creating inequities in the viability and productivity of natural resources), have given new emphasis to a third discourse associated with food and its meaning: that of food as a basic *human right*. In this discourse, dating back to the United Nations' Universal Declaration of Human Rights of 1948 and now institutionalized in the position of the UN's Special Rapporteur on the Right to Food (www2.ohchr.org/english/issues/food/overview.htm), food is considered, first and foremost, as a basic human need, a need that fulfils the most basic physiological, cultural and safety needs of any human being (UN, 1948). In the hierarchy of human needs as proposed by Maslow (1943), food (along with clothing and shelter) is placed at the bottom of the pyramid of needs for life, providing the foundation for our most essential needs. The denial of the right to food is therefore highly contentious. Populations perceiving that their rights have been infringed upon may well mobilize and protest in an effort to have their voices heard and rights reinstated. In cases where institutions are unresponsive, protests can become violent as populations resort to extra-legal or more aggressive behaviours in an effort to ensure their needs are met. As John Burton (1990) has argued, aggressions and violent conflicts can be seen as the direct result of some institutions and social norms being incompatible with inherent human needs.

While there are potential synergies among the three discourses on food and the disparate values associated with food, each perspective has given rise to its own set of institutions, actors and goals, as well as distinct expectations for addressing conflict. These institutions and actors are described below. The absence of a set of institutions that encompasses all three perspectives on food indicates that the risk of food-related violence may well increase in the future, as the pressures of globalization and GEC aggravate existing tensions in vulnerable and food insecure regions.

Institutional failures and food violence

Violence associated with food can erupt between countries at an international level (Box 17.2, p257), at a national level often as a result of actions of a government against its people (Box 17.3, p258), among distinct populations at a sub-national level (Box 17.4, p259), and locally within a population group where individuals are struggling over access to scarce resources. In each case, interventions to reduce the risk of violence depend in part on adequate institutional arrangements at international, national and sub-national levels that can legitimize the intervention and allow for conflict.

Trade institutions and conflict

At the supranational level, the WTO and associated institutions formally structure the food supply chain and food distribution, and contain mechanisms for addressing conflict associated with the international trade of food. National governments are also heavily involved in the regulation of food supply and access, and most formally subscribe to the free trade principles embodied in the WTO. Nevertheless, in practice, the governance of food systems is the result of a dynamic negotiation of interests and power in relation to food system resources, both within countries and between countries. This negotiation reveals unresolved tensions in characterizing food in a simplified sense as a tradable commodity. On the one hand, the formal institutional arrangements currently tend to promote the idea that markets are the most efficient means of achieving food security and adequate distribution of food globally. On the other hand, in practice, government policy has often tended towards nationalistic goals of protecting consumer and producer markets, particularly in industrialized nations. For example, while developing nations removed non-tariff barriers and reduced tariffs under agreements with the International Monetary Fund and World Bank, as well as under WTO agreements, industrialized nations, such as the USA, were able to lobby for ambiguous language in the same WTO agreements that allowed them to increase agricultural subsidies on export crops such as maize (Gonzalez, 2004).

GEC will enhance the imperative for nation states to take action to ensure adequate food supplies and productive resources, including land, water and energy, for their populations (see Box 17.2). Some of these actions, such as those pertaining to commodity trade, are governed by existing international trade and investment policy, and will potentially raise concerns over the balance of protectionism and free trade interests. Other actions, such as 'land grabbing' described above, are currently taking place in an institutional vacuum.

In the domain of food product trade, the tension between protectionist policies in relation to food security and environmental goals and the general movement towards free-trade principles have been partially addressed in the Uruguay Round WTO agreement through the use of the 'green box' designation. This designation permits some non-trade distorting subsidies in agriculture in order to address food security shocks, poverty concerns and environmental objectives. The designation is, however, highly contested and it is the responsibility of each country to argue for why their food security or environmental goals cannot be satisfied through free-trade mechanisms. In the current round of agricultural negotiations, additional mechanisms have been proposed to enable developing nations to enact policies to protect food security and the environment. Among these is a proposal for a 'development box' that would permit trade interventions to secure basic commodities for food security purposes and potentially protect smallholder producers from market shocks. Nevertheless, there has been no consensus among developing nations on the advantages of including such a 'box' in the current negotiations, and ultimately whether achieving food security requires enhanced trade or protectionist policies is heavily disputed (WTO, 2004).

Box 17.2 *The food crisis of 2007–8:*
A failure of supranational institutions?

At an international level, the food crisis of 2007–8 illustrates the inadequacy of current systems of international governance to address unexpected global shocks to the food system. Many countries reacted to rising food prices and domestic political unrest by implementing protectionist policies such as banning exports (Mitchell, 2008). These policies were met with significant opposition in trading partners, and the retaliatory measures implemented in some cases exacerbated food insecurity. For example, in response to citizen protests over rising prices, Egypt put a ban on the exports of rice to protect their domestic consumers and China froze the price of food staples (Tenenbaum, 2008). In response to increasing wheat prices, India restricted rice exports with the idea of keeping rice stocks in the domestic market to allow for dietary substitution of rice for wheat. Vietnam and Indonesia followed suit, and in an unforeseen consequence of these policies, the international markets experienced further shocks, which drove prices higher and led to further protests in many nations (Brahmbhatt and Christiaensen, 2008). While the food crisis was a global phenomenon, driven by coincident economic, environmental and political factors (see Chapters 1 and 7), no international organization was prepared to cope with the crisis and coordinate national responses to avoid aggravating food insecurity. The complexity of the origins of the crisis implied that the violence resulting from rising food prices was not a direct concern of humanitarian organizations, and trade agreements and policy were not equipped to address such emergent shocks and national responses. In the absence of other organizational formats, a special High Level Task Force on the Global Food Crisis was organized by the UN in 2008 to develop a strategy for international response to the crisis (Murphy and Paasch, 2009).

The lack of agreement on how to address the concerns illustrates the problems associated with relying on trade policy to effectively resolve conflicting goals and meanings associated with food. In effect, in the midst of the most recent food crisis, the WTO spokesperson Lamy declared that the organization was not an effective means for addressing the short-term crisis, although ultimately open markets and freer trade would lead to greater market stability and ultimately food security for vulnerable nations (Lamy, 2008). Recently, the WTO declared that 'Trade is the transmission belt that allows food to move from the land of the plenty to the land of the few' (Lynn, 2009). In contrast, other groups of stakeholders argue that the current institutions governing food trade are, in fact, the source of conflict. The US Working Group on the Food Crisis (a consortium of non-governmental actors) argued, for example, to the G8 that a re-evaluation was needed of the utility of policies such as strategic grain reserves and stockholding, antitrust enforcement, as well as guarantees for 'rights to food and build healthy local and regional food systems that foster social, ecological and economic justice' (US Working Group on the Food Crisis, 2009).

Box 17.3 *Zimbabwe, sovereignty and domestic food crises*

Recent events in diverse contexts illustrate the inadequacies of institutional arrangements at national and international levels for addressing the complex nature of food-related violence, especially when exacerbated by GEC. In 2008, the food security situation in Zimbabwe was severe: 45 per cent of the population was reportedly food insecure. Institutional failures played a significant role in the crisis. A series of inappropriate and inadequate policies and incentive structures had precipitated socio-economic collapse, with severe implications for food availability and access. According to media reports, in the face of political unrest over rising hunger in 2007, the government intervened directly in food prices, setting off uncontrolled inflation and the diversion of food into the black market, where it became largely inaccessible to most households (Raath, 2007). The inflation rate alone, which was over 2 million per cent in July 2008, was devastating to consumers. The profound and prolonged civil conflict between the ruling party and opposition parties precipitated the dissolution of any moral contract the government had to provide for the population. The 2007 rainy season was characterized by inconsistent rainfall, a pattern that led to alternate droughts and floods across the country (FAO/GIEWS, 2008). In January 2008, 95 per cent of the country received 150 per cent of average rainfall, disrupting farm activities and causing flooding, soil leaching and erosion (IFRC, 2008). This situation is typical of anticipated GEC-induced changes in climate variation and extreme events. The combined effect of economic collapse, political conflict and market failures, coupled with climatic shocks, led to widespread malnutrition with many child and adult deaths, a situation further exacerbated by a cholera epidemic affecting over 70,000 people at its height in early 2009 that resulted in over 3700 deaths (WHO, 2009). In this context, food aid has been used as a political weapon against opposition supporters (Zimbabwe Peace Project, 2008). The food shortages spurred the migration of hungry populations to neighbouring South Africa. Hunger relief agencies and humanitarian organizations aiming to protect the population from starvation have found their activities crippled by having to work through official government channels and distribution networks (Zimbabwe Peace Project, 2008), which have manipulated food supplies for political end.

Humanitarian organizations and conflict

Emergent food crises have traditionally been addressed through bilateral and international humanitarian organizations such as the World Food Programme (WFP) and Famine Early Warning Systems Network (FEWS NET; see Chapter 6), as well as through national food safety nets and poverty alleviation programs. WFP provides emergency food relief as well as works to prevent future hunger with a vision that all should have access to food to provide for a healthy lifestyle and through focusing on food security analysis, food procurement, nutrition and logistics (WFP, 2009). The material food aid provided by the USA through the Food for Peace Act (PL 480) is one product of the substantial agricultural subsidies allocated to US producers, resulting in an

---

Box 17.4 *Palm oil in Indonesia*

Indonesia's palm oil industry expanded dramatically over the period 1967–90, with annual crude palm-oil production increasing 12 per cent, and the area under palm oil expanding to 20 times what it was in the 1960s. The expansion has been driven by the massive increase in demand by the international food industry for palm oil, and has contributed to deforestation and the displacement of small rural communities, exacerbating social conflict. Much of the palm-oil development has been managed by large international conglomerates. Reformists and local communities have been demanding to play a larger role in local development and, in some cases, have attempted to reclaim their land from the large palm oil conglomerates. Violence toward conglomerate representatives in the form of burning plantations, offices and processing plants was not uncommon in the late 1990s. Looting and violence against workers on the plantations was also on the upswing. The violence has caused some plantations to reduce their planting targets or close down their operations (Casson, 1999). The last several years of high oil prices have once again triggered palm oil expansion. Confusion regarding forest boundaries and land tenure has made the conflict worse. Along with concern regarding deforestation, locals are protesting the low wages paid to them by the palm oil plantations (Reuters, 2009).

---

excessive production of basic commodity crops. PL 480 allows the USA to use this excess production as international food aid, albeit at considerable cost both in terms of food shipments and associated environmental externalities, and in terms of the effects on the receiving country's domestic production and food import markets (Director International Affairs and Trade, 2007).

It remains to be seen whether such emergency relief services are adequate for dealing with the complexity in cause and scope of food emergencies such as the world experienced in 2007–8 or, for that matter, the type of security disturbances that are anticipated under climate change scenarios (Purvis and Busby, 2004). Barrett and Maxwell (2006) argue that the 'existing international mechanisms governing food aid are dysfunctional and outdated' (Barrett and Maxwell, 2006). They claim that the Food Aid Convention, of which only donor countries are members, is inadequate for ensuring that food aid achieves the humanitarian and development objectives for which it is ostentatiously intended. Emergency relief organizations are necessary, and play critical roles in mobilizing resources to avoid human food catastrophes. Food relief programmes are not, however, designed to ensure that the political, social and economic contexts for food distribution and governance are robust in the face of changing natural resource availability and environmental shocks. Many of the drivers of food conflict are deeply embedded in specific modes of social organization and development paths, and demand flexible yet systematic means of addressing these drivers at multiple levels of government (as are described in greater detail below).

Human rights institutions and conflict

The FAO defined the right to food as 'a human right, inherent in all people, to have regular, permanent and unrestricted access, either directly or by means of financial purchases, to quantitatively and qualitatively adequate and sufficient food corresponding to the cultural traditions of people to which the consumer belongs, and which ensures a physical and mental, individual and collective fulfilling and dignified life free of fear' (Ziegler, 2002). Food as a human right has a relatively long history in the institutional frameworks of the United Nations (UN, 1948), including the establishment of a Special Rapporteur on Rights to Food. The recognition of food as a human right by the United Nations does not, however, necessarily mean that effective conflict resolution mechanisms will be available at the international or national and subnational levels. Nevertheless, non-governmental and civil society groups have created highly visible global activist networks, which have pressured national governments and international organizations to address food rights concerns in national and international policy (Murphy and Paasch, 2009). Such groups argue for recognition of individual freedoms associated with productive assets (seeds, land, water, fishing grounds), as well as for framing food security concerns such as issues of distribution and access, cultural sovereignty and human dignity (Menezes, 2001). The efforts of these groups can be seen in terms of the UN Covenant on Economic, Social and Cultural Rights, the African Charter on Human Rights (which has been extended to include the right to food by the African Commission on Human and People's Rights), and the American Convention on Human Rights in the area of Economic, Social and Cultural Rights.

To date, there have been only a limited number of cases in which these institutions have been used to address conflict associated with food rights. Two cases where the right to food has been recognized by regional courts included the case of 'Social and Economic Rights Action Centre and the Centre for Economic and Social Rights (SERAC) v Nigeria (2001)' and the case of the 'Indigenous community Yaka Aye v Paraguay (2005)'. In the first case, the African Commission on Human and People's Rights found the Nigerian military government guilty of, among other infringements, violating the right to food implicit in the African Charter by: (a) destroying and contaminating crops; and (b) failing to respect and protect the Ogoni people from this infringement by non-state actors. The Commission ordered various actions to be taken by the government. Nevertheless, the Commission lacks an effective mechanism for enforcing this decision. In the latter case, the Intra-American Court of Human Rights reaffirmed its wide interpretation of the right to life to include the right to adequate food as set out by the Protocol of San Salvador. The Court held that the state of Paraguay had an obligation to adopt measures to ensure a dignified life (including the right to food) for high-risk, vulnerable groups and ordered the government to submit a report on measures that it would implement in a year's time.

Although the progressive nature of these decisions is unquestionable, the lack of effective enforcement mechanisms means that there is often no concrete improvement in ensuring rights that have been agreed to in charters. In addition to lack of enforcement, bringing cases to international tribunals can be cumbersome and costly for participants. While some national governments who recognize rights to food (e.g. Congo, Ecuador, Haiti, South Africa) may have institutional structures to accommodate and address conflict arising over rights violation, it remains to be seen how such institutions will manage emerging conflicts aggravated by environmental change and how such institutions will be reconciled with potentially contradictory rules and norms associated with trade regimes and policy.

Emerging and alternative systems of governance

In the absence of formal governance structures to address food-related conflict and GEC, other actors from civil society and the private sector are mobilizing to reconcile conflicts on their own terms. There are cases in which private sector actors are participating in sustainable agriculture and food system initiatives, in recognition of emerging issues of resource scarcity and competition associated with issues such as global warming. For example, the World Dairy Associations and Companies (including representatives of the dairy industry from both developing and industrial nations) recently demanded proactive policy change to recognize the role of dairy production in social and environmental sustainability (Reuters, 2009). Other industries are participating in round-table talks with the FAO and other international organizations to discuss formalizing and coordinating actions regarding food production, distribution and processing. One example of such collaboration is the Sustainable Agricultural Initiative Platform (www.saiplatform.org), in which companies such as Danone, Kraft, McDonalds, Nestle, Sara Lee and others are discussing ways in which the food industry can work together to achieve common objectives concerning sustainability, including economic, social and ecological dimensions. While neither violence reduction nor 'rights' per se are explicit objectives of this group, it does offer a model of voluntary private sector governance. The role of non-state actors in the food system is discussed more generally in Chapter 18.

Other long-established initiatives attempt to avoid structural violence through formal rules and principles of transparency and accountability in food trade. The World Fair Trade Organization and Fair Trade Federation, and the associated certifying agencies (Fairtrade Labelling Organization, Transfair USA and Transfair Canada), require members to, among other things, establish social and environmental accountability in trade relationships, provide safe working conditions and engage in trade that sustains and enhances producers' livelihoods. These are voluntary organizations and thus cannot address conflict that arises among non-participating parties, or that fall outside the scope of the specific trade partnerships that they regulate. Although some scholars argue that by 'commodifying' the social and environmental attributes of production these conventions do not represent true alternatives to the dominant trade regime (Taylor, 2005; Jaffee and Howard, 2009), fair trade potentially offers a means of addressing conflicts associated with GEC that may arise among trading partners. Assisting coffee farmers to adapt to climate change, for example, has become a theme of fair-trade coffee importers and retailers such as Green Mountain Coffee and Starbucks.

Some civil society groups such as the International Network for Economic, Social and Cultural Rights (ESCR-Net) and the food sovereignty movement are taking action to modify existing institutions to be more conducive to protecting rights. ESCR-Net has created a working group on trade, investment, finance and human rights that is a vocal advocate in WTO negotiations for a human-rights-based approach to global trade. ESCR has also established a database of case law in which rights violations, including 'the right to food security', have been defended and addressed in national or international courts (ESCR-Net, 2009). In another case, the International NGO/CSO Planning Committee (IPC) defines itself as a global network of non-governmental and civil society groups concerned with the diversity of rights associated with food production, distribution and consumption. The IPC provides a forum to communicate the interests of its network to the appropriate international organizations (e.g. the FAO, the World Bank, the WTO) with the aim of getting a rights-based perspective on food incorporated into international agreements, and having such perspectives reflected in development practice (e.g. the Millennium Development Goals).

Together with consumer action groups, organizations such as these are reshaping some elements of the food system. Their actions provide examples of possible structures for food system governance that reflect the networks that are consolidating around shared social interests and values, food production, consumption and trade. These emergent networks of social actors provide instructive insights into alternative means of organizing food systems according to the relationship of actors across scales of time and space. It will be increasingly important to draw from these innovations and instructive examples in considering what forms of governance are needed to reduce the risk of violence in the face of increasingly integrated markets and unprecedented rates of environmental change.

A new structure for food system governance to reduce violence

The existing systems of governance, while working relatively well to enhance food commodity trade and to motivate production, have shown to be less well equipped to address changes in the food system that provoke violence. This failure can only be exacerbated by the pressures resulting from GEC. The experience reviewed above suggests that while there are lessons to be learned from current institutions and practices, the structure and functions of food governance to reduce the risk of violence will require definition and thus this is a critical area of new research.

Governance can be defined as:

> the sum of the formal and informal rule systems and actor-networks at all levels of human society that are set up in order to influence the co-evolution of human and natural systems in a way that secures the sustainable development of human society (Biermann, 2007).

To avoid the potential for increased food violence, any new form of governance must address the underlying tensions between different perspectives and resulting institutions related to food and its diverse value in society: food as *commodity*, food as *basic need* and food as a part of *ecosystem services* (Figure 17.1). The diverse values and meanings of food will persist; what is needed is a democratic forum in which alternative perspectives on food and its value to society have equal footing, and rules and norms are established to compensate for the differential power and political interests associated with different perspectives on food values. Currently the lack of recognition of alternative views on food and its value allows the single and narrow perspective of 'food as commodity' to dominate policy and social transactions in relation to food and the environment. A reconciliation of a broader spectrum of values associated with food should be considered the foundation of any new arrangement of food system governance. The reconciliation of diverse perspectives on food meaning requires a close examination of how governance should be structured, who should participate in governance, what aspects of the food system require coordination and/or regulation and specifically how conflicts can be resolved at the diverse scales and levels at which they manifest. Here the authors echo scholars and practitioners who are arguing that a rights-based approach to food system offers the greatest potential for reconciliation of diverse meanings and values associated with our complex food system, as well as the greatest opportunity to reduce the risk of violence (e.g. Anderson, 2008; Murphy and Paasch, 2009).

Some of the emerging research questions needed to address the challenges associated with achieving the goal of improved food security while reducing the risk of violence in food systems are described below. Those aspects of governance that specifically relate to the causes, consequences and resolution of issues of violence and food insecurity are addressed. Many of the questions described in this section directly contribute to the analytical themes identified as key issues of concern in the international research project 'Earth System Governance' (Biermann, 2007): architecture, agency, adaptiveness, accountability, and allocation and access.

How should any new form of governance be structured in terms of specific institutional arrangements and shared principles?

It is not yet apparent whether an entirely new food governance system is required, or whether existing governance systems such as the WTO can be modified (as many NGOs are now advocating) to address the need for reconciling diverse perspectives on food meanings, and the rising risk of conflict associated with food. In the 2009 World Summit on Food Security, the member countries of the FAO agreed to a reform of the Committee on World Food Security (CFS) that begins to address some of the challenges associated with the risk of violence in food systems. For example, under the reforms, the CFS would create a more inclusive space for discussing global food security policy, emphasize linkages among countries, regions and local governance, and promote greater accountability in relation to food security and nutrition goals. Promoting the use of 'Voluntary Guidelines to Support the Progressive Realization of the Right to Adequate Food in the Context of National Food Security' in member countries is part of the reform programme. While the membership of the CFS would only include international agencies with explicit mandates in food security and nutrition, the WTO and international finance organizations would be invited as formal participants in CFS discussions (FAO, 2009c). The reforms do not include any discussion of enforcement or conflict resolution measures, but an emphasis on rights-based policy is evident. It is perhaps encouraging to note that over the last few years the UN has been discussing concrete ways of reconciling the WTO with the perspective of food as a right, as enshrined in the *Universal Declaration of Human Rights* (De Schutter, undated, 2009). In addition to concerns with food trade, most transactions in land, water and energy currently fall outside of any specific institutional arrangements associated with food systems. The potentially serious implications of these transactions for food security and the risk of violence in food systems add weight to the argument that the global architecture for food governance needs to be reconsidered.

In order to reduce the risk of violence, which actors can and should participate in governing the food system? Which actors have the responsibility and legitimacy to play active roles in conflict resolution at different spatial levels?

As described above, it is evident that the world food system is being reconfigured through both the actions of 'authoritative actors' operating within legitimate and representative contexts, as well as the autonomous actions of different social, political and economic groups in the food system (including the actions of the most vulnerable) to ensure their own economic objectives and immediate food requirements. 'Authoritative actors' are not, however, independent: they are responding to pressures from constituencies and interest groups. The domination of 'food as commodity' perspectives in global food systems suggests that corporate actors and agribusiness interests have considerable power and influence in current food systems (see Chapter 18). Thus underlying many situations of conflict is the lack of equitable participation by affected local populations in designing food and food resource transactions at national, regional and supranational levels. Any new governance system will need to create mechanisms that enable participation of these diverse actors, and in doing so, legitimize their distinct perspectives on the meaning of food. Those organizations and actors with the authority to resolve conflict will also need the legitimacy, mandate and standing to do so. Critical research questions relate to the participation of different actors in food system governance, how conflicts over resources and access are resolved, and whose rights take precedence in contexts of conflict.

How can institutions designed to diminish the risk of violence associated with food be made sufficiently flexible and adaptive to effectively deal with shocks and surprises, while also addressing the underlying causes associated with food violence?

The inability of the WTO and other international institutions to effectively intervene to reduce violent outcomes associated with the recent spike in commodity prices illustrates the importance of *adaptiveness* as a core attribute of food governance aiming to reduce the risk of violence. The pace of market and environmental change will require institutions to manage uncertainty and surprise, including the emergence of conflict at different spatial levels. Currently food security crises and social unrest are addressed in a reactionary manner, through emergency measures and disaster institutions, and reversion to clumsy domestic price and trade controls. New governance structures will need to have an embedded capacity to consider scenarios of sudden shocks and change, respond to crises as they arrive, and have sufficient flexibility and authority to alter practices and norms when circumstances require it. More insight is needed into how governance and institutional arrangements either constrain or enhance the flexibility of decision-making, and thus exacerbate or reduce the risk of conflict.

How can institutions be designed to have sufficient legitimacy and accountability in relation to a broad spectrum of food system actors such that they can become effective agents for reducing and resolving conflict associated with food security?

Current institutional arrangements are not equally accountable to all the actors that are affected by change in food systems. While complete equity is an unreasonable expectation of any institutional arrangement, fairness is a reasonable goal. The recent food crisis illustrates the ramifications for governments who are perceived as accountable for food insecurity, resulting primarily from volatility in global trade and financial systems. International trade institutions are currently accountable primarily to active participants in trade transactions, which may or may not include the needs and rights of food insecure populations. In other domains, private sector actors in food systems are increasingly being held accountable by consumer groups for their actions in the food system, particularly in relation to fair trade, labour practices and food safety, all of which increasingly speak to the human rights associated with food supply chains. Some of these practices are now being codified in international agreements about fair trade and/or ethical business standards. For example, the upcoming ISO26000, focusing on corporate social responsibility, is due out in 2010. While a voluntary standard, its development speaks for the growing belief that there is a need for both public and private organizations to act in a more socially responsible manner (ISO, 2009). New areas of research will need to address how and if these practices have successfully mitigated issues of conflict and violence in food systems, and whether these practices offer lessons that can be used to construct more formal mechanisms for achieving food security. Of particular concern are those populations most vulnerable to food insecurity (e.g. those in the global south, as well as those within countries without access to political power and representation). How to enhance the accountability of food system governance is thus a critical area of future research. These questions, together with what the role of formal government should be, are discussed further in Chapter 18.

In what ways can greater equity in resource allocation and improved access to food reduce the risk of violence?

Underlying all concerns of violence and food systems are issues of *allocation* and *access*. The allocation and access to land, water and energy critical for food production, and to food itself by society at large, are ultimately political concerns with strong ethical and moral dimensions. The moral and ethical aspects of resource distribution are not addressed in current international trade agreements. While human rights law raises these issues as concerns, it has not yet been used extensively to intervene in allocation decisions. Conflict also emerges in cases where populations have limited access to information about food quality and safety and the origins of their food, as well as lack information about or participation in the development of technologies that can enhance their food security. New systems of food governance must not only address issues of access and allocation of food production resources such as water, land, labour and energy, but also access to markets, to decision-making processes and to knowledge and technology. A critical area of research is how to design food system governance to anticipate both gradual and sudden changes in resource availability, while enhancing access in regions where food resources are already scarce and vulnerability to violence is high.

Priorities for a food system aiming to reduce risks of violence

The challenge of creating a system of food system governance that reduces the risk of violence is large and, given the pace of environmental change and increased risk of more frequent or severe environmental and economic shocks, the challenge is also urgent. Cutting across all the elements of governance that are described above are three priority concerns.

First, a rights-based approach to food governance immediately raises difficult questions concerning *whose rights?* And *whose rights matter most, under what conditions?* In other words, there is a need to explicitly address which ordering of rights (to food, to profits, to natural/environmental resources) will best reconcile the disparate perspectives associated with the values associated with food and will best provide the overall rights framework within which new governance mechanisms can be designed.

Second, the globalized economy now requires that any governance system addresses transactions and processes occurring at supranational levels. This international governance environment will likely need to promote food security through direct incentives, the harmonization of diverse institutions associated with conflict, food trade, food security and the environment (e.g.

aligning food aid with the MDGs and the policy objectives espoused in trade agreements), and through the provision of a recourse mechanism in cases of any violation of the agreed ordering of food rights.

Third, there is a need for the design and implementation of food governance systems that respect conceptualizations of national sovereignty and the diversity of viewpoints on food and its meaning among nations and districts and communities within nations. Such national and sub-national systems will need to be well articulated with those at the supranational level to avoid countervailing effects (and see Chapter 13). They will have to reflect the particular rights-ordering of relevance to a country, and will need to have the flexibility to accommodate distinct cultural needs and values associated with food and the diversity of actors that participate in food systems that are largely governed by national-level institutions. These systems will also need to be adaptive, as values and needs change with the evolving challenges in food system governance and food security. It will be increasingly important that there are in-country mechanisms to address the drivers of violence and resolve conflict once it has emerged, as well as mechanisms that allow specific populations to find recourse for rights violations outside their immediate political domain should such recourse be required.

Conclusion

The confluence of GEC, tightly integrated and consolidated markets and limited resources for food production is giving rise to an increased risk of violence associated with food access, availability and utilization. Underlying the rising risk of violence in food systems are different values and meanings associated with food, and the way these meanings have become embedded in the institutions that currently govern food systems. Of particular concern is the inadequacy of institutions characterizing 'food as commodity' for addressing the diverse issues that have emerged around the concept of 'food as right'. Reconciling counteracting discourses in food system governance is thus a core concern in mitigating the risk of violence. Such governance systems will need to view food security as embedded in the larger domains of human security and human rights. Food security, from this perspective, is not only freedom from want, from fear and from direct forms of food-related violence, but also the freedom to make food choices and to actively pursue individual and communal food sovereignty.

References

Anderson, K. and W. Martin (eds) (2005) *Agricultural Trade Reform and the DOHA Development Round*, New York, Pargrave Macmillan and The World Bank

Anderson, M. D. (2008) 'Rights-based food systems and the goals of food systems reform', *Agriculture and Human Values*, 25, 593–608

Atkins, P. and I. Bowler (2000) *Food in Society*, London, Arnold Press

Barrett, C. B. and D. G. Maxwell (2006) 'Towards a global food aid compact', *Food Policy*, 31, 2, 105–18

Bello, W. F. (2009) *The Food Wars*, London, Verso

Biermann, F. (2007) '"Earth system governance" as a crosscutting theme of global change research', *Global Environmental Change*, 17, 3–4, 326–37

Blanchard, T. and M. T. Blanchard (2007) Retail concentration, food deserts and food-disadvantaged communities in rural America. In Hinrichs, C. C. and T. Lyson (eds) *Remaking the North American Food System*, Lincoln, University of Nebraska Press

Brahmbhatt, M. and L. Christiaensen (2008) *Rising Food Prices in East Asia: Challenges and Policy Options*, Mimeo, World Bank

Bruinsma, J. (ed) (2003) *World Agriculture: Towards 2015/2030, A FAO Perspective*, London, Earthscan

Burke, M. B., E. Miguel, S. Satyanath, J. A. Dykema and D. B. Lobell (2009) 'Warming increases the risk of civil war in Africa', *Proceedings of the National Academy of Sciences*, 106, 49, 20670–674

Burton, J. (ed) (1990) *Conflict: Human Needs Theory*, New York, St Martin's Press

Buttel, F. H. (1997) Some observations on agro-food change and the future of agricultural sustainability movements. In Goodman, D. and M. J. Watts (eds) *Globalizing Food: Agrarian Questions and Global Restructuring*, New York, Routledge

Casson, A. (1999) *The Hesitant Boom: Indonesia's Oil Palm Sub-Sector in an Era of Economic Crisis and Political Change*, Bogor, Indonesia, Program on the Underlying Causes of Deforestation

Cornia, G. A. (1999) *Liberalization, Globalization and Income Distribution*, Helsinki, UNU/WIDER

Davis, C. G., C. Y. Thomas and W. A. Amponsah (2001) 'Globalization and poverty: Lessons from the theory and practice of food security', *American Journal of Agricultural Economics*, 83, 714–21

Davis, M. (2001) *Late Victorian Holocausts: El Nino Famines and the Making of the Third World*, London, Verso Books

De Schutter, O., The right to food. www.srfood.org/index.php/en/right-to-food (accessed: undated)

DeSchutter, O. (2009) *Seed Policies and the Right to Food: Enhancing Agrobiodiversity and Encouraging Innovation*, United Nations General Assembly, Sixty-Fourth Session, New York

Director International Affairs and Trade (2007) *Foreign Assistance: Various Challenges Limit the Efficiency and Effectiveness of US Food Aid*, Washington, DC, Committee of Foreign Affairs, Subcommittee on Africa and Global Health, House of Representatives

Dyer, G. (2008) *Climate Wars*, Toronto, Random House

ESCR-Net, Caselaw Database. www.escr-net.org/caselaw/ (accessed: 26 October 2009)

FAO (2003) *World Agriculture Towards 2015/2030. An FAO Perspective*, Rome, FAO

FAO (2009a) *The State of Food Insecurity in the World 2009: Economic Crises – Impacts and Lessons Learned*, Rome, FAO

FAO (2009b) *The State of Agricultural Commodity Markets (SACM): High Food Prices and the Food Crisis – Experiences and Lessons Learned*, Rome, FAO

FAO (2009c) *Committee on World Food Security, thirty-fifth session: Reform of the Committee on World Food Security*, Rome, FAO

FAO/GIEWS, Zimbabwe 2007/08 Agricultural Season Update. www.fao.org/giews/english/shortnews/zimbabwe080410.htm (accessed: 1 May 2010)

Galtung, J. (1969) 'Violence, peace, and peace research', *Journal of Peace Research*, 6, 3, 167–91

Gledhill, J. (1995) *Neoliberalism, Transnationalization and Rural Poverty*, Boulder, Westview Press

Gonzalez, C. G. (2004) 'Trade liberalization, food security and the environment: The neoliberal threat to sustainable rural development', *Transnational Law and Contemporary Problems*, 14, 2, 419–98

Goodman, D. and M. J. Watts (eds) (1997) *Globalising Food: Agrarian Questions and Global Restructuring*, New York, Routledge

Gordon, L. J., G. D. Peterson and E. M. Bennett (2008) 'Agricultural modifications of hydrological flows create ecological surprises', *Trends in Ecology and Evolution*, 23, 4, 211–19

GRAIN, Seized: The 2008 landgrab for food and financial security. www.grain.org/briefings/?id=212 (accessed: undated)

Gregory, P. J., J. S. I. Ingram, R. Andersson, R. A. Betts, V. Brovkin, T. N. Chase, P. R. Grace, A. J. Gray, N. Hamilton, T. B. Hardy, S. M. Howden, A. Jenkins, M. Meybeck, M. Olsson, I. Ortiz-Monasterio, C. A. Palm, T. W. Payn, M. Rummukainen, R. E. Schulze, M. Thiem, C. Valentin and M. J. Wilkinson (2002) 'Environmental consequences of alternative practices for intensifying crop production', *Agriculture, Ecosystems & Environment*, 88, 3, 279–90

Hauge, W. and T. Ellingsen (1998) 'Beyond environmental scarcity: Causal pathways to conflict', *Journal of Peace Research*, 35, 3, 299–317

Hawkes, C. (2006) 'Uneven dietary development: linking the policies and processes of globalization with the nutrition transition, obesity and diet-related chronic diseases', *Globalization and Health*, 2, 1, 4

Homer-Dixon, T. F. (1999) *Environment, Scarcity, and Violence*, Princeton, NJ, Princeton University Press

IFRC (2008) Zimbabwe: Food Insecurity. Emergency Appeal no. MDRZW003 GLIDE JT-2008-000097-ZWE Periodical; Zimbabwe: Food Insecurity. Emergency Appeal no. MDRZW003 GLIDE JT-2008-000097-ZWE International Federation of Red Cross and Red Crescent Societies. www.ifrc.org/docs/appeals/08/MDRZW 00301.pdf (accessed: 1 May 2010)

ISO (2009) ISO 26000: Social Responsibility, Geneva, ISO

Jaffee, D. and P. Howard (2009) Corporate cooptation of organic and fair trade standards. *Agriculture and Human Values*, DOI 10.1007/s10460-009-9231-8, Dordrecht, Springer, Netherlands

Lamy, P. (2008) *The Doha Round can be part of the answer to the food crisis*, High-Level Conference on World Food Security, Rome

Le Billon, P. (2001) 'The political ecology of war: natural resources and armed conflicts', *Political Geography*, 20, 5, 561–84

Liechenko, R. and K. O'Brien (2008) *Environmental Change and Globalization: Double Exposure*, Oxford, Oxford University Press

Lynn, J. (2009) WTO's 'Lamy: Give Doha negotiators more flexibility', *The Guardian*, 28 September 2009

MA (2005a) *Ecosystems and Human Well-being: Synthesis*, Washington, DC, Island Press

MA (2005b) *Millennium Ecosystem Assessment*, Washington, DC, Island Press

Maslow, A. (1943) 'A theory of human motivation', *Psychological Review*, 50, 370–96

McKie, R. (2007) 'Climate wars threaten billions', *The Observer*, 7 November 2009

McMichael, P. (1994) *The Global Restructuring of Agro-Food Systems*, Ithaca, Cornell University Press

Menezes, F. (2001) 'Food sovereignty: A vital requirement for food security in the context of globalization', *Development*, 44, 4, 29–33

Milonavic (2003) 'The two faces of globalization: Against globalization as we know it', *World Development*, 31, 667–83

Mitchell, D. (2008) A Note on Rising Food Prices. *Policy Research Working Paper 4682*, Washington, DC, The World Bank

Murphy, S. and A. Paasch (eds) (2009) *The Global Food Challenge: Towards a Human Rights Approach to Trade and Investment Policies*, Minneapolis, IATP

Nelson, D. (2009) 'India joins "neo-colonial" rush for Africa's land and labour', *Telegraph*, 28 June 2009

Paarlberg, R. (2000) 'Genetically modified crops in developing countries: Promise or peril?', *Environment*, 42, 1, 19

Peluso, N. L. and M. Watts (eds) (2001) *Violent Environments*, Ithaca, NY, Cornell University Press

Pingali, P. (2007) *Agricultural Growth and Economic Development: A View Through the Globalization Lens*, Rome, FAO

Purvis, N. and J. Busby (2004) *The Security Implications of Climate Change for the UN System*, Washington, DC, Brookings Institute

Raath, J. (2007) 'Violence looms as Zimbabwe runs out of food – except for the elite', *The Times*, 6 September 2007

Reuters (2009) 'Worldwide dairy industry to sign global declaration on climate change', *Reuters*, 18 September 2009

Rosegrant, M. W., X. Cai and S. A. Cline (2002) *World Water and Food to 2025: Dealing with Scarcity*, Washington, DC, International Food Policy Research Institute

Ryal, J. and M. Pflanz (2009) 'Land rental deal collapses after backlash against "colonialism": The world's largest deal to lease African farmland to rich countries to grow their own crops has collapsed amid accusations of "neo-colonialism"', *The Daily Telegraph*, 15 January 2009, p16

SANNA/Reuters, Yemen Water Crisis Threatens Population. www.alarabiya.net/articles/2009/08/30/83384.html (accessed: 26 October 2009)

Seto, K. C., R. de Groot, S. Bringezu, K. Erb, T. Graedel, N. Ramankutty, A. Reenberg, O. Schmitz and D. Skole (In press) Land: Stocks, flows, and prospects. In Graedel, T. and E. van der Voet (eds) *Measuring Sustainability*, Cambridge, MA, MIT Press

Shiva, V. (1991) 'The green revolution in the Punjab', *The Ecologist*, 21, 2, 57–62

Spitz, P. (1978) 'Silent violence: Famine and inequality', *International Social Science Journal*, 30, 4, 867–92

Staff (2009) 'Sale of agricultural land to foreign investors: LHC calls report from federal government', *Daily Times* (online), 15 September 2009

Suhrke, A. and S. Hazarika (1993) *Pressure Points: Environmental Degradation, Migration and Conflict*, Cambridge, MA; Toronto, Ont., American Academy of Arts and Sciences; Peace and Conflict Studies Program, University College, University of Toronto

Taylor, P. L. (2005) 'In the market but not of it: Fair Trade coffee and Forest Stewardship Council Certification as market-based social change', *World Development*, 33, 1, 129–47

Tenenbaum, D. (2008) 'Food vs. fuel: Diversion of crops could cause more hunger', *Environmental Health Perspectives*, 116, A254–57

Theisen, O. M. (2008) 'Blood or soil? Resource scarcity and internal armed conflict revisited', *Journal of Peace Research*, 45, 801–18

Tilman, D., K. G. Cassman, P. A. Matson, R. Naylor and S. Polasky (2002) 'Agricultural sustainability and intensive production practices', *Nature*, 418, 6898, 671–77

Tobar, H. (2008) 'Mexican Farmers Protest NAFTA', *Los Angeles Times*, 3 January 2008

Turner II, B. L., E. F. Lambin and A. Reenberg (2007) 'The emergence of land change science for global environmental change and sustainability', *PNAS*, 104, 52, 20666–71

UN, Universal Declaration of Human Rights. www.un.org/en/documents/udhr (accessed: 26 October 2009)

US Working Group on the Food Crisis, Call to Action on the World Food Crisis. www.usfoodcrisisgroup.org (accessed: 26 October 2009)

Valdes, A. a. J. Z. (1980) *Agricultural Protection in OECD Countries: Its Cost to Less-Developed Countries*, Washington, DC, International Food Policy Research Institute

Vidal, J. (2009) 'International: Fears for the world's poor countries as the rich grab land to grow food: UN sounds warning after 300 hectares bought up: G8 leaders to discuss "neo-colonialism"', *The Guardian*, 4 July 2009, p24

von Braun, J. and R. Meinzen-Dick (2009) *Land Grabbing? by Foreign Investors in Developing Countries: Risks and Opportunities*, Washington, DC, International Food Policy Research Institute

Wade, R. (2004) 'Is globalization reducing poverty and inequality?', *World Development*, 32, 4, 567–89

Watts, M. and D. Goodman (1997) Agrarian questions: Global appetite, local metabolism: Nature culture and industry in *fin-de-siecle* agro-food systems. In Goodman, D. and M. Watts (eds) *Globalising Food: Agrarian Questions and Global Restructuring*, London, Routledge

WBGU (2008) *Climate Change as a Security Risk,* German Advisory Council on Global Change (WBGU), London and Sterling, VA, Earthscan

Welzer, H. (2008) *Klimakriege: wofür im 21', Jahrhundert Getötet Wird*, Hamburg, Fischer Verlag

WFP, WFP Executive Director visits Cyclone Nargis victims; Calls for greater support, http://one.wfp.org/english/?ModuleID=137&Key=2863 (accessed: 26 October 2009)

WFP, Our Work. www.wfp.org/our-work (accessed: 26 October 2009)

WHO (2009) Periodical, Cholera in Zimbabwe – update 2. Global Alert and Response (GAR) World Health Organization. www.who.int/csr/don/2009_02_20/en/index.html (accessed: 1 May 2010)

World Bank (2001) *World Bank Development Report 2000/2001: Attacking Poverty*, New York, World Bank

Wrigley, N. (2002) '"Food deserts" in British cities: Policy context and research priorities', *Urban Studies*, 39, 11, 2029–40

Wrigley, N., D. Warm and B. Margetts (2003) 'Deprivation, diet and food-retail access: Findings from the Leeds "food deserts" study', *Environment and Planning A,* 35, 1, 151–88

WTO, Agricultural Negotiations: Backgrounder – The issues and where are we now.www.wto.org/english/tratop_e/agric_e/negs_bkgrnd00_contents_e.htm (accessed: 26 October 2009)

Ziegler, J. (2002) *Report of the Special Rapporteur of the Commission on Human Rights on the Right to Food*, New York, United Nations

Zimbabwe Peace Project, Zimbabwe Peace Project Information. www.freedom-house.org/uploads/zpp_hrviol_report_0809.pdf (accessed: 26 October 2009)

18
Governance Beyond the State: Non-state Actors and Food Systems

Rutger Schilpzand, Diana Liverman, David Tecklin,
Ronald Gordon, Laura Pereira, Miriam Saxl
and Keith Wiebe

Introduction

One of the most significant trends in both food systems and environment is the increasing importance of private sector and non-governmental actors in governance – the formal and informal rules, institutions and practices that guide the management of food and environment within a complex network of governments, organizations and citizens (Biermann, 2007). Rather than a shift from public oversight and investment to the purely private sphere, emerging forms of governance are often a hybrid that combines government action with that of business and non-government organizations (NGOs). Examples of this occur when the private sector develops emission reduction projects within a trading framework established by UN intergovernmental processes, or when corporations adopt certification systems for sustainable or fair trade foods that have been developed in collaboration with NGOs.

This chapter discusses the governance of food and environmental systems by non-state actors through a series of case studies that illustrate the agency of the private sector and civil society. It covers a wide range of governance mechanisms, and the intersections with decisions taken by governments, when it comes to new ways of adapting food system governance to global environmental change (GEC) that is already taking place. Because research and policy often overlook the importance of non-state actors, a future research agenda should pay attention to their roles and impact on both food and environmental governance.

Who are the key non-state actors in the governance of food systems and food security? Corporations are clearly extremely important, with large-scale producers, processors and retailers of food increasingly dominating food markets and taking responsibility for rules that range from food safety to trade to sustainability (Marsden et al, 2008). Figure 18.1 shows the increasing sales of

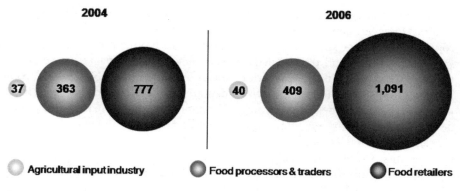

Source: von Braun, 2007

Figure 18.1 *A 'corporate view' of the world food system: sales of top ten companies (in billions of US dollars) in 2004 and 2006*

top global companies active in the food system. Between 2004 and 2006 the sales of food retailers increased by a disproportionately large amount compared to the sales of food processors and of companies in the food input industry (von Braun, 2007).

Such corporations are joined in influencing the global food system by a multitude of additional actors, including farming and fishing trade organizations, large processors and retailers, consumer organizations, energy and transportation corporations, insurance and banking, research companies, and water companies and irrigation cooperatives (Meinzen-Dick, 2007; Oosterveer, 2007; Marsden et al, 2008; Clapp and Fuchs, 2009a; Gereffi and Christian, 2009). Small-scale producers are also very important, with millions of farmers, herders and fishers making decisions and innovating in ways that affect consumers and the environment, and joining cooperatives and networks in order to gain power to negotiate with governments and business. Non-governmental organizations with power in the food system include humanitarian and social justice organizations with interests in food security, environmental groups concerned with ecological sustainability, consumer groups concerned with health and pricing, and indigenous and community organizations promoting self-determination and protection of local food systems. Consumers and the media have been especially significant in campaigns for food safety.

A growing body of work is pointing to the emerging, and important, role in governing environmental change action of non-state actors, such as multinational energy corporations, cities and non-national government actors, carbon traders and off-setters, environmental NGOs, certifying bodies, and research networks (Newell, 2000; Betsill and Bulkeley, 2006; Bumpus and Liverman, 2008; Andonova et al, 2009). Of central importance are those private sector firms whose operations have a direct impact on the global environment. Among these are major emitters of greenhouse gases (GHGs) such as energy and cement companies, private landowners (whose use of land affects a number of GHG budgets and which also affects biodiversity), and industrial producers whose use of chemicals can also perturb global biogeochemical cycles and affect water quality and ecosystem health (Newell, 2000). However, new sets of private sector actors have emerged around environmental markets such as those for carbon, including project developers, verifiers, investors and traders. Many other actors have appeared in response to concerns over climate adaptation, including specialized NGOs and consulting groups. Furthermore, environmental NGOs have become relatively powerful in environmental governance through providing research and policy designs, influencing the media, business, public and governments, implementing environmental conservation projects, and participating in international negotiations (Corell and Betsill, 2001).

Many organizations are involved in both food and environmental governance (Table 18.1). Private sector actors operating within the context of government policies have long dominated the food system, but the challenges posed by GEC transcend the traditional decision-making horizons of both private actors and governments. As a result, the roles of actors across the food system are being redefined. These many intersections also produce contradictions or tradeoffs when, for example, investments in renewable energy for decarbonization, such as biofuels, threaten food supplies or biodiversity, or when campaigns to reduce carbon footprints through reducing food miles and

Table 18.1 *Some examples of non-state actors and their roles in food and environment*

Non-state actor	Food system role	Environmental role and impacts
Farmer/fisher/herder	Produces food Jobs	Energy, land, water, seeds/fish/livestock waste, pollution, biodiversity
Food processing and manufacturing	Processes, preserves and packages food Branding, marketing Jobs	Energy, water, waste, pollution
Transportation sector	Moves and stores food Jobs	Energy, land, pollution
Retailers	Sells food Jobs	Energy, land, water, waste, pollution
Consumers	Processes and eats food	Energy, water, waste, pollution
Finance and insurance	Provides credit, commodity trading, insurance	Carbon trading, hazard insurance
Environmental NGO	Monitors and advocates for organics and supply chain standards, etc.	Influence practices to improve environmental management
Social justice NGO	Advocate for fair trade and for justice to local organizations in the South	Shift in power distribution and value attribution
Humanitarian NGO	Distributes food	Influences water use and quality, agricultural practices

Note: Others: Consultancies, certification and standard organizations, business networks, scientists and universities, international organizations, hybrids of NGO–private–government.

imports create problems for overseas fair-trade producers who rely on exports for their livelihoods.

Shifts in governance and non-state actors

From a governance perspective, GEC and food systems share many traits. Both are highly diffuse and fragmented since objects of governance involve decision-making across a wide array of contexts and actors at multiple spatial levels (Auer, 2000). The growing literatures on governance of GEC and food systems indicate that this has been deeply shaped by several broad trends associated with recent patterns in economic globalization, including: (i) the diminishing (if unevenly so) regulatory authority of nation-states and the tendency for them to shift into facilitative roles; (ii) conversely the growing authority and 'regulatory' role of large corporations, particularly through supply chain management and private contracting, which is also often described as 'private rulemaking'; (iii) the spread of corporate social responsibility (CSR) doctrine and practices, as well as an explosion of public–private or social–private alliances; (iv) a parallel growth in the role of social but particularly of environmental NGOs at all levels of governance; and (v) the emergence of global networks as a key cross-cutting organizational form, and the way in which global supply chains have become the focus of regulatory efforts. While these trends have heterogeneous affects depending on the context, the general picture is one of increasing interdependence between public and private forms of authority, the multiplicity of organizational forms engaged simultaneously in governance, and decentralization in how the key institutions – or 'rules of the game' – take shape and are enforced. Simultaneous with this flattening and fragmentation of regulations, multinational corporations have emerged as increasingly powerful and dominant actors (Clapp and Fuchs, 2009a).

The shift in regulatory authority from states to corporations
Much recent work has discussed the weakening of the regulatory state and its retooling to assume a more facilitative rather than direct regulatory role following the spread and deepening of neoliberal policies of free trade, reduced government and privatization in the 1980s (Harvey, 2005; Liverman and Vilas, 2006). Global trade regimes have constrained state authority to regulate investment and trade flows (Dicken, 2007). This has been further accompanied by a wave of privatizations that have transferred the direct control over significant areas of production and distribution of goods and services from the state to the private sector – although this has not been a unilinear process and reversals have occurred, particularly in the case of water infrastructure (Conca, 2006). At the same time, states have adopted more facilitative roles through, for example, the large number of non-binding guidelines issued by the EU for agri-food systems (Marsden et al, 2008), the endorsement of private certification systems (Gulbrandsen, 2005; Gulbrandsen, 2006), and support for industry self-regulation.

Transnational corporations (TNCs) have increasingly become a central actor in global environmental governance, outstripping the analytical frameworks that have traditionally been applied to policy-making and international regimes (Newell, 2000; Levy and Newell, 2005). Through their control over industrial GHG emissions, their active role in deforestation (and sometimes reforestation), and their substantial financial, technological and organizational resources, businesses are powerful engines of change in GEC issues. Within the global food system the market role of TNCs has grown such that the top ten companies control at least a quarter of the global market in seeds, pesticides, food retailing and food manufacturing (Clapp and Fuchs, 2009b, citing ETC Group 2005; and see Figure 18.1). Moreover, their influence extends beyond the historical role of lobbying governments and multilateral institutions, as TNCs increasingly have replaced those institutions in research and knowledge dissemination, as well as the definition of norms and rules. As corporations have grown more dominant and more visible, they have also increasingly become the targets of social mobilization (Klein, 2000; Schurman, 2004; Seidman, 2007; King, 2008), yet relatively little is known about the determinants and longer-term outcomes of such mobilization.

Businesses are starting to play important direct and indirect roles in environmental governance, including the management of GHGs and climate impacts in food systems. Major companies such as DuPont and Unilever are publicly disclosing their GHG profiles, voluntarily reporting and aiming to manage them (Hoffman, 2005; Okereke, 2007). Some are even lobbying for a stronger regulatory landscape in which to make new, profitable, green investments; the insurance and finance sectors, for example, are calling upon governments to act to reduce risks from climate change, including those to food and agriculture (Blyth et al, 2007; Deutsche Bank, 2007; Mills, 2009). In general, business policy positions have emphasized the promotion of market mechanisms (such as emissions trading and offsets), incentives for clean technology investment, and learning-by-doing approaches to carbon management and reduction. Importantly, beyond the indirect lobbying role of businesses, those with major market power (such as Walmart or DuPont) now have the ability to more directly shift environment and food policies up and down global supply chains (Burch and Lawrence, 2007; Minx et al, 2008).

Business has also engaged in self-regulation and reformed practices with regard to food security and environment as a result of the growth of the CSR movement and concern for the triple bottom line of profit, social and environmental responsibility (Blowfield and Murray, 2008; Epstein, 2008). CSR in the food system has been dominated by concerns about food safety, animal welfare, fair trade, human rights and social access, as well as environmental impacts (Hughes et al, 2007; Blowfield and Murray, 2008; Spence and Bourlakis, 2009). Some of the turn to CSR may be a strategy to gain competitive and early-mover advantages, and many studies highlight that this is often associated with efforts to move regulation in particularly favourable directions. Examples include energy intensity commitments from energy companies with emerging capacity in low-carbon-generating technologies and the largest food retailers who have committed to expand the sale of organic produce or the sourcing of sustainably produced seafood. Some evidence also shows CSR as a strategy to attract and maintain young talent, particularly in the finance and high technology sectors. Nonetheless, CSR drivers clearly also exceed such factors and are rooted in wider societal pressures and concern for the environment, which has been documented in particular in terms of pressures from investors and consumers (Emel, 2002; Bartley, 2007; Okereke, 2007; Newell, 2008).

These trends towards direct business roles in governance have been characterized as manifestations of growing private authority and the associated rise of private rulemaking (Cutler et al, 1999; Pierre, 2000; Cashore et al, 2004). Examples of private rule-making include industry-wide standards (Angel et al, 2007), process certification such as the well-known ISO systems, social and environmental certifications such as the Forest Stewardship Council and Marine Stewardship Council (MSC), and fair-trade certification for products like coffee and chocolate. This context of dispersed, privately controlled and polycentric (see Ostrom, 2005) rule-making and enforcement contrasts with traditional political analysis focused on the state control over processes of enforceable rule-making and thus raises many new research questions for GEC and food governance.

NGOs and governance

NGOs have been described as important actors in shaping governance through several complementary roles, including influencing governments and international regimes, as leaders in cultural change and in instilling new norms; as innovators and implementers of social and environmental management techniques or technologies; and as leaders in 'civil regulation' (Liverman, 2004; Newell, 2005; Conca, 2006). In their role as a social and environmental lobby, many of the largest NGOs channel influence through a formal consultative role in the United Nations Framework Convention on Climate Change (UNFCCC), as well as in several international regimes related to the global food system such as the Convention on International Trade in Endangered Species of Wild Fauna and Flora (CITES), and the Convention on Biological Diversity. Such efforts with particular relevance to food governance extend also to international regimes where NGOs have no formal role, such as the World Trade Organization (WTO). Underlying most such work, however, are efforts to further the growth of certain social and environmental values and project these through the news media. Particularly in Europe, the large environmental and social NGOs (such as the World Wide Fund for Nature (WWF) and Oxfam) are some of the most widely recognized entities and considered to have more credibility than government or business by large proportions of the population. They thus exercise a powerful role in projecting norms and shaping the policy agenda (Conca, 2006).

The range of NGO approaches to food systems and GEC underlies the diversity and size of the sector. Many organizations that are deeply rooted in churches or labour and civic organizations, such as Christian Aid, the Catholic Relief Services and Oxfam, are associated in the public eye with humanitarian aid and development projects but are now also actively involved with multiple initiatives in both environment and food system governance. Oxfam, for example, has long been involved in promoting greater equity in food systems through fair trade – supporting producers, raising awareness, lobbying companies and governments, and contributing to the development of certification and labelling schemes (Renard, 2003; Nicholls and Opal, 2005). Starting with handicrafts sold in charity stores, fair trade expanded to food products, especially coffee, chocolate, wine and tropical fruit, and to sales through major supermarkets worth more than US$3 billion worldwide. These changes were partly a response to civil society concerns about fair trade. Humanitarian NGOs are also participating in the development of strategies for adaptation to climate change, including the development of rules for allocating funds.

Humanitarian NGOs have an increasingly important role in international food relief and policies on food security, and governments often rely on NGOs to distribute aid and to expand the overall funding for food programmes (Natsios, 1995; Tvedt, 1998; Paarlberg, 2002). NGOs are also important advocates for the food insecure, using the media and their supporters to pressure governments in responding to food crises. They can also play a role in monitoring emerging problems, including those related to climate and environment, such as in the case of the assistance which NGOs have offered in famine early warning.

Although the organizational models, objectives and methods of environmental NGOs have proliferated, those active in GEC and food systems remain relatively concentrated in the global North. Understanding the geographies of NGOs has become highly relevant to governance research. Some environmental organizations like WWF have used relatively decentralized models of organization to develop a nearly global presence. Others like the Nature Conservancy have consolidated enormous fundraising and technical capacities within an essentially corporate structure and US base. Many environmental NGOs have a strong focus on standard-based approaches to reducing production impacts. WWF has been influential in shaping food governance through promotion of standards and best practices for fisheries, several of the most traded aquaculture species, potatoes, palm oil and soybeans, as well as many other lesser known products worldwide (see, for example, Clay, 2004). It has also been active in climate change policy at the international regime level through participation in the UNFCCC meeting of the parties, through partnerships with business which in the US-based scheme alone are claimed to add up to reductions in 14 million tonnes of carbon emissions annually, and has published a steady flow of reports, scientific articles and books on the impacts of GEC (WWF, 2009).

The role of NGOs in protest and as watchdogs and critics of government and corporations is well-known. What is less recognized are the diverse ways in which NGOs have developed new forms of partnerships with the private sector and acted as 'hinges' or key linkages between actors in global networks (Newell, 2008; Perez-Aleman and Sandilands, 2008). This has increasingly led to the emergence of hybrid organization models, in which commercial incentives and social goals are related and mutually reinforcing. NGOs have also either designed or participated in crucial ways in the development of new standards, codes of conduct and certification systems across a variety of sectors. These tendencies have begun to break down the often significant barriers between the NGOs with different priorities (e.g. biodiversity conservation, local economies and foods, animal welfare) and those dividing environmental NGOs from the classic consumer organizations, producing an amalgam of strategies and forms of collaboration. An important aspect is how these different strategies interact in practice. They can sometimes be coordinated but often are more 'spontaneous': watchdog NGOs like Greenpeace put pressure on an issue by campaigning in public opinion, while dialogue NGOs like Oxfam or WWF discuss agreements towards a solution of the issue (see Box 11.5).

Political and commercial networks of non-state actors

The immense development of networks in food and GEC makes this a crucial area for exploring current debates about the significance and operation of network approaches versus more traditional hierarchical and territorially bounded forms of governance. The role of transnational activist networks in shaping policy agendas as diverse as tropical deforestation, dam building, chemicals and genetically modified organisms has long been documented (Keck and Sikkink, 1998; Schurman, 2004; Conca, 2006). More recent trends in environmentally oriented networks include those established by major cities as a means of influencing the international climate regime (Betsill and Bulkeley, 2006). Parallel to these are well-established trade associations for grain, seed and dairy (www.agrifood.net/contacts.htm), whose role as defenders of sectoral interests is well understood. However, these are increasingly complemented by hybrid groups such as the Roundtable on Sustainable Palm Oil (RSPO), producer organizations such as the International Federation of Agricultural Producers, and many food governance networks including fair trade and certification groups (such as the Fairtrade Labelling Organizations). Moreover, many businesses now work through networks like the World Business Council for Sustainable Development, Global Compact and the Climate Group to promote their vision of politically feasible and profitable ways towards greater sustainability (e.g. reducing greenhouse-gas emissions within the international climate regime).

Arguably, however, from the point of view of food systems governance, the most important type of network to emerge is the global supply chain of large retail and marketing corporations (Gereffi et al, 2001, 2005; Appelbaum and Lichtenstein, 2006; Marsden et al, 2008; Gereffi and Christian, 2009). Economic globalization throughout much of the 20th century was driven by the expansion of commodity chains for manufacturing and extractive industries, particularly the fossil fuels sector. Although the term 'supply chain' is only a few decades old (Appelbaum and Lichtenstein, 2006), in that period, global supply chains (GSCs) organized by mega-retailers and brander-marketers have emerged as arguably the world's dominant organizational force (Gereffi and Christian, 2009) and this is particularly true of the food sector (Oosterveer, 2007; Marsden et al, 2008). GSCs are defined as a 'network of firms that contributes both inbound and outbound products and services along an industry value chain' (Miles and Snow, 2007). Much of the world's commerce now flows through these global supply chains, which while quite different from other forms of networks, share similar challenges of securing coordination and cooperation across organizations that are not hierarchically dependent (Bair, 2008).

Supply/food/commodity chains link producers of food to consumers, often through complex and multinational connections, several steps of processing and distribution, and with many other actors interacting with different stages in the chain (e.g. banks finance and insurance). In many cases the environmental impacts are greatest at the beginning of supply chains, with water and energy heavily used in production and less at later stages. Consumers often feel disconnected from farmers, especially where heavily processed food combines multiple producers and a variety of additives or preservatives. NGOs have learned to target the whole supply chain in promoting environmental and social responsibility, looking at GHG emissions or labour practices at each step in the food system and pushing for sustainability across the whole system. Retailers who feel vulnerable to market or environment-driven variations in production may try to manage and control the supply chains through contract farming or vertical integration of operations that includes production, processing and retail.

Major processors and retailers began to develop private quality standards and accounting systems in the 1980s, when quality control in food production became increasingly complicated. At this stage, some governments left producers largely in de facto control and confined themselves to 'meta-control'. Liability concerns were often a driving force, particularly following major food risk scares. Once systems were established internally, they were later enlarged to cover suppliers as a means of influencing both the quality of products and the security of supply. More recently, the term 'Wal-Mart effect' (Fishman, 2006) has been used to describe the use of market power to reshape entire supply chains, and some researchers look to this as a major force for the greening of industry and note that incorporation of environmental and social concerns into private contracting within supply chains has become ubiquitous (Vandenbergh, 2007). In recognition of this fact GSCs have become a focus of regulatory efforts for both climate change and food systems. NGOs are centring demands on corporate supply-chain management, seeking commitments to abide by standards and codes, adherence to a third-party certification system or to make overall performance improvements. Supply chain management has thus become the terrain on which large corporations engage with civil society, but it is also an area by which corporations seek new competitive advantages. On the one hand, this is to safeguard their own supplies in terms of quality, quantity and price in the face of increasing scarcity of raw materials and strong price fluctuations (Clay, 2004); on the other hand, and beyond the traditional strategies of price and quality, as they seek to distinguish themselves in terms of new consumer values such as food provenance or low-carbon impact (Morgan et al, 2006; Marsden et al, 2008). Part of the explosion in private contracting is due to the convergence of private supply chain management and public regulation around reducing food (and other product) risks (Marsden et al, 2008), but the extent to which public interests and public regulatory efforts will converge with private efforts in broader environmental areas such as GHG emissions is very much an open question.

Non-state actors in action

A series of case studies illustrates some of the ways in which non-state actors are contributing to the environmental governance of food systems and create new solutions for GEC. These case studies also suggest some of the emerging research questions around the roles of different actors and the relations between them. Case studies have been selected to cover different activities of the food system including producing, processing, distributing and retailing food, as well as cross-cutting governance issues such as product labelling and certification in supply chains, and financing.

Producing food

Farmers, farm families and farmer cooperatives are some of the more important agents in the food system, providing for themselves and the market, and finding creative solutions to problems in the absence of government interventions. Farmers have experimented and innovated ever since early peoples initiated the process of plant and animal domestication, breeding for desired traits and finding ways to adapt to extreme or variable environments through technologies such as terracing, irrigation and social networks. In some cases governmental interference or new governance models were indispensible (e.g. water management). In other cases, the intervention of governments may even have suppressed the adaptability, resilience and entrepreneurship of farmers as in the case of colonial governance of African agriculture dismissing African traditional practices (Richards, 1985), or when government programmes promoted monocultures that increased vulnerability (see Chapter 19). Government intervention and support has, of course, provided many food system benefits, ranging from safety nets for the poor to the distribution of technology innovation and environmental protection, through agricultural extension and various other support and incentive systems. But in many countries economic crisis and restructuring have undermined government support systems, leaving farmers once again to develop their own responses to environmental and economic challenges. Many farms are business enterprises and thus farm experiments and innovations can be considered private rather than public forms of governance.

The literature on farmer innovation includes case studies from both developed and developing countries. For example, case studies of farm experiments in animal husbandry, water and soil management, and agroforestry in Africa, show how farmers improved agricultural yields and livelihood security (Reij and Waters-Bayer, 2001). Innovations in land management, such as the control of soil erosion, can also originate in farmer experimentation and include improvements in water harvesting, irrigation (see Box 18.1) and organic matter management (Chambers et al, 1989; Critchley, 2000). Box 18.2 gives a case study from larger-scale agribusiness.

Box 18.1 *Innovation in irrigation by small farmers in southern India (adapted from Buechler and Mekala, 2005)*

Most of India's irrigation depends on groundwater, but groundwater supplies are being depleted due to diminished recharge, as precipitation becomes more erratic, droughts more frequent, and increased withdrawals occur for growing urban populations. In the semi-arid and drought-prone Telangan region of Andhra Pradesh in south India, increased groundwater pumping for the city of Hyderabad for domestic and industrial uses has at least tripled since 1980. Most of this water is returned untreated to the Musi River, thereby increasing the volume and regularity of flow, even while it becomes more polluted. Thus, the state's inability to regulate groundwater withdrawals in the face of climate change and increased demand, as well as the inability of public investment in water treatment to keep pace with increased urban demand, has resulted in new patterns of water availability and quality.

Small-scale rice farmers downstream of Hyderabad have responded to these changes with a variety of adaptations and innovations. Traditional controls exercised by water user associations have been relaxed for river water, thereby devolving decision-making to individual farmers who have substituted surface water for more expensive groundwater, which requires drilling and pumping. Irrigated areas have been expanded but the higher salinity and nitrogen content of the wastewater increased susceptibility to pests and diseases and reduced rice yields. Yet a new set of farmer-led innovations have followed, including a switch to rice varieties that are more tolerant of higher salinity, reducing fertilizer application to compensate for the higher nitrogen levels in wastewater, increasing spacing between seedlings to improve grain formation and reduce the risk of pests, and (in some cases) switching from rice to hardier fodder crops that are becoming increasingly profitable as the dairy industry expands in response to rising urban demand.

Innovations have also involved changes in water management, including more precise intermittent irrigation (depending on crop needs, precipitation and pest conditions), mixing of groundwater and wastewater (to balance quality and cost considerations), and irrigating alternately with groundwater and wastewater at different stages of crop production (as the different stages vary in sensitivity to water quality). Because irrigation with groundwater is more costly, farmers who relied on these latter strategies tended to be wealthier and of higher caste, and to have larger landholdings (2–3ha) than the average in the study area (0.5ha).

The study found that farmers continuously experiment, adapt and innovate to increase their incomes and food security in response to changing demographic, economic and environmental conditions. Informal communication between farmers and observation of practices by others were the primary means by which innovations were disseminated. The way in which farmer practices emerged without an enabling policy framework underlies the many spontaneous adaptations of individual producers. The authors also point to a reversal of knowledge flows

– that is, that these farmers' innovations did not result from expert advice or extension services, but from farmer experimentation, communication and observation. These findings point to the need for greater recognition and dissemination of local innovations entailing improved linkages between local populations, researchers, managers, development workers and policy-makers.

Box 18.2 *The Peanut Company of Australia's response to climate change (derived from www.pca.com)*

The Peanut Company of Australia (PCA) provides an example of a farm business making a major investment to relocate its production in response to changing precipitation patterns associated with climate change. In 2007, PCA paid AUS$9 million for 11,700ha near Katherine, Northern Territory (NT), 3000km northwest of its traditional peanut-growing areas around Kingaroy, Queensland.

'The move north was prompted by traditional peanut-growing areas in Queensland becoming increasingly dry. A foothold in the NT will secure PCA's peanut supply.' PCA had been exploring options to shift its production since the mid-1980s as a risk-mitigation strategy, so they were prepared when its traditional growing areas in southern Queensland became increasingly dry over the past five to ten years. In 2007, PCA's Queensland crop was so poor it had to import peanuts to meet domestic orders: 'Katherine's annual rainfall is 960 millimetres, nearly all of which falls in the wet season between November and March. It's a tantalising figure for an industry being belted by drought.' (However, note that the Australian Government Grains Research and Development Corporation website says 'Between 1500 and 1800 millimetres of rain falls on PCA's Katherine farms between November and April'.)

Over the next four to five years, PCA plans to invest a further $20 million and eventually farm 4000ha of their 11,700ha holding, the remainder being left in virgin state. By 2012, PCA expects to harvest 13,500 tonnes of peanuts annually, one-third of its total production, from this operation. PCA will produce the same varieties in NT as it grows in Queensland, using no-till techniques, precision application of fertilizers and pesticides, and centre-pivot irrigation to draw water from underground aquifers that are filled by wet-season rains and hold water as close as 30m to the surface. After initial drying on-farm, the peanuts are trucked 3000km to Kingaroy for further processing.

This type of large-scale investment has generated both support and concerns from other actors in the food system. The NT government welcomes the jobs and economic benefits such investment brings, and local land values have doubled in the past 18 months, driven by an influx of buyers from the south and east. But WWF questions the 'false view that the north is going to get uniformly wetter and better for farming and is going to substitute for drought in the south. It's not what the science says.' WWF also raises concerns about impacts on local fishing, tourism and indigenous communities, and local horticultural producers are concerned about increased withdrawals from aquifers. The NT Minister of Agriculture says scientists will play a critical role in understanding how much water is available in order to ensure that expanded production remains environmentally safe.

Processing food

As noted above, food processing and retailing (see below) have become the most concentrated stages in the value chain, with only a few processing and retailing companies relative to the number of primary producers and consumers at either end. In terms of governance, processors' strategic role in the food system and high level of market power is complemented by growing liability for food safety in the wake of food scares such as BSE. At the same time, concerns with supply under changing market and environmental conditions reinforce their engagement in governance. An example is given in Box 18.3.

Box 18.3 *Tiger Brands and global environmental change (derived from an interview conducted by Laura Pereira with Bongiwe Njobe, Johannesburg, former Director General of Agriculture in South Africa, and current executive director of corporate sustainability for Tiger Brands, September 2009)*

Tiger Brands is a food processing and manufacturing company for grains, beverages and meat, based in South Africa, with commercial interests in other African countries. It has recognized the need for responses to the impacts of GEC from both an adaptation and mitigation perspective. In South Africa the power of the major retailers and processors is key for ensuring food security under GEC impacts in the food system because this is the space in which major decisions with regard to product development, procurement and, more controversially, prices, are determined. Their annual report admits that they are a heavy user of non-renewable energy (coal and petroleum), but commits to focusing on energy efficiency and disclosure of CO_2 emissions. They also note their role as a water user in drought-prone South Africa and a corporate response promoting improved water use efficiency. They explicitly mention the risk of climate change to fruit, vegetable, wheat, maize and sugar production in South Africa over the next 30 years. Predictions for South African agriculture under climate change in 2030 show a major decrease in staple crops like maize, but relatively little impact on 'traditional crops' like sorghum (Lobell et al, 2008).

Tiger Brands' response to these projections has included a focus on product development that is adaptive to changing climatic conditions. This is exemplified by the product *Morvite*, which is an instant sorghum-based breakfast cereal. This is thought to be a means of contributing to food security while meeting the company's commercial interests. This product is already very popular in non-domestic markets (e.g. Kenya), but is still to become mainstream in South Africa, where it is still seen as a nutrition supplement in areas where people are affected by disease and poverty or in disaster relief situations. In June 2009, it was used in a disaster relief programme for flood victims in Namibia. There is therefore a need for consumer awareness around innovative products like these that could make a significant contribution to food security under GEC and, in particular, climate change. For the company, such product development complements a socially responsible investment programme through their 'Unite Against Hunger' programme, which provides food to more than 100,000 orphans and vulnerable children in collaboration with NGOs.

Distributing food

The distribution of food contributes to GEC through GHG emissions and other pollution from transport. Food distribution is also vulnerable to GEC (Chapter 7), especially to supply disruptions during extreme events and disasters. Events such as Hurricane Katrina, the Asian tsunami and the Haiti earthquake illustrate the serious disruption of food distribution systems in the aftermath of disaster, compounded by the challenges of getting food relief to the worst affected people and places (Kovacs and Spens, 2007; Douglas, 2009). The role of both the private sector and NGOs in disaster logistics is extremely important, as they are often asked to distribute government relief and to be the first responder in many areas. Some have suggested that the retail sector's response to Katrina was more successful than that of government agencies, with companies like Walmart and Home Depot mobilizing to send many hundreds of truckloads of supplies into hard-hit areas, including free food and prescription drugs (Horwitz, 2009; Stewart et al, 2009).

Regarding GHG emissions, one of the most interesting developments in food system governance is the surge in interest in 'food miles' as a measure of the environmental impact of the food system. Food miles are simply the number of miles to transport a product from the producer to the consumer and are a surrogate for energy use and GHG emissions from transporting a product (Coley et al, 2009). Campaigns by NGOs in Europe have focused on reducing food miles through local purchasing and have been enabled by food labels that identify the source of products and even whether they travelled by air (see Box 18.4). Thus both NGOs and the private sector are engaging in this new approach to governing the environmental impacts and food. But tensions are also arising where, for example, in New Zealand concerns have been raised about their ability to participate in international food trade given their isolated location. However, studies based on a full life-cycle assessment suggest that relatively lower energy use within the New Zealand production system means that food miles are not always a good indicator of GEC impacts (Saunders and Barber, 2007). In the UK, contradictions have emerged over the ways in which a focus on food miles could undermine attempts to promote development in poor regions of the world through fair trade (MacGregor and Vorley, 2006; Morgan, 2008).

Retailing food

The retail sector is a very powerful actor in food systems and has become deeply engaged in environmental governance in relation to climate change, forestry, fisheries and water (Baldwin, 2009). In the UK some of the largest food retailers (such as Marks and Spencer, Waitrose, Sainsbury's and Tesco) have taken significant steps to respond to consumer and NGO pressure, anticipate government regulation, and gain competitive advantage by greening their supply chains and operations (Box 18.4). They are also beginning to show interest in Fairtrade and locally produced foods to varying extents. As major private sector actors in the UK food system and with much of consumer expenditure allocated to the food retail sector each year, these actors have considerable power to shape food and environmental governance both domestically and in those regions where they operate internationally. The common ingredients in this greening have included: a continuing expansion of environmental labelling to facilitate consumer choice; the introduction of new environmental standards to supply chain management and operations to complement existing quality and safety oriented standards; more GHG-friendly energy use (solar panels, energy conservation, biofuels); and the adoption of sustainability reporting (using the Global Reporting Initiative model).

Box 18.4 *A case study of Marks and Spencer and Tesco*

Two of the UK's largest retailers – Tesco (the largest with almost £60 billion revenue in 2009) and Marks and Spencer (with £10 billion) – recently announced major corporate commitments to the environment, and to the reduction of GHG emissions and other environmental impacts (Baldwin, 2009). This follows announcements of similar initiatives by the US-based Walmart corporation – now the world's largest corporation, with more than 8400 retail units under 55 different banners in 15 countries and US$405 billion in annual sales (2010). Given its role as a market leader, Walmart's new policies to reduce its environmental impact by committing to zero waste, 100 per cent renewable energy (it is one of the largest private users of electricity in the USA) (Gunther, 2006), sourcing from sustainable fisheries (Box 18.5) and a packaging initiative, can be seen as an important context for all retail approaches to environmental governance.

Beginning in about 2003, Marks and Spencer set out to address environmental concerns more comprehensively through a set of standards for fruit and vegetables that included low pesticide use, fair trade, non-GM and food safety, as well as pricing organics at the same level as comparable conventionally produced foods. They decided to only purchase fish with a clear origin in sustainable fisheries, according to guidance from the MSC, and set a standard for produce from fish farms. Operational efficiencies were made in energy, refrigerants and the logistics of transport, and reduced packaging in stores and for transport. In 2007 they launched 'Plan A', with 100 commitments to social, environmental and ethical practices, supported by a £200 million investment, but have admitted that meeting these during the economic downturn is challenging, although much of

---

the investment paid off in lower waste and energy use (Marks and Spencer, 2009). Plan A is designed to: achieve carbon neutrality through energy conservation, investing in bio-diesel and renewable energy; set new standards in ethical retail by integrating fair-trade and organic produce; end all waste to landfill sites by making packaging purely from sustainable sources and recyclable; increase sustainable and local sourcing; help staff and customers to lead a healthier life; and only sell fish from sustainable sources.

Tesco also announced a set of commitments, though these are reduced in scope and more focused on climate change. It has taken steps to reduce carbon emissions from their operations (by 50 per cent from 2006 to 2020), drive down emissions in their supply chain, and alter consumer behaviour through information, a reward card scheme and selling energy-efficient products (Tesco, 2009). They have also announced and introduced waste reduction schemes and recycling, and claim to monitor the sustainability of products associated with palm oil, forests and fisheries. Tesco's best known environmental initiative is probably their announcement of plans to place a carbon label on their products, so that consumers can choose to purchase goods with lower GHG emissions in the supply chain. It has also taken steps to 'green' its operations and supply chain by establishing a Sustainable Consumption Research Institute, and promising to provide consumer information in the form of carbon labels on its products.

The scope of private governance of the environmental impacts of the retail sector, especially the potential to manage environmental impacts along supply chains, is clearly very large. The effectiveness of these actions is not yet clear and some have questioned the real impacts and sincerity of the commitments with calls of 'greenwashing' and claims that efforts are insignificant (Simms, 2007; Dale, 2008). There are also concerns that sustainability efforts will be reduced as a result of economic bottom line concerns (Carrigan and De Pelsmacker, 2009). The long-term legitimacy of these actions will probably depend on careful and independent monitoring and audit (Laufer, 2003).

Labelling and certifying food

One of the most evident shifts from public to private rule-making can be seen in the growth of certification and labelling schemes for both food and environment. Of course, labelling and standards have a long governmental history and in part continue to be the purview of governments and multilateral institutions, as evidenced by the continuing conflicts around the UN's *Codex Alimentarius* food standards guidelines (Smythe, 2009). Non-state certification and labelling schemes have, however, greatly expanded since the 1990s (Marsden et al, 2008; Perez-Aleman and Sandilands, 2008). The rise of standards is linked in recent work to the supremacy of retailers in the food sector who have adopted a range of quality and safety standards. Some authors argue, however, that this contributes to the consolidation and concentration of the food industry and constitute barriers to entry for new producers, particularly in developing countries. Such standards include those of, for instance,

Europe Gap, the Global Food Safety Initiative and the International Federation of Organic Agriculture Movements (Marsden et al, 2008; Clapp and Fuchs, 2009a; Fuchs et al, 2009).

Environmental and social certification schemes are closely related to labelling, though they may never generate a consumer label and instead function within supply chains. Following innovation of the organic agriculture certifications in the USA and UK, the approach has since spread importantly to forestry, and from there to fisheries, aquaculture, wine, and even to such questionable terrain as bottled water and hydropower. As environmental governance shifts to increasingly market-based mechanisms (Lemos and Agrawal, 2006), the Forest Stewardship Council (FSC) has become the inspiration and institutional model for dozens of other certification efforts, including the Marine Stewardship Council (Box 18.5). Recently, such approaches have been translated to climate with the emergence of the Voluntary Carbon Standard certification system, and a variety of labelling approaches whose goal is to make transparent the GHG emissions associated with products and companies.

Box 18.5 *A case study of WWF and the Marine Stewardship Council (derived from www.msc.org)*

In 1997 the World Wide Fund for Nature (WWF) partnered with Unilever, then the largest buyer of frozen fish, to launch the Marine Stewardship Council (MSC) as the world's first certification body for fisheries. The MSC is particularly interesting as a dynamic initiative that is intermediate between the more established certification system of the Forest Stewardship Council (FSC, beginning in 1992), and the still incipient certification efforts in aquaculture and other sectors.

The MSC closely follows the FSC recipe in the general approach to rule-making, but only loosely does so in internal governance and organization. Most fundamentally, the MSC differs in terms of its hybrid public–private design, which recognizes that fisheries management (where it exists) relies on state-regulated regimes. MSC certification seeks to ensure that these regimes meet its three principles of: (i) maintaining sustainable fish stocks; (ii) minimizing environmental impact; and (iii) effective management.

The MSC has expanded rapidly since its first certification in 2000, now certifying 56 fisheries, 37 species and spanning the territorial waters of 14 countries. It has also certified the chains of custody of 900 companies. Importantly, unlike other sectors with multiple competing certifying initiatives, MSC has emerged as the only global certifier for seafood (Gulbrandsen, 2009). The MSC estimates that 6 million tonnes of seafood were either certified or under assessment in 2009, amounting to nearly 7 per cent of the world's total wild harvest, and with a market value of US$1.5 billion annually.

It also manifests a hybrid NGO–private sector character, both in its emergence as a strategic partnership between WWF and Unilever, in its current internal governance which draws from seafood corporations and NGOs, and in the structure of its funding. To a greater extent than other certification schemes, it has focused on large retailers and global supply chain management. This approach, commonly known as the 'Wal-mart effect', relies on the market power of large buyers to push change down commodity chains through the adoption of purchasing policies (Fishman, 2006). A steadily increasing number of large retailers have made commitments to support MSC and source-certified seafood, though this has not carried over to Asia, which is the world's largest seafood market.

Despite this rapid advance, the MSC and its founders have faced criticisms on the grounds of legitimacy/credibility, accountability, and of effectiveness in meeting ecological goals. The MSC's emergence under the auspices of a single corporation has from the beginning given rise to criticism that it lacks transparency and accountability (Constance and Bonanno, 2000), and the perception of its dependence on large retailers has been a source of scepticism among some producers (Kaiser and Edwards-Jones, 2006). Others have questioned whether the formal structure of accountability established results in substantive accountability to stakeholders (Gulbrandsen, 2008).

The MSC demonstrates some of the potential and tradeoffs associated with the use of market-based private rule-making as environmental and food governance. Notwithstanding criticisms, its ability within a decade to establish minimal legitimacy, secure resources and expand to cover 7 per cent of the world's seafood market shows the extraordinary power of non-state actors to develop and impose new rules of the game for global food systems.

Critics of carbon-labelling initiatives argue that it places too much weight on consumer ability to understand the label, take the time to make comparison, and feel that their actions are effective. It also relies on uncertain science (Green and Capell, 2008; Lillywhite and Collier, 2009) and may also negatively impact imports from poor countries based on incomplete life-cycle analysis (Brenton et al, 2008). It would be better, the critics argue, for the private sector and for government to ensure that all products have lower carbon footprints (Dale, 2007; Specter, 2008).

This raises the important question: as food systems increasingly face stresses associated with GEC, what are the potential contributions and limitations of certification and labelling schemes to the governance of food systems and GEC, contrasted with a legally enforced 'bottom line'? A better understanding of the conditions and design characteristics that shape the emergence and performance of such private institution building is a major challenge for future research.

Financing the food system

The finance and insurance sectors are often overlooked in studies of both food and environmental management, although they are essential to food security and are increasingly relevant to environmental governance. In the food system, farmers, manufacturers, retailers and consumers all take advantage of credit for the purchase of seeds, energy and commodities, and may insure themselves against natural disaster and other crises. Interest and insurance costs can be a significant share of expenditures. Three key trends at the intersection of finance, food and environmental governance are the growth of socially and environmentally responsible investing, the potential of insurance as an adaptation to climate change in the food system and the use of microcredit to assist poor farmers and women in the developing world. These are important areas for future research as they have considerable potential to foster sustainability.

Socially and environmentally responsible investment practices are developing from international level to that of the local high street bank. At the international level the Equator Principles are influencing lending by multilateral (e.g. World Bank) and many of the largest private banks (e.g. Citigroup, JP Morgan Chase and Barclays). The ten principles include requirements for social and environmental assessment, compliance with host country environmental regulations, impact on health and pollution, consultation and disclosure, and independent review and monitoring (www.equator-principles.com). The principles apply only to project-level credits over US$10 million and are thus highly relevant to energy projects, but are mostly inapplicable to the food system. A study by Scholtens and Dam (2007) found that the principles are mainly adopted by larger financial institutions, at some cost, but are generally supported by shareholders because of concerns about reputational risks should issues emerge about the environmental or human rights impacts of projects.

Pension funds were some of the first finance sector actors to promote social and environmental investment criteria by offering 'green' portfolios to their members and by evaluating the environmental and social performance of the companies in which they invested. These were promoted by shareholders and NGOs following earlier activism around issues such as tobacco and anti-apartheid (Guay et al, 2004). Concerns about climate change (in particular) – its impacts, the potential for government regulation of GHG emissions, opportunities in new environmental markets and ethical responsibility – have resulted in major banks and investors using climate risk as criteria for investments. Several new risk-ranking indicators, such as the Dow Jones DJSI and FTSE4GOOD, have been developed to guide investors towards companies who are seen as taking climate change seriously by controlling emissions, or developing clean energy or adaptation technologies (Deutsche Bank, 2010). Table 18.2 suggests opportunities for investors in relation to climate risks, including several relating to food and agriculture systems such as desalination, irrigation, insurance and seed technology.

Investment companies have joined the Institutional Investors Group on Climate Change and are reporting climate risks to the Carbon Disclosure Project and are making climate-related investments in clean energy, water and agribusiness and in reducing their own emissions (Deutsche Bank, 2010).

Both banks and investment agencies increasingly take part in multistakeholder initiatives to bring certain sustainability initiatives into practice. Examples are the participation of Rabobank in global initiatives such as MSC (fish), FSC (wood), RSPO (palm oil) and the Round Table on Responsible

Table 18.2 *Investment opportunities in response to climate change*

Cleaner Energy	Environmental Resources	Energy and Material Efficiency	Environmental Services
Power Generation • Solar • Wind • Clean coal (capture, sequestration, infrastructure) • Other clean power generation (geothermal, hydro, methane capture) • Increased efficiency • Fuel switch: gas, biomass • Nuclear	**Water** • Desalination purification • Wastewater treatment • Distribution and management	**Advanced Materials** • Advanced coatings • Lightweight substitutes • Solvents and biodegradables	**Environmental Protection** • Land conservation • Environmental restoration • Timberland • Forestry • Sea defences
Clean Tech Innovation • Infrastructure management • Supply chain management	**Agriculture** • Irrigation innovation • Clean pesticides • Consumer food purity • Seeds and breeding technologies	**Building Efficiency** • Building management, including data centre management • Heating and cooling systems • Lighting systems • Insulation and materials • Micro generation micro CHP	**Business Services** • Insurance • Logistics • Green focused banking • Microfinance • Consultancy/advisory • Intellecual property
Transport • Emissions reduction • Propulsion systems • Battery technology	**Waste Management** • Recycling • Toxin management • Waste to energy • Land remediation	**Power Grid Efficiency** • Transmission (including smart grids) • Distribution (including home area networks, smart devices and meters) • Storage: Batteries, CA, flywheels, pump storage • Infrastructure • Energy management systems	**Carbon credit developers**
Sustainable Biofuels • Bio-diesel • Ethanol		**Enabling Technologies** • Lasers • Others	

Source: Adapted from Deutsche Bank, 2010

Soy. Among the 400 members of RSPO, there are seven banks and finance institutions. Ethical investment organizations are also becoming increasingly important. The credit and bank crisis of 2008–9 stressed the importance of their long-term sustainability approach. The most important examples of companies and organizations, explicitly working with ethical investments, include Triodos Bank, and the umbrellas Global Alliance for Banking on Values and Eurosif.

Private sector insurance has a long tradition of managing environmental risk in the food system, providing insurance coverage to farmers, industries, transport companies and even consumers, to assist them in surviving environmental and economic variation. The insurance sector was one of the earliest private sector actors to become concerned about the impacts of climate change on infrastructure and food systems (Dlugolecki, 1992; Leggett, 1993). Insurance groups and researchers have collaborated to argue for the significant role that private sector insurance can play within the international climate change adaptation regime (Paterson, 2001; Mills, 2005; Linnerooth-Bayer and Mechler, 2006). Pilot programmes are making insurance available to herders and farmers in some of the poorer regions of Africa (Hellmuth et al, 2009). In Kenya's Marsabit district, the International Livestock Research Institute (ILRI), microfinance pioneer Equity Bank and African insurance provider UAP Insurance Ltd have partnered to offer insurance to pastoralists (CGIAR, undated). New proposals for index-based insurance instruments – designed to provide cover in relation to an average measure such as yield or rainfall and to avoid some of the difficulties of farm-level verification and information asymmetries of traditional insurance – have promise but are limited by design challenges and information gaps (Barrett et al, 2007).

One of the most famous examples of innovations in financing development in poorer countries is that of the Grameen Bank in Bangladesh (Box 18.6), which inspired a worldwide microfinance movement. Microfinance is the delivery of loans to those living in poverty so that they can engage in productive activities, build assets and maintain consumption, filling gaps left by state and private sector finance programmes that often overlook them.

Following the GB initiative, microfinancing initiatives have been implemented in many developing countries, by commercial players and NGOs, often in cooperation with governments and international development organizations and aimed at farmers, other small-scale entrepreneurs as well as consumers. Many of these have performed with mixed results, exceptions being Bolivia's BancoSol, the Bank Rakyat of Indonesia and the GB itself (Bhatt and Shui-Yan, 2001). Schicks (2007) compared the microfinancing strategies of the BancoSol in Bolivia and the GB in Bangladesh and concluded that, despite differences in approaches and target clients of different degrees of poverty, both initiatives provide a valuable service that contributes to the economic development needs of the poorest of the poor.

Clarke (2009) observed that microfinance can be considered environmentally 'green' since it promotes and facilitates recipients' investment in sustainable economic ventures, thus obviating the need for them to pillage their surrounding natural resources for food and shelter. Having acknowledged the

> ### Box 18.6 *A case study of Grameen Bank*
> ### *(derived from www.grameen-info.org)*
>
> The Grameen Bank (GB) began as a project in 1976 by Professor Muhammad Yunus (who later received the Nobel prize), with objectives of extending banking services to poor men and women, eliminating their exploitation by money lenders and creating opportunities for self-employment among the extensive unemployed population in rural Bangladesh. The Grameen Bank specifically targets women and is now is an established bank with majority ownership (90 per cent) by its target clients and the remainder by the government. By virtue of its target market and operating strategies this bank established itself as an alternative to the 'traditional' institutional arrangements within financial markets. The GB shuns many of the characteristics of the traditional banks by, inter alia: (i) going to the doorsteps of clients and potential clients to deliver its services rather than requiring clients to visit the bank; (ii) embracing the philosophy that access to credit is a basic human right, and shunning the traditional requirement of collateral from its clients in favour of a dependence on peer pressure and community support to encourage a strong and healthy repayment response; (iii) monitoring the socio-economic circumstances of its clients' families (children's education, housing, sanitation, access to clean water), and their capacity to respond to and cope with disasters and emergency situations; and (iv) establishing both economic and socioeconomic corporate goals by awarding commendations to its branches for the achievement of both economic performance and socio economic targets. (Level of profits and loan disbursements are among the former, while education of clients' children and elevation of clients from poverty are some of the latter.)
>
> As of January 2010, the GB services over 8 million borrowers, 97 per cent of whom are women. Through its 2558 branches throughout the country it currently serves all of the villages in Bangladesh, and has been recognized for its positive impact on improving the welfare and livelihoods of those living in poverty in the country through many documented studies and reports. For example, the analysis of Khandker et al (1995) suggested that the GB makes a definite contribution to the economic development of the country through a significant positive effect on the wages of men and children. They demonstrate that the mean men's wage in GB villages was higher than that in villages without a GB programme. This led to the conclusion that the GB alleviates poverty on a sustainable basis in Bangladesh (World Bank, 1995).

strong link between poverty and the natural environment, Goldsworthy (2008) discusses and suggests the adoption of a development paradigm that proactively uses environmentally sensitive microfinance policies in the pursuit of poverty alleviation strategies. Her empirical research in Uganda demonstrates the strong potential of such an approach.

There is a clear role for microfinancing in poverty alleviation, while establishing the strong link between the circumstances of poverty and those of environmental security. One appealing transition is into the evolution of development strategies that promote microfinancing with a focus on environmentally sustainable or 'green' activities (Hammill et al, 2008). This will build on the novel governance arrangements embodied within microfinancing strategies, with an enhancement of the bottom-up approach therein. Such a paradigm enhances both food and environmental security.

In addition to microcredit, microinsurance is now developing as a tool to help poor consumers and entrepreneurs to overcome adversity. This can become a potent tool for small-scale farmers, facing increasing uncertainty of harvest and income as a consequence of climate change.

Conclusions

This chapter has outlined a wide variety of ways in which actors beyond government are involved in the governance of the food system as related to GEC and has identified a number of exciting new policy options and research priorities. The authors believe that the role of business, NGOs and farmers in sustaining food systems in the face of GEC is critical. However, despite the accumulation of research to date, much work remains to be done to better understand the relative effectiveness of different governance approaches, including CSR, certification schemes and standards for supply chain management. Given the potential tradeoffs and conflicts that may emerge between decentralized governance approaches, the ways in which conflicts are resolved and coordination achieved where necessary is also a promising area for further research.

Moreover, non-state actions take place within a framework of government policies, at national and international levels, that can either enable or constrain the responses of non-state actors. In many cases it is partnerships between government, the private sector and NGOs that have developed new forms of governance such as certification schemes. In others the private sector and NGOs are filling gaps left by the withdrawal, inadequacy or absence of state programmes. Non-state actors are sometimes acting to shape or anticipate government policies in their own interests and although private sector and NGO actions can be significant, government policies and regulation are often needed to ensure that less responsible companies and consumers respond to environmental and other concerns. The diverse forms of connection between the state and non-state actors, as well as how the outcomes they produce shift under different conditions, remain central issues for future research.

References

Andonova, L. B., M. M. Betsill and H. Bulkeley (2009) 'Transnational climate governance', *Global Environmental Politics*, 9, 2, 52–73

Angel, D. P., T. Hamilton and M. T. Huber (2007) 'Global environmental standards for industry', *Annual Review of Environment and Resources*, 32, 1, 295–316

Appelbaum, R. and N. Lichtenstein (2006) 'A new world of retail supremacy: Supply chains and workers' chains in the age of Wal-Mart', *International Labor and Working-Class History*, 70, 1, 106–25

Auer, M. R. (2000) 'Who participates in global environmental governance? Partial answers from international relations theory', *Policy Sciences,* 33, 2, 155–80

Bair, J. (2008) 'Analysing global economic organization: embedded networks and global chains compared', *Economy & Society,* 37, 3, 339–64

Baldwin, C. (ed) (2009) *Sustainability in the Food Industry,* Oxford, Wiley-Blackwell

Barrett, C., B. Barnett, M. Carter, S. Chantarat, J. Hansen, A. Mude, D. Osgood, J. Skees, C. Turvey and M. Ward (2007) Poverty traps and climate risk: Limitations and opportunities of index-based risk financing. IRI Technical Report 07-02, Beltsville MD, Integrity Research Institute

Bartley, T. (2007) 'Institutional emergence in an era of globalization: The rise of transnational private regulation of labor and environmental conditions', *American Journal of Sociology,* 113, 2, 297–351

Betsill, M. M. and H. Bulkeley (2006) 'Cities and the multilevel governance of global climate change', *Global Governance,* 12, 2, 141–59

Bhatt, N. and T. Shui-Yan (2001) 'Delivering microfinance in developing countries: Controversies and policy perspectives', *Policy Studies Journal,* 29, 2, 319

Biermann, F. (2007) '"Earth system governance" as a crosscutting theme of global change research', *Global Environmental Change,* 17, 3–4, 326–37

Blowfield, M. and A. Murray (2008) *Corporate Responsibility: A Critical Introduction,* New York, Oxford University Press

Blyth, W., R. Bradley, D. Bunn, C. Clarke, T. Wilson and M. Yang (2007) 'Investment risks under uncertain climate change policy', *Energy Policy,* 35, 11, 5766–73

Brenton, P., G. Edwards-Jones and M. Jensen (2008) 'Carbon labelling and low income country exports: An issues paper', *MPRA Paper,* 8971

Buechler, S. and G. D. Mekala (2005) 'Local responses to water resource degradation in India: Groundwater farmer innovations and the reversal of knowledge flows', *The Journal of Environment and Development,* 14, 4, 410–38

Bumpus, A. G. and D. M. Liverman (2008) 'Accumulation by decarbonization and the governance of carbon offsets', *Economic Geography,* 84, 2, 127–55

Burch, D. and G. Lawrence (2007) *Supermarkets and Agri-food Supply Chains: Transformations in the Production and Consumption of Foods,* Cheltenham, Edward Elgar Publishing

Carrigan, M. and P. De Pelsmacker (2009) 'Will ethical consumers sustain their values in the global credit crunch?', *International Marketing Review,* 26, 6

Cashore, B. W., G. Auld and D. Newsom (2004) *Governing Through Markets: Forest Certification and the Emergence of Non-state Authority,* New Haven, Yale University Press

CGIAR, A New Option for Managing Climate Risk in Africa. www.cgiarclimate-change.wordpress.com (accessed: undated)

Chambers, R., A. Pacey and L. A. Thrupp (eds) (1989) *Farmer First: Farmer Innovation and Agricultural Research,* London, Intermediate Technology Publication

Clapp, J. and D. A. Fuchs (eds) (2009a) *Corporate Power in Global Agrifood Governance,* Cambridge, MA, MIT Press

Clapp, J. and D. A. Fuchs (2009b) Agrifood corporations, global governance and sustainability: A framework for analysis. In Clapp, J. and D. A. Fuchs (eds) *Corporate Power in Global Agrifood Governance,* Cambridge, MA, MIT Press

Clarke, E. (2009) 'Microfinance: A little can mean a lot to the environment', *Christian Science Monitor,* 101, 57

Clay, J. W. (2004) *World Agriculture and the Environment: A Commodity-by-Commodity Guide to Impacts and Practices,* Washington, DC, Island Press

Coley, D., M. Howard and M. Winter (2009) 'Local food, food miles and carbon emissions: A comparison of farm shop and mass distribution approaches', *Food Policy*, 24, 150–55

Conca, K. (2006) *Governing Water: Contentious Transnational Politics and Global Institution Building*, Cambridge, MA, MIT Press

Constance, D. H. and A. Bonanno (2000) 'Regulating the global fisheries: The World Wildlife Fund, Unilever, and the Marine Stewardship Council', *Agriculture and Human Values*, 17, 2, 125–39

Corell, E. and M. Betsill (2001) 'A comparative look at NGO influence in international environmental negotiations: desertification and climate change', *Global Environmental Politics*, 1, 4, 86–107

Critchley, W. R. S. (2000) 'Inquiry, initiative and inventiveness: Farmer innovators in East Africa', *Physics and Chemistry of the Earth, Part B: Hydrology, Oceans and Atmosphere*, 25, 3, 285–88

Cutler, A. C., V. Haufler and T. Porter (1999) *Private Authority and International Affairs*, Albany, State University of New York Press

Dale, G. (2007) '"On the menu or at the table": corporations and climate change', *International Socialism*, 116, 117–38

Dale, G. (2008) '"Green shift": an analysis of corporate responses to climate change', *International Journal of Management Concepts and Philosophy*, 3, 2, 134–55

Deutsche Bank (2007) *Investing in Climate Change. An Asset Management Perspective*, Sydney, Frankfurt, Deutsche Asset Management Press

Deutsche Bank (2010) *Investing in Climate Change 2010: A Strategic Asset Allocation Perspective*, New York, Deutsche Bank Climate Change Advisors

Dicken, P. (2007) *Global Shift: Mapping the Changing Contours of the World Economy*, New York, Guilford Press

Dlugolecki, A. (1992) 'Insurance implications of climatic change', *The Geneva Papers on Risk and Insurance – Issues and Practice*, 17, 3, 393–405

Douglas, I. (2009) 'Climate change, flooding and food security in south Asia', *Food Security*, 1, 2, 127–36

Emel, J. (2002) 'An inquiry into the green disciplining of capital', *Environment and Planning A*, 34, 827–43

Epstein, M. (2008) *Making Sustainability Work: Best Practices in Managing and Measuring Corporate Social, Environmental and Economic Impacts*, Sheffield, Greenleaf

Fishman, C. (2006) *The Wal-Mart Effect: How the World's Most Powerful Company Really Works – And how it's Transforming the American Economy*, New York, Penguin

Fuchs, D. A., A. Kalfagianni and M. Arentsen (2009) Retail power, private standards, and sustainability in the global food system. In Clapp, J. and D. A. Fuchs (eds) *Corporate Power in Global Agrifood Governance*, Cambridge, MA, MIT Press

Gereffi, G. and M. Christian (2009) 'The impacts of Wal-Mart: The rise and consequences of the world's dominant retailer', *Annual Review of Sociology*, 35, 1, 573–91

Gereffi, G., J. Humphrey and T. Sturgeon (2005) 'The governance of global value chains', *Review of International Political Economy*, 12, 78–104

Gereffi, G., J. Humphrey, R. Kaplinsky and T. J. Sturgeon (2001) 'Introduction: Globalisation, value chains and development', *IDS Bulletin-Institute of Development Studies*, 32, 3, 1–8

Goldsworthy, H. (2008) Financing a tragedy of the commons? Microfinance and sustainable environmental development in Uganda. *ISA's 49th Annual Convention, Bridging Multiple Divides*, San Francisco, CA

Green, H. and K. Capell (2008) 'Carbon confusion', *Business Week*, 17 March 2008, 4075, 52–55

Guay, T., J. Doh and G. Sinclair (2004) 'Non-governmental organizations, shareholder activism, and socially responsible investments: ethical, strategic, and governance implications', *Journal of Business Ethics*, 52, 1, 125–39

Gulbrandsen, L. H. (2005) 'Mark of sustainability? – Challenges for fishery and forestry eco-labeling', *Environment*, 47, 5, 8–23

Gulbrandsen, L. H. (2006) 'Creating markets for eco-labelling: are consumers insignificant?', *International Journal of Consumer Studies*, 30, 477–89

Gulbrandsen, L. H. (2008) 'Accountability arrangements in non-state standards organizations: Instrumental design and imitation', *Organization*, 15, 4, 563–83

Gulbrandsen, L. H. (2009) 'The emergence and effectiveness of the Marine Stewardship Council', *Marine Policy*, 33, 4, 654–60

Gunther, M. (2006) 'The green machine', *Fortune Magazine*, 154, 42–57

Hammill, A., R. Matthew and E. McCarter (2008) 'Microfinance and climate change adaptation', *IDS Bulletin*, 39, 4

Harvey, D. (2005) *A Brief History of Neoliberalism*, Oxford/New York, Oxford University Press

Hellmuth, M., D. Osgood, U. Hess, A. Moorhead and H. Bhojwani (2009) *Index Insurance and Climate Risk: Prospects for Development and Disaster Management*, Columbia University, New York, International Research Institute for Climate and Society

Hoffman, A. (2005) 'Climate change strategy: The business logic behind voluntary greenhouse gas reductions', *California Management Review*, 47, 3, 21–46

Horwitz, S. (2009) 'Wal-Mart to the rescue: Private enterprise's response to hurricane Katrina', *The Independent Review*, 13, 4

Hughes, A., M. Buttle and N. Wrigley (2007) 'Organisational geographies of corporate responsibility: a UK-US comparison of retailers' ethical trading initiatives', *Journal of Economic Geography*, 7, 4, 491

Kaiser, M. J. and G. Edwards-Jones (2006) 'The role of ecolabeling in fisheries management and conservation', *Conservation Biology*, 20, 2, 392–98

Keck, M. E. and K. Sikkink (1998) *Activists Beyond Borders: Advocacy Networks in International Politics*, Ithaca, NY, Cornell University Press

Khandker, S. R., B. Khalily and Z. Khan (1995) Grameen Bank: Performance and sustainability. *World Bank – Discussion Paper 306*, Washington, DC, World Bank

King, B. G. (2008) 'A political mediation model of corporate response to social movement activism', *Administrative Science Quarterly*, 53, 3, 395–421

Klein, N. (2000) *No Space, No Choice, No Jobs, No Logo: Taking Aim at the Brand Bullies*, New York, Picador

Kovacs, G. and K. Spens (2007) 'Humanitarian logistics in disaster relief operations', *International Journal of Physical Distribution & Logistics Management*, 37, 2, 99–114

Laufer, W. (2003) 'Social accountability and corporate greenwashing', *Journal of Business Ethics*, 43, 3, 253–61

Leggett, J. (1993) 'Climate change and the insurance industry', *European Environment*, 3, 3, 3–8

Lemos, M. C. and A. Agrawal (2006) 'Environmental governance', *Annual Review of Environment and Natural Resources*, 31, 297–325

Levy, D. L. and P. Newell (2005) *The Business of Global Environmental Governance*, Cambridge, MA, MIT Press

Lillywhite, R. and R. Collier (2009) 'Why carbon footprinting (and carbon labelling) only tells half the story', *Aspects of Applied Biology*, 95, 73–78

Linnerooth-Bayer, J. and R. Mechler (2006) 'Insurance for assisting adaptation to climate change in developing countries: a proposed strategy', *Climate Policy*, 6, 6, 621–36

Liverman, D. (2004) 'Who governs, at what scale and at what price? Geography, environmental governance, and the commodification of nature', *Annals of the Association of American Geographers*, 94, 4, 734–38

Liverman, D. M. and S. Vilas (2006) 'Neoliberalism and the environment in Latin America', *Annual Review of Environment and Resources*, 31, 1, 327–63

Lobell, D. B., M. B. Burke, C. Tebaldi, M. D. Mastrandrea, W. P. Falcon and R. L. Naylor (2008) 'Prioritizing climate change adaptation needs for food security in 2030', *Science*, 319, 5863, 607–10

MacGregor, J. and B. Vorley (2006) *Fair Miles? The Concept of 'Food Miles' Through a Sustainable Development Lens*, London, International Institute for Environment and Development

Marks and Spencer, Annual Report 2009. www.annualreport.marksandspencer.com (accessed: undated)

Marsden, L., T. Marsden, R. Lee, A. Flynn and S. Thankappan (2008) *The New Regulation and Governance of Food: Beyond the Food Crisis?*, Abingdon, Routledge

Meinzen-Dick, R. (2007) 'Beyond panaceas in water institutions', *PNAS*, 104, 15200–205

Miles, R. E. and C. C. Snow (2007) 'Organization theory and supply chain management: An evolving research perspective', *Journal of Operations Management*, 25, 2, 459–63

Mills, E. (2005) 'Insurance in a climate of change', *Science*, 309, 5737, 1040

Mills, E. (2009) 'A global review of insurance industry responses to climate change', *The Geneva Papers on Risk and Insurance – Issues and Practice*, 34, 3, 323–59

Minx, J., G. Peters, T. Wiedmann and J. Barrett (2008) *GHG Emissions in the Global Supply Chain of Food Products*, Seville, Input Output and Environment

Morgan, K. (2008) Local and green vs global and fair: The new geopolitics of care. *Working Paper Series*, Cardiff, The Centre for Business Relationships, Accountability, Sustainability and Society

Morgan, K., T. Marsden and J. Murdoch (2006) *Worlds of Food: Place, Power and Provenance in the Food Chain*, Oxford/New York, Oxford University Press

Natsios, A. (1995) 'NGOs and the UN system in complex humanitarian emergencies: conflict or cooperation?', *Third World Quarterly*, 16, 3, 405–19

Newell, P. (2000) *Climate for Change: Non-State Actors and the Global Politics of the Greenhouse*, Cambridge, UK, Cambridge University Press

Newell, P. (2005) 'Citizenship, accountability and community: the limits of the CSR agenda', *International Affairs*, 81, 3

Newell, P. (2008) 'Civil society, corporate accountability and the politics of climate change', *Global Environmental Politics*, 8, 3, 122–53

Nicholls, A. and C. Opal (2005) *Fair Trade: Market-Driven Ethical Consumption*, London, Sage

Okereke, C. (2007) 'An exploration of motivations, drivers and barriers to carbon management: The UK FTSE 100', *European Management Journal*, 25, 6, 475–86

Oosterveer, P. (2007) *Global Governance of Food Production and Consumption: Issues and Challenges*, Cheltenham/Northampton, MA, Elgar

Ostrom, E. (2005) *Understanding Institutional Diversity*, Princeton, New Jersey, Princeton University Press

Paarlberg, R. L. (2002) Governance and food security in an age of globalization. *Food, Agriculture and the Environment Discussion Papers*, Washington, DC, International Food Policy Research Institute

Paterson, M. (2001) 'Risky business: insurance companies in global warming politics', *Global Environmental Politics*, 1, 4, 18–42

Perez-Aleman, P. and M. Sandilands (2008) 'Building value at the top and the bottom of the global supply chain: MNC-NGO partnerships', *California Management Review*, 51, 1, 24–49

Pierre, J. (2000) *Debating Governance*, Oxford/New York, Oxford University Press

Reij, C. and A. Waters-Bayer (2001) *Farmer Innovation in Africa: A Source of Inspiration for Agricultural Development*, London, Earthscan

Renard, M. (2003) 'Fair trade: quality, market and conventions', *Journal of Rural Studies*, 19, 1, 87–96

Richards, P. (1985) *Indigenous Agricultural Revolution: Ecology and Food Production in West Africa*, London, Vintage

Saunders, C. and A. Barber (2007) 'Carbon footprints and food miles: global trends and market issues', *New Zealand Science Review*, 64, 54

Schicks, J. (2007) 'Developmental impact and coexistence of sustainable and charitable microfinance institutions: Analysing BancoSol and Grameen Bank', *European Journal of Development Research*, 19, 4, 551–68

Scholtens, B. and L. Dam (2007) 'Banking on the equator. Are banks that adopted the equator principles different from non-adopters?', *World Development*, 35, 8, 1307–28

Schurman, R. (2004) 'Fighting "Frankenfoods": Industry opportunity structures and the efficacy of the anti-biotech movement in Western Europe', *Social Problems*, 51, 2, 243–68

Seidman, G. (2007) *Beyond the Boycott: Labor Rights, Human Rights and Transnational Activism*, New York, Russell Sage Foundation

Simms, A. (2007) *Tescopoly: How One Shop Came Out on Top and Why it Matters*, London, Constable

Smythe, E. (2009) In whose interests? Transparency and accountability in the global governance of food: Agribusiness, the Codex Alimentarius, and the World Trade Organization. In Clapp, J. and D. A. Fuchs (eds) *Corporate Power in Global Agrifood Governance*, Cambridge, MA, MIT Press

Specter, M. (2008) 'Big foot', *The New Yorker*, 25

Spence, L. and M. Bourlakis (2009) 'The evolution from corporate social responsibility to supply chain responsibility: the case of Waitrose', *Supply Chain Management: An International Journal*, 14, 4, 291–302

Stewart, G., R. Kolluru and M. Smith (2009) 'Leveraging public-private partnerships to improve community resilience in times of disaster', *International Journal of Physical Distribution & Logistics Management*, 39

Tesco, Corporate Responsibility – Environment webpage. www.tescoplc.com/plc/corporate_responsibility_09/environment/ (accessed: 2009)

Tvedt, T. (1998) *Angels of Mercy Or Development Diplomats?: NGOs & Foreign Aid*, Trenton, NJ, Africa World Press

Vandenbergh, M. (2007) 'The new Wal-Mart effect: The role of private contracting in global governance', *UCLA Law Review*, 54, 913

von Braun, J. (2007) *The world food situation: new driving forces and required actions*, IFPRI's Biannual Overview of the World Food Situation, CGIAR Annual General Meeting, Beijing

World Bank (1995) Grameen Bank: Performance and sustainability. In Khandker, S. R., Z. H. Khan and M. A. Baqui Khalily (eds) *World Bank Discussion Paper No. 306*, Washington, DC, World Bank

WWF (2009) *Climate Change: Confronting a global challenge*, Washington, DC, World Wildlife Fund

19
Green Food Systems for 9 Billion

Michael Obersteiner, Mark Stafford Smith, Claudia Hiepe, Mike Brklacich and Winston Rudder

Introduction

Satisfying the food security needs of 9 billion people while having a smaller environmental footprint are both key goals of the international and development agendas. At present these twin goals are insufficiently linked and at times apparently conflicting in practice. A food systems approach (see Chapter 2) helps explicate the complex interactions across time and space, and across different sectors and societal goals, which can cause this apparent conflict. To meet both goals, this must be resolved at a range of spatial and temporal levels, albeit with sensitivity to transition issues and to ancillary goals such as equity among regions.

Although there has historically been more than enough food to provide adequate nutrition for the global population, hunger has always been a part of everyday life for some people, for some, if not all, of the time. In addition to the underlying growth in demand from the growing global population, food security has also been affected by shifting diets, especially the growing demand for livestock-based products as economies expand, particularly in Asia. The long-term consequences of changing diets have also become evident (e.g. obesity problems), and for the first time in modern history average life expectancy for children in North America may fall short of the life expectancies of their parents (Olshansky et al, 2005). More recently, these trends have also become evident in some developing countries, where food insecurity results from both insufficient and excessive, or unbalanced, diets, in different parts of their populations. Despite the wide range of technical and policy responses, food insecurity persists for about 1 billion people today and the prospects of increasing food production by 50 per cent and feeding about 9 billion people by 2050 remain globally challenging, particularly in the face of global environmental change (GEC) (FAO, 2006, 2009a; Johnson et al, 2007; Chapter 1).

The complementary, simultaneous challenge of feeding 9 billion is to reduce the environmental footprint of food systems. Recent evidence suggests that 60 per cent (15 of 24) of key ecosystem services have been adversely impacted by human activity (MA, 2005) and that some key thresholds in earth system functioning may have been passed (Rockström et al, 2009; Chapter 1). The reclaiming of ecosystem services at the global level is central to several international environmental agreements, including the Framework Convention on Climate Change and its Kyoto Protocol, the Biodiversity Convention and the Convention to Combat Desertification. Proposals such as lowering the carbon footprint of agriculture and other food systems activities (McKinsey & Co., 2009) show considerable unrealized promise but these must be evaluated in the larger context of other GEC issues, of livelihoods and equity of all actors in the food system, and interactions with other sectors that are linked to food systems such as energy and water.

The twin goals of achieving food security for 9 billion people in 2050 while reducing environmental footprints are routinely included in the rhetoric of global agendas, but the achievement of these goals is threatened by weak coordination between often independent and localized efforts to reduce food insecurity and environmental footprints (see Chapter 13). Our understanding of these interplays remains underdeveloped, and key challenges include improving the understanding of interactions between food systems operating at local through regional levels with global food requirements; addressing mismatches between the capacity of current institutions to manage for both food and environmental outcomes; and creating an approach to responding to these issues that is sophisticated and nuanced but not so complex as to be unachievable.

This chapter explores examples of the interactions between the food security and environmental outcomes of food systems in order to set a research agenda to help deliver environmentally sound food systems that will satisfy the needs of the global population in 2050.

Interactions between food security and environmental footprints

In general, policy and management activities related to food systems can contribute towards the two goals of enhancing food or environmental outcomes in various ways (see Figure 19.1). The quadrants of Figure 19.1 represent cases where activities contribute to neither goal, help one goal but hinder the other, or, most preferably, contribute to both. The figure is intended to spark discussion about the interplay between food security and environmental outcomes in an aggregate sense, so the axes only approximately embrace all aspects of each for illustrative purposes, rather than being quantitatively precise. In particular, it aims to highlight how activities taken because of a perceived benefit at a given spatial or temporal level, or in only one system (domain), may not in fact deliver the same value when analysed more comprehensively; understanding the interrelations between scales and levels is crucial (see Chapters 2 and 13).

Some activities that are related or have different implications when analysed at different spatial and temporal levels are linked by pale lines. Examples in the different quadrants include:

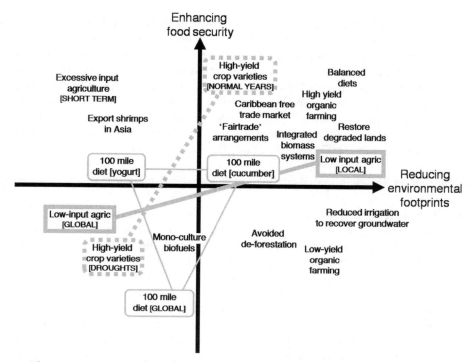

Figure 19.1 *Examples of activities from various components of the food systems at various levels, plotted to show their approximate contribution to food security and environmental outcomes*

- *Fail to deliver to either goal (bottom left)*: mono-culture biofuel plantations on good agricultural land in regions with low food security may appear to provide an income for the local people, but in fact probably neither improve environmental footprints (being aimed at energy rather than environmental outcomes) nor enhance food security (if payments are delayed and at risk in shocks on the global carbon market; or food production is displaced) (e.g. Mendoza, 2007; Altieri, 2009).
- *Enhance food security but not environmental outcomes (top left)*: intensive shrimp production systems in Asia that export to global markets are an example of a food system activity that can temporarily enhance local incomes and hence food security, but which causes local environmental damage and are subject to global economic fluctuations such that further environmental damage ensues when producers cannot afford to look after their land (Belton and Little, 2008).
- *Improve environmental but not food security (bottom right)*: large-scale policies supporting extensive systems and/or lower input organic farming, especially in the developed world, which are accompanied with lower yields, may provide local or regional environmental benefits but fail to deliver sufficient food (Halberg et al, 2006).

- *Deliver to both goals (top right)*: intensive organic farming systems have the potential to deliver both food security and improved environments (by maintaining high production levels with fewer pesticides and artificial fertilizers). Similarly, a move to diets with less red meat may improve food security (by reducing obesity-related health problems and allowing land, where suitable, to be re-allocated from grazing and producing feed for livestock to producing food for direct human consumption), while improving environmental outcomes (e.g. by reducing methane emissions) (Johnson et al, 2007; Hamm, 2008; Dixon et al, 2009).

Other examples, highlighted by links in Figure 19.1, include:

- The current '100-mile diet' movement in Europe, and similar 'food miles' concepts (e.g. Watkiss, 2009; and see Chapter 18), work well at a local level for some products but not for others. For example, it is good for vegetables such as cucumbers, which can be grown widely and are relatively expensive to transport; however, it is often environmentally bad in the case of yogurt, as it means pasturing cows on suboptimal lands where local production would lead to higher total overall life-cycle emissions. In an ever-more urbanizing world, if such a goal was implemented for everyone globally, it would almost certainly reduce overall food security since some people would not have access to enough food; and some regions would be forced to grow foods to which they are not environmentally suited.
- Low input agriculture if practised in only a few places can deliver environmental benefits with no great impacts on food security overall; however, if this was practised globally, food production per unit area cultivated would drop and to compensate, large areas of 'new' land would be converted to this form of agriculture to supply demand, with big impacts on global environments (Gregory et al, 2002).
- High yielding cultivars in variable environments may greatly improve food production in 'average' years (notwithstanding possible higher water use), but may have greater yield reductions in drought years, in which case not only is food security reduced but natural resources are degraded.
- Environmental restoration activities, such as closing off irrigation systems in order to allow depleted groundwater systems to recover, may have great environmental benefits but reduce food production in the meantime; however, this is an issue of timeframes, since the environmental impacts would eventually have affected food production anyway.

These examples highlight that it is important to analyse the contribution of different food system activities to food and environmental outcomes at the appropriate level in time and space, in the appropriate context, and while being cognizant of the interactions with other domains such as energy and human livelihoods. These types of issues are now explored in more detail.

Principles for evaluation of synergies and tradeoffs

The choices to adapt food systems to improve both food and environmental outcomes are highly contextual and often location-specific. Here three specific issues are explored where actions that currently seem to lead to deleterious tradeoffs between food security and environmental concerns need to be re-evaluated.

In general, strategies will only rarely be robust along all dimensions, so that good judgement and choices about tradeoffs and synergies between and within the dimensions of food and environmental outcomes are often needed. Concurring with Tilman et al (2002), sustainable agriculture is defined as practices that meet current and future societal needs for food and fibre, for ecosystem services and for healthy lives, and that do so by maximizing the net benefit to society when all costs and benefits of the practices are considered. Focus is given to the integration of costs and benefits along spatial and temporal scales and spill over to other sectors.

Global implications of local actions

Many current adaptations in food systems, which appear to be environmentally beneficial at the local level, might however turn out to be disadvantageous in terms of both food security and environmental performance in wider landscape or global food system contexts. For example, while de-intensifying agricultural production can reduce deleterious environmental impacts locally, applied as a global strategy would lead to more deforestation and expansion of agriculture into higher-risk areas if yields decline, in turn leading to an overall higher environmental impact and reducing food security (Halberg et al, 2006; Herrero et al, 2009; Melillo et al, 2009). Conversely, solutions assessed only at a global level, such as global, unsustainable, intensification of agricultural management systems (Richardson et al, 2009), could lead to tremendous environmental degradation in a few globally competitive production regions, hence causing further food insecurity and dependency in food importing regions. Indeed, such simplistic global solutions to enhancing food and environmental security may cause new vulnerabilities, in particular to price shocks given the enhanced dependence on food imports in many countries under such proposals. Because all land-based resources (land, water, soil, biodiversity, etc.) are finite, clever portfolios of locally adapted food systems have to be constructed with the main aim of minimizing land-use expansion at the global level.

Tradeoffs between short- and long-term objectives

Transitioning food systems to a better state might involve short-term degradation of some dimensions of food and environmental security. For example, while a direct, universal switch from subsistence farming to high-yield organic farming may be desirable on first principles, it is not possible due to lack of knowledge and technological resources, nutrient supply and markets for premium products, and would probably necessitate an intermediate stage that may be less environmentally benign. Thus, transitioning strategies through management practices that are environmentally less desirable may enable long-term adaptation towards superior strategies, through learning and building of sufficient human and technological capital. Recognition of this possibility should generate potential solutions.

Similarly, strategies that aim at a global concentration of industrial high-input farming will eventually lead to soil fertility decline and thus only delay the environmental impacts arising from relocating agricultural production and expanding into natural ecosystems elsewhere. On the other hand, such a strategy, applied transiently, might buy time for the development of new technologies to increase yields and reverse environmental degradation, such as promised by the biochar community (Lehmann, 2007). Thus pathways to transition matter, and understanding the acceptability of transient tradeoffs is critically important.

System boundaries for green food systems

The potential of agriculture to impact environmental security is substantial, both within agri-food systems and beyond, and both positively and negatively. In feeding the world, the provision of environmental services thus needs to be appraised from a total systems perspective. Agriculture is a major driver of deforestation, so that intensification of agriculture will continue to help avoid further deforestation. However, there are tradeoffs across system boundaries: intensification is likely to have negative impacts on net greenhouse gas emissions, water consumption and local biodiversity, although the overall balance of environmental services at a landscape level might be superior and more importantly the impacts per unit production might be less than with less intensive systems. Conversely, policies that avoid deforestation through forcing agriculture to switch to more costly production systems will increase food prices with consequences for food security. Yet farmers' incomes might benefit from structurally higher prices, thus improving their livelihoods and production systems (Ewing and Msangi, 2009), provided that higher-level policy settings (e.g. trade liberalization, etc.) are effective.

The logic around such complex tradeoffs across system boundaries can equally be applied to the currently controversial issue of biofuels and biomass use (Searchinger et al, 2009). Integrated biomass systems (which include renewable energy systems) increase the level of economic activity in rural areas and diversify the economy, thus increasing overall resilience and income levels; smallholder grower schemes are more likely to lead to a better integration of biofuel cash crops into a wider spectrum of cropping systems, whereas large-scale monoculture biofuel production is more likely to have negative impacts on food and environmental security.

In short, an integrated systems view is required, with clear bounds to the system, which accounts for interactions (both positive and negative) in time and space. Given the goals of feeding a growing global population with a lower environmental footprint into the future, these bounds really need to be set by all actors in the food system (see Chapters 2 and 18), and at global level across domains which include food, water and energy systems. Monitoring and evaluation at global level and over long time periods will be necessary.

However, as noted above, within such analyses, attention needs to be paid to transient effects in time and local effects in space, as either of these may help to design pathways to transition, and which may be vital for meeting ancillary goals such as equity and resilience of livelihoods (see Chapters 17 and 20).

Quo vadis?

The foregoing analysis leads to the conclusion that the weak links between food security and environmental agendas need strengthening by concentrating on actions that aim to move outcomes from the left and bottom quadrants of Figure 19.2 to the top right, while being cognizant of axes not represented on this graph (e.g. livelihoods, equity and human security). Future analysis needs to focus much more clearly on this issue, while taking cross-scale and sector interactions and the interplay among strategies clearly into consideration. There are many legitimately competing issues concerning food, environment, social and economic outcomes at local and regional levels, as illustrated by some of the competing models in Figure 19.1. However, the goal of feeding 9 billion people with less environmental impact is a global goal, and the analysis needs to proceed at the global level to resolve what appear to be intractable tradeoffs when the system is analysed partially. This is not to say that cross-level effects relating to inter-regional equity and developing regional resilience should not be considered, but that their resolution at lower level(s) should be judged for their global-level impacts to the changes schematized in Figure 19.2. Progress can only be made efficiently and effectively towards low-carbon and environmentally sound food systems when local actions are globally consistent. Comprehensive and systematic assessment tools to judge global consistency of local actions can either be imposed by regulatory arrangements or internalized in the design of sound food systems. Food and environmental security will then be either the outcome of a (self-)imposed good governance scheme or the emergent property of a sound food system.

Simply stated, this is an enterprise of immense complexity with cascading sets of tradeoffs and interactions that could paralyse decision-making. However, the foregoing analysis highlights how different areas of policy can help address the needed changes in a somewhat disaggregated way. Figure 19.2 illustrates this point with some specific actions that could contribute to this change, which are not hugely interdependent, such as the topical issue of ensuring that all climate mitigation instruments pay close attention to their interactions with adaptation. We now explore these actions as case studies to identify some lessons for future approaches to decision-making and research.

Carbon mitigation taking better account of adaptation

A specific and topical example refers to the international climate negotiations, which are heavily dominated by developing regulations, markets and incentives to reduce net CO_2 emissions, both globally and at national levels (Watson et al, 2000). At present, if successful, these arrangements will slow climate change, but potential negative implications on food security are often not considered. Large areas of monoculture forest plantation, for example, may store

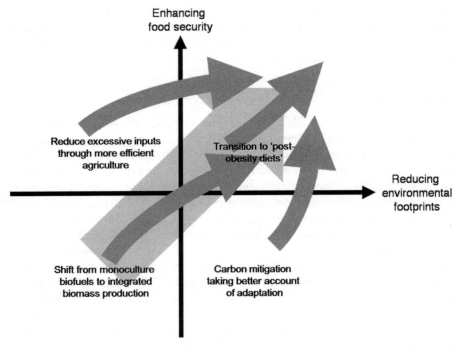

Figure 19.2 *Examples of activities that could be pursued reasonably independently (individual foreground arrows) under the common vision of moving global food systems towards the upper right quadrant (large background arrow)*

carbon but displace food production, as may a major move into biofuel production on high-quality agricultural land (Searchinger et al, 2009). More subtle effects may be felt through the socio-economic system if carbon sequestration activities reduce the diversity or productivity of agricultural regions in ways that make their inhabitants less resilient to shocks in energy or food prices (Janssen et al, 2009).

In the future, mitigation instruments should explicitly consider synergies and tradeoffs with food security and strive for multiple benefits rather than perverse outcomes (Tubiello et al, 2008; FAO, 2009b). For example, markets could be devised which incorporate the effects of large-scale mitigation on food system resilience, thus bringing this consideration explicitly into a market-based instrument. At national level, decision-makers could ensure that agricultural agencies, as well as environmental and energy agencies, are involved in the design of national carbon tax or cap-and-trade systems. Carbon pricing will impact all stages of the food system, whether producing, processing, transporting or consuming food (see also below). Research is needed to help design mitigation instruments that deliver such multiple benefits.

Reduce excessive inputs through more efficient agriculture

A major thrust of the green revolution to increase food production was to encourage farmers to increase their inputs of irrigation water, chemical fertilizers and pesticides. While this has had enormous success in enhancing productivity, the cost in terms of an overused water cycle (Molden, 2007), nitrogen leakage and eutrophication around coasts globally (Duce et al, 2008), and unintended impacts of excess pesticides and herbicides (Freemark and Boutin, 1995), has become apparent. As a result, particularly in developed nations, various forms of more efficient agriculture have emerged, all based on the concept of increased input–use efficiency, where the 'input' can be germplasm, nutrients, agrochemicals, water, energy and/or knowledge. Examples include drip irrigation and closed irrigation canals to waste less water, precision agriculture emphasizing more accurate management of fertilizer to meet plant needs in years of different rainfall, and organic farming to reduce other environmental side effects. These all aim to maintain – or even increase – productivity at the same time as reducing environmental impacts. However, much developing world agriculture still faces the need to greatly increase production but has hardly made the first transition to agriculture that is inefficient but effective in terms of increasing production, let alone the second transition to be both efficient *and* effective (Pingali, 2007).

The risk that, in order to meet food security goals, large areas of developing-world agriculture will make only the first transition, indicates the urgent need to set in place processes to encourage these regions to skip the first transition and go directly to the second as much as possible. However, this leapfrogging is capital-intensive in financial, built, human and even social terms, so may not be feasible everywhere (López, 1995). It requires targeted investment in knowledge and training, including investments in off-farm activities, as well as the provision of capital investment to facilitate the transition to high technological support where needed (Reardon and Timmer, 2007). Research can contribute to this process by systematically identifying the forms of efficient agriculture that would be most appropriate in different regions, and ensuring that the locally appropriate technologies and cultivars are available and accessible at reasonable cost to permit their wide implementation. There are also ramifications for other food system activities in terms of processing, transport and consumer preferences for some food types, as exemplified in Chapter 18.

Transition to 'post-obesity diets'

Recent years have shown that access to cheap food has led to an explosion of obesity-related problems in developed nations. This is slowly leading to public awareness campaigns and community movements for more healthy diets (e.g. in Europe; Barling et al, 2009); a trend towards 'post-obesity diets'. As much as these diets tend to emphasize less red meat (closely aligned to methane emissions and the relatively inefficient use of land to grow grain for feed-lotting), less processed food, less packaging and lower food miles, and less consumption overall, this trend could contribute to environmental goals across the whole food system, as well as enhanced food security in terms of lower total demand (Gerbens-Leenes and Nonhebel, 2002; Eshel and Martin, 2009).

Although obesity-related problems are now also emerging in developing countries, the trend towards 'post-obesity diets' is still in its infancy in developed nations, while countries in transition are rapidly developing the underlying health problems, even among poorer members of society and particularly in cities (Frazão et al, 2008). There is thus a need to promote the transition to 'post-obesity diets' worldwide, and especially in countries where some or all of the population no longer face day-to-day food security problems (McMichael, 2007). Research is needed to support this process and validate the implications of this move for improved food and environmental outcomes, as well as to understand subtle differences in implementation which may be important. As global food prices can be expected to fall, a full food systems analysis of the implications for small farmers in developing nations is also needed to determine whether such moves might cause unexpected negative income impacts and consequently on their livelihoods.

Shift from monoculture biofuels to integrated biomass systems

Large monoculture biofuel plantations that have been established in developing countries in recent years have been criticized for their potential impacts on both environmental and food outcomes. These occur through negative impacts on food prices, land, water and biodiversity, as well as net greenhouse gas emissions if land-use change is taken into account (Tan et al, 2009). Recent attempts to define criteria for 'sustainable' biofuels partially address these concerns such as those elaborated by the Roundtable for Sustainable Biofuels. It has been argued that well-designed biofuel production systems could help to mitigate climate change, restore degraded land and contribute to local food security through improved access to markets and income (Janssen et al, 2009). Second-generation biofuels may also reduce the competition for land; however, they may favour monoculture systems even more, and their technology-intensive processing needs may restrict smallholder involvement (Havlik et al, forthcoming).

To improve joint food and environmental outcomes, yields need to be sustained and further improved, while at the same time the risks reduced of large-scale crop and forage losses due to the spread of pests and diseases (Holmann and Peck, 2002). This points to a need to move from large-scale monocultures of narrow genetic variance to more diverse integrated biomass systems. Institutional arrangements (e.g. cooperatives) that can involve smallholders, but at the same time guarantee an 'economy of scale' to stay attractive for foreign investors, are one avenue that could be explored. Careful analyses are needed for least developed countries to determine whether it is in their interests to jump on the 'biofuels bandwagon'. Bio-energy, when viewed more broadly (not just liquid, but also biogas and heat), includes integrated crop-energy systems that provide many benefits for smallholder food security and the environment, such as rural electrification, soil carbon sequestration and waste management (Buchholz et al, 2009).

Keys to pathways towards green food systems for 9 billion

There is no single or optimal pathway to creating food security for 9 billion people with low environmental impact. Work is needed on many alternative pathways that can contribute to the vision outlined in Figure 19.2, considering interplay among food system components and actors as well as interactions with many other sectors. However, there are some common actions and capabilities that are required to support the set of transition pathways: (i) governance and institutions (see Chapter 18); (ii) food choices and diets; and (iii) food system management (see Chapter 2). The interplay among these actions and capabilities across levels is crucial; the challenge is to have all three components simultaneously striving to move towards achieving food security, while reducing the environmental footprint of food systems.

Food choices and diets in society

Ultimately food production systems respond to changes in food demand, and this in turn is driven by the combination of growing population, economic growth, national and international food policies, as well as societal choices and consumer preferences related to diet. Here the population growth and minimum diet needs to avoid hunger is taken as given, so that the key to changing net demand lies with dietary choices among those that can make such choices. Such changes are likely to emerge as a result of the combined effects of education and awareness, regulation and peer-group pressure, with an important role for non-state actors (see Chapter 18). Urban populations are becoming increasingly detached from rural food production systems, and are hence unaware of the consequences of their food choices that routinely occur at a considerable distance from the urban consumption point. Improved awareness about these impacts need to be underpinned by better information about the impacts of 'bad' diets on both personal and public health locally, and on environmental impacts and food shortages elsewhere in the world. At the same time there is emerging evidence that the public health consequences of diets have a systemic underlying driver in societal inequality (e.g. Wilkinson and Pickett, 2009), which would need to be addressed by much wider policy instruments.

Challenges for national and regional decision-makers include delivering the appropriate public awareness campaigns, coupled with supportive regulations such as food labelling. These actions need underpinning by better research on the environmental footprint of different foods within their whole food systems (such as extended life-cycle analyses), which help determine whether communications concepts such as food miles are too simplistic, and if so replace them with meaningful measures of environmental and social impacts. At the same time, governmental exchanges at a global level are needed to assist developing nations to skip as quickly as possible over the highest environmental-impact diets, led by good examples in the developed nations. Research is also needed on how behavioural change in this area can be promoted.

In short, the needs are to facilitate, educate and regulate to encourage low-carbon diets based on low-carbon food systems; develop and gain acceptability of better food-system-wide measures of net environmental and social impacts of different foods; and transfer these lessons, through example and support, as fast as possible to developing communities so that they can avoid the worst health and environmental aspects of food system transition.

Food institutions and governance decisions

The food security and environmental agendas are usually the preserve of different institutions at global (e.g. Food and Agriculture Organization of the United Nations (FAO), the United Nations Environment Program and global environmental conventions), national (primary production as compared to environmental conservation government departments) and even local levels. Not all policy can be collapsed into a single organization at any level, so integration must be obtained by appropriate interorganizational (e.g. working groups) or transdepartmental (e.g. cross-cutting institutions such as emissions trading schemes) arrangements. At the global level, governments must require these institutions to collaborate on the global goal of feeding 9 billion with lower environmental impacts. Research can support decision-makers at all levels by identifying where such governance arrangements are needed and what forms may work best, as well as by providing the integrative data and assessments that inform action within these arrangements (Barling et al, 2003).

There is also a need for food system institutions to be better coordinated across scales and levels and along food system value chains. In fact there are various examples of this developing: the Caribbean Community (CARICOM) Single Market Economy is slowly coordinating trade within that region (see Chapter 14), while the European Union aims to do similarly at the European level; both aim to help these regions move towards the upper right quadrant of Figure 19.2. Chapter 18 identifies multiple ways that this coordination is already being explored by the non-state sector, often integrating local to global levels. This style of coordination is needed in other regions of the globe and needs to be extended to the global level. To be effective, all spatial levels need functional multistakeholder forums as well as monitoring systems on how regional outcomes contribute to – or detract from – global-level improvements. A global treaty on sustainable food systems for 9 billion may well need exploring, in the spirit of developing global-level adaptation actions in the face of GEC. This might involve a mechanism to ensure that carbon financing transferred to developing countries is done with a view to promoting food security outcomes, as well as reduced environmental impacts. In this case and others, research must deliver better understanding of the cross-scale, cross-level and intersectoral tradeoffs within the context of the global goal, to assist decision-makers in linking food security and environmental outcomes. For example, Liu and Savenije (2008) pointed to the benefits of virtual water transport via food trading for the south of China to the north, compared to the planned deviation of major rivers for irrigation.

In short, there is a need to create institutional arrangements which coordinate between food and environmental interests at all spatial and temporal levels, in particular considering whether this deserves a treaty on sustainable food systems at the global level (and see Chapters 17 and 18); and further develop open and universally shared quantitative decision support tools and biophysical and socio-economic observing systems of food systems across spatial levels up to global, which enable decision-makers to explore and learn from integrated food, social, economic and environmental outcomes at all levels with a consistent understanding of emergent full-system implications.

Food systems management and technologies

Notwithstanding the importance of food demand and governance, there is still a need for the continued delivery of technical solutions for running efficient and effective food systems. Both existing technologies that are not widely spread, and new innovative technologies, must play a major role in improving food security and environmental outcomes while responding to new challenges. Research is also needed to place these technologies firmly in the entire food system context so as to identify the synergies and tradeoffs among technologies, as well as potential risks associated with them. In this regard, research can contribute improved cropping systems that are specifically targeted at delivering resilient, low-impact production in the face of GEC, while exploring the value of traditional systems in this endeavour (Coward, 1977; Magombeyi and Taigbenu, 2008); it can work on-farm to collaboratively affirm or transform local 'best practices' into locally suited options for addressing the joint goal, including local traditional knowledge in the mix as appropriate (Evenson, 2001). It can also deliver monitoring and analysis techniques which help more rapid learning cycles throughout the food system value chain (Horton and Mackay, 2003; Hazell, 2010).

The experience of the green revolution showed that a focus on the singular goal of increased production, primarily at a local level, could cumulate up to a broadly successful global outcome, although with costs to ancillary values. Today the challenge for research is to ensure that the goals of a new food system revolution encompass the broad socio-environmental footprints of the system, across a range of scales and levels on them, and in relation to other GEC issues which were not well articulated at the time of the original green revolution. This will likely necessitate novel institutional mechanisms to facilitate technological development and adoption.

In short, the needs are to create a forward-looking assessment of technical options that may be needed across the food system to meet the twin challenges of feeding more people with less environmental impact; and deliver a systems understanding of the nested network of global food systems within which new technologies and practices can be holistically assessed against their economic, social and environmental outcomes.

In conclusion, there is an urgent need for a systems view to be taken to develop food systems able to meet the joint goals of feeding over 9 billion people over the coming decades while reducing environmental impact. Synergies from the interactions between food, environmental and social factors need to be identified and translated into socio-economic realities. Crucial in this endeavour is the consistency of local management solutions in a global context. Complementary actions can be taken relatively independently, providing that they are framed within a common and comprehensive global analysis.

A number of clear challenges for societal action and supportive research emerge from this appraisal, helping to set an urgent agenda for the next decade.

References

Altieri, M. A. (2009) 'The ecological impacts of large-scale agrofuel monoculture production systems in the Americas', *Bulletin of Science Technology Society*, 29, 3, 236–44

Barling, D., T. Lang and M. Caraher (2003) Joined-up food policy? The trials of governance, public policy and the food system. In Dowler, E. and C. J. Finer (eds) *The Welfare of Food: Rights and Responsibilities in a Changing World*, Oxford, Blackwell

Barling, D., T. Lang and G. Rayner (2009) Current trends in food retailing and consumption and key choices facing society. In Rabbinge, R. and A. Linnemann (eds) *European Food Systems in a Changing World*, Wageningen, European Science Foundation/COST Forward Look

Belton, B. E. N. and D. Little (2008) 'The development of aquaculture in central Thailand: Domestic demand versus export-led production', *Journal of Agrarian Change*, 8, 1, 123–43

Buchholz, T., E. Rametsteiner, T. A. Volk and V. A. Luzadis (2009) 'Multi criteria analysis for bioenergy systems assessments', *Energy Policy*, 37, 2, 484–95

Coward, E. W. (1977) 'Irrigation management alternatives: Themes from indigenous irrigation systems', *Agricultural Administration*, 4, 3, 223–37

Dixon, J. M., K. J. Donati, L. L. Pike and L. Hattersley (2009) Functional foods and urban agriculture: two responses to climate change-related food insecurity. *NSW Public Health Bulletin 20*, Sydney, NSW Government

Duce, R. A., J. La Roche, K. Altieri, K. R. Arrigo, A. R. Baker, D. G. Capone, S. Cornell, F. Dentener, J. Galloway, R. S. Ganeshram, R. J. Geider, T. Jickells, M. M. Kuypers, R. Langlois, P. S. Liss, S. M. Liu, J. J. Middelburg, C. M. Moore, S. Nickovic, A. Oschlies, T. Pedersen, J. Prospero, R. Schlitzer, S. Seitzinger, L. L. Sorensen, M. Uematsu, O. Ulloa, M. Voss, B. Ward and L. Zamora (2008) 'Impacts of atmospheric anthropogenic nitrogen on the open ocean', *Science*, 320, 5878, 893–97

Eshel, G. and P. A. Martin (2009) 'Geophysics and nutritional science: Toward a novel, unified paradigm', *American Journal of Clinical Nutrition*, 89, 5, 1710S–16S

Evenson, R.E. (2001) 'Economic impacts of agricultural research and extension', *Handbooks in Economics*, 18, 1A, 573–628

Ewing, M. and S. Msangi (2009) 'Biofuels production in developing countries: Assessing tradeoffs in welfare and food security', *Environmental Science & Policy*, 12, 4, 520–28

FAO (2006) *World Agriculture: Towards 2015/30*, Rome, FAO

FAO (2009a) Global Agriculture Towards 2050. *High Level Expert Forum Issues Paper*, Rome, FAO

FAO (2009b) *Food Security and Agricultural Mitigation in Developing Countries: Options for Capturing Synergies*, Rome, FAO

Frazão, E., B. Meade and A. Regmi (2008) 'Converging patterns in global food consumption and food delivery systems', *Amber Waves*, 6, 22–29

Freemark, K. and C. Boutin (1995) 'Impacts of agricultural herbicide use on terrestrial wildlife in temperate landscapes: A review with special reference to North America', *Agriculture, Ecosystems & Environment*, 52, 2–3, 67–91

Gerbens-Leenes, P. W. and S. Nonhebel (2002) 'Consumption patterns and their effects on land required for food', *Ecological Economics*, 42, 1–2, 185–99

Gregory, P. J., J. S. I. Ingram, R. Andersson, R. A. Betts, V. Brovkin, T. N. Chase, P. R. Grace, A. J. Gray, N. Hamilton, T. B. Hardy, S. M. Howden, A. Jenkins, M. Meybeck, M. Olsson, I. Ortiz-Monasterio, C. A. Palm, T. W. Payn, M. Rummukainen, R. E. Schulze, M. Thiem, C. Valentin and M. J. Wilkinson (2002) 'Environmental consequences of alternative practices for intensifying crop production', *Agriculture, Ecosystems & Environment*, 88, 3, 279–90

Halberg, N., T. B. Sulser, H. Høgh Jensen, M. W. Rosegrant and M. T. Knudsen (2006) The impacts of organic farming on food security in a regional and global perspective. In Halberg, N., H. F. Alrøe, M. T. Knudsen and E. S. Kristensen (eds) *Global Development of Organic Agriculture: Challenges and Prospects*, Wallingford, CABI Publishing

Hamm, M. W. (2008) 'Linking sustainable agriculture and public health: Opportunities for realizing multiple goals', *Journal of Hunger & Environmental Nutrition*, 3, 169–85

Havlik, P., U. A. Schneider, E. Schmid, H. Böttcher, R. Skalský, K. Aoki, S. de Cara, G. Kindermann, F. Leduc, I. McCallum, A. Mosnier, T. Sauer and M. Obersteiner (forthcoming) 'Global land-use implications of first and second generation biofuel targets', *Energy Policy*, forthcoming

Hazell, P. B. (2010) Chapter 68: An assessment of the impact of agricultural research in South Asia since the Green Revolution. In Evenson, R. and P. Pingali (eds) *Handbook of Agricultural Economics: Agricultural Development: Farm Policies and Regional Development*, Amsterdam, Elsevier

Herrero, M., P. K. Thornton, P. Gerber and R. S. Reid (2009) 'Livestock, livelihoods and the environment: Understanding the trade-offs', *Current Opinion in Environmental Sustainability*, 1, 2, 111–20

Holmann, F. and Peck, D. C. (2002) 'Economic damage caused by spittlebugs (Homoptera: Cercopidae) in Colombia: a first approximation of impact on animal production in Brachiaria decumbens pastures', *Neotropical Entomology*, 31, 2

Horton, D. and Mackay, R. (2003) 'Using evaluation to enhance institutional learning and change: recent experiences with agricultural research and development', *Agricultural Systems*, 78, 2, 127–42

Janssen, R., D. Rutz, P. Helm, J. Woods and Rocio-Diaz-Chavez (2009) *Bioenergy for sustainable development in Africa – Environmental and Social Aspects*, COMPETE. www.compete-bioafrica.net/index.html (accessed: undated)

Johnson, J. M. F., A. J. Franzluebbers, S. L. Weyers and D. C. Reicosky (2007) 'Agricultural opportunities to mitigate greenhouse gas emissions', *Environmental Pollution*, 150, 1, 107–24

Lehmann, J. (2007) 'Bio-energy in the black', *Frontiers in Ecology and the Environment*, 5, 7, 381–87

Liu, J. and H. H. G. Savenije (2008) 'Time to break the silence around virtual-water imports', *Nature*, 453, 7195, 587

López, R. (1995) 'Synergy and investment efficiency effects of trade and labor market distortions', *European Economic Review*, 39, 7, 1321–44

MA (2005) *Ecosystems and Human Well-being: Synthesis*, Washington, DC, Island Press

Magombeyi, M. S. and Taigbenu, A. E. (2008) 'Crop yield risk analysis and mitigation of smallholder farmers at quaternary catchment level: Case study of B72A in Olifants river basin, South Africa', *Physics and Chemistry of the Earth, Parts A/B/C*, 33, 8–13, 744–56

McKinsey & Co. (2009) *Pathways to a Low-Carbon Economy*, New York, McKinsey & Co.

McMichael, P. (2007) Feeding the world: agriculture, development and ecology. In Leys, C. and L. Panitch (eds) *Socialist Register 2007: The Ecological Challenge*, New York, Monthly Review Press

Melillo, J. M., A. C. Gurgel, D. W. Kicklighter, J. M. Reilly, T. W. Cronin, B. S. Felzer, S. Paltsev, C. A. Schlosser, A. P. Sokolov and X. Wang (2009) Unintended Environmental Consequences of a Global Biofuels Program. *MIT Joint Program on the Science and Policy of Global Change*, Cambridge, MA, MIT

Mendoza, T. C. (2007) 'Are biofuels really beneficial for humanity?', *Philippine Journal of Crop Science*, 32, 85–100

Molden, D. (ed) (2007) *Water for Food, Water for Life: A Comprehensive Assessment of Water Management in Agriculture*, London and Colombo, Earthscan and International Water Management Institute

Olshansky, S. J., D. J. Passaro, R. C. Hershow, J. Layden, B. A. Carnes, J. Brody, L. Hayflick, R. N. Butler, D. B. Allison and D. S. Ludwig (2005) 'A potential decline in life expectancy in the United States in the 21st Century', *New England Journal of Medicine*, 352, 11, 1138–45

Pingali, P. (2007) Chapter 54: Agricultural mechanization: Adoption patterns and economic impact. In Evenson, R. and P. Pingali (eds) *Handbook of Agricultural Economics: Agricultural Development: Farmers, Farm Production and Farm Markets*, Amsterdam, Elsevier

Reardon, T. and C. Timmer (2007) Chapter 55: Transformation of markets for agricultural output in developing countries since 1950: how has thinking changed? In Evenson, R. and P. Pingali (eds) *Handbook of Agricultural Economics: Agricultural Development: Farmers, Farm Production and Farm Markets*, Amsterdam, Elsevier

Richardson, K., W. Steffen, H. J. Schellnhuber, J. Alcamo, T. Barker, D. M. Kammen, R. Leemans, D. Liverman, M. Munasinghe, N. Stern and O. Wæver (2009) *Synthesis Report: Climate Change, Global Risks, Challenges & Decisions*, Copenhagen, Climate Congress

Rockström, J., W. Steffen, K. Noone, A. F. Persson, I. Stuart Chapin, E. F. Lambin, T. M. Lenton, M. Scheffer, C. Folke, H. J. Schellnhuber, B. Nykvist, C. A. de Wit, T. Hughes, S. van der Leeuw, H. Rodhe, S. Sorlin, P. K. Snyder, R. Constanza, U. Svendin, M. Falkenmark, L. Karlberg, R. W. Corell, V. J. Fabry, J. Hansen, B. Walker, D. Liverman, K. Richardson, P. Crutzen and J. A. Foley (2009) 'A safe operating space for humanity', *Nature*, 461, 472–75

Searchinger, T. D., S. P. Hamburg, J. Melillo, W. Chameides, P. Havlik, D. M. Kammen, G. E. Likens, R. N. Lubowski, M. Obersteiner, M. Oppenheimer, G. Philip Robertson, W. H. Schlesinger and G. David Tilman (2009) 'Fixing a Critical Climate Accounting Error', *Science*, 326, 5952, 527–28

Tan, K. T., K. T. Lee, A. R. Mohamed and S. Bhatia (2009) 'Palm oil: Addressing issues and towards sustainable development', *Renewable and Sustainable Energy Reviews*, 13, 2, 420–27

Tilman, D., K. G. Cassman, P. A. Matson, R. Naylor and S. Polasky (2002) 'Agricultural sustainability and intensive production practices', *Nature*, 418, 6898, 671–77

Tubiello, F., J. Schmidhuber, M. Howden, P. G. Neofotis, S. Park, E. Fernandes and D. Thapa (2008) 'Climate Change Response Strategies for Agriculture: Challenges and Opportunities for the 21st Century', *Agriculture and Rural Development Discussion Paper 42*, Washington, World Bank

Watkiss, P. (2009) Current trends in distribution and packaging. In Rabbinge, R. and A. Linnemann (eds) *European Food Systems in a Changing World*, Wageningen, European Science Foundation/COST Forward Look

Watson, R. T., I. R. Nobel, A. Bolin, N. H. Ravindranath, D. J. Verardo and D. J. Dokken (2000) Land Use, Land-Use Change and Forestry, *IPCC Special Report on Land Use, Land-Use Change and Forestry*, Geneva, IPCC

Wilkinson, R. and K. Pickett (2009) *The Spirit Level: Why More Equal Societies Almost Always do Better*, London, Allen Lane

20
Surprises and Possibilities

Alison Misselhorn, Andrew Challinor, Philip Thornton,
James W. Jones, Rüdiger Schaldach and
Veronique Plocq-Fichelet

What are surprises in the context of GEC and food systems, and why are they important?

Changes in food systems can occur that are not anticipated. These 'surprises' are important when they have significant consequences for food security – whether positive or negative. Some examples include the public acceptance or otherwise of genetically modified crop varieties or other technological advances, or food price spikes. Multiple changes can occur in more than one aspect of a food system, can interact to yield further change or may occur through 'feedback' mechanisms. Changes may be in a socio-economic or environmental driver (or both) (see Figure 2.1a), or in a food system activity or outcome (see Figure 2.1b). They become a surprise due to lack of monitoring or not being predicted, or when their consequences are not expected, such as those that might lead to violence or conflict (see Chapter 17). This chapter discusses global environmental change (GEC) and food system 'surprises'. Of particular interest is their role in helping to look ahead – not only for studying and understanding, but also for helping decision-making for better managing food systems.

Figure 20.1 illustrates the scope of the chapter. The starting point is a 'surprise' of some kind. This may be either a bolt out of the blue (e.g. an unpredicted weather event, or the food price spike of 2008) or a predicted event that has surprising consequences (e.g. the introduction of mobile phone technology in developing countries, spawning mobile phone-based environmental and market information and banking systems). The common characteristic of such events is that they have unanticipated elements – either their very *occurrence* is unanticipated or their *impacts* are unanticipated (even if the event itself occurs quite frequently), or a combination of the two. Figure 20.1 maps surprises in two-dimensional space, one defined by their degree of 'anticipatability' and the other by the nature and extent of the impacts that ensue. Impacts may be positive or negative, small or large. In the case of a nasty surprise (i.e. unanticipated with negative impacts, mapped in the lower

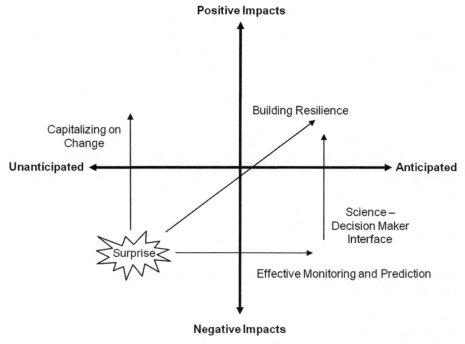

Figure 20.1 *Plotting surprises against their potential impacts and avenues of response*

left quadrant of Figure 20.1), two things might be done: one is to make similar kinds of events in the future less unanticipated (less of a shock); the other is to take some kind of action that can help to reduce the negative impacts of the event (either of the surprise itself or of a similar surprise in the future).

In order to move the impacts of 'surprises' from being potentially negative to being neutral or even positive, building adaptive capacity into food systems is a prerequisite. To increase the 'anticipatability' of an event, however, a clearer or more complete understanding of the processes that are operating is likely to be necessary, possibly in the form of a model and/or better monitoring of key indicators from appropriate baseline data. This monitoring and prediction may then allow movement to higher levels of anticipatability (i.e. from the bottom left-hand to the bottom right-hand quadrant in Figure 20.1). Effective monitoring and prediction mechanisms – including but not limited to robust and credible measures of risk and vulnerability (Adger, 2006) – mean actors in food systems (individuals, community-based organizations, non-government organizations, government bodies, private sector and academia) are in a better position to mitigate negative outcomes and/or capitalize on change. However, their anticipatory and predictive capacities are only enhanced if there is engagement, communication and exchange of knowledge and information between actors; it has been clearly articulated, for example, that 'there is a need to bridge the gap between scientific and local knowledge in order to create projects capable of withstanding stronger natural hazards' (Blanco, 2006). A further expression of a surprise rooted in the links or feedbacks *between* two or more environmental or social food system elements is that of outcomes of policy decisions or technologies that have positive impacts on one 'part' of a food system but unforeseen impacts in other areas as discussed in Chapter 19.

This chapter's sections are mapped against Figure 20.1 as follows: section 2 discusses different forms of surprises, as well as their impacts; section 3 deals with resilience in food systems; section 4 deals with what has been or can be done to prepare for surprises, and discusses the tools and information systems that are needed to better address uncertainty and risk in the development of resilient food systems; section 5 discusses considerations and priorities for research and decision-making, and the interface between the two; section 6 reviews the role of technology in generating, but also in anticipating and preparing for, surprises; and, finally, section 7 looks ahead to considerations for science and decision-makers into the future.

What kinds of surprises (and possibilities) affect food systems?

Part II of this book has outlined many forms of vulnerability in food systems across their ecological and social components, and in their linkages between the two. Surprises also have a number of features that assist us in characterizing or defining vulnerability. Surprises may manifest in ecological or environmental changes (e.g. an unanticipated drought or flood), or in social, political or economic changes (e.g. the 2007–8 food price crisis). The two types might of course be linked. Chapter 7 defines a number of 'generic' characteristics of drivers of food system vulnerability: the time frames over which drivers arise and their pace, magnitude and extent of impact; their level of predictability; the extent to which they are understood or recognized by knowledge systems;[1] their level of reversibility (including possible thresholds and tipping points); and the extent to which there are possibilities for adaptation. These characteristics clearly also have profound implications for the likelihood of a vulnerability outcome being anticipated – or the likelihood of surprise impacts. Chapter 7 also addresses some key drivers of food system vulnerability and provides a number of examples of food system vulnerabilities, including some that have come as surprises. (Some of the issues raised in Chapter 7 around climate, ecosystem services, 'knowledge systems' and predictability are of particular interest in the context of food system surprises and are briefly reconsidered in the discussions that follow.) Finally, human responses to GEC in the production component of food systems can also have unintended or poorly thought through consequences, which can come as surprises. An example noted in Chapter 7 is the impact on groundwater and aquifer reserves of high-yielding crops in many parts of the world, including Mexico, the USA, China, India and the Middle East (Brown, 2005), which, in turn, detrimentally impact on crop productivity and hence regional food production.

Climate

The viability of food systems depends heavily on environmental, notably climatic, conditions. Extreme events such as heat waves, storms, floods and drought are often felt as surprises (even though they are a 'normal' part of the climatology of any place). They can have devastating impacts on local as well as regional food systems, depending on their extent and the ability of people to cope. Climate change per se is no longer unanticipated, but despite the tremendous body of research on climate change, numerous uncertainties in environmental responses and their impacts on food systems remain, raising issues as to whether the past research focus needs to be changed.

Changes in the severity and frequency of extreme events and the effects of high-impact but low-probability events lead to surprises. These could have significant impacts on a range of food system activities, including food storage and distribution infrastructure as well as food production (Gregory et al, 2005).

Ecosystems and their services

One way in which ecosystem services can be compromised is through tipping points (see Chapter 5) and thresholds being reached, which can lead to negative surprises. An environmental and/or social manifestation of a surprise, or of its impacts, does not necessarily uncover causal factors, and these may be rooted in the links or feedbacks *between* two or more environmental or social food system elements or perturbations. An example might be that of increasing levels of HIV in a region coinciding with a climate shift towards one more suitable for vectors of malaria. (Research in southern Africa indicates that HIV-1 may increase malaria incidence by as much as 28 per cent and deaths by as much as 114 per cent (Korenromp et al, 2005).) Because of the complexity of the potential feedbacks involved, such a situation and its potentially far-reaching impacts on human health and food systems may be unpredictable. Thresholds or tipping points being reached in, for example, population dynamics, reduced biodiversity, food web complexity, diversity within functional groups, as well as decreases in the size of organisms (Simon, 2009), may lead to substantive shifts in ecosystem services and food systems. A system tipping point was arguably reached, for example, in fish stocks in the Northern Atlantic during 1992, when the collapse of cod stocks off the east coast of Newfoundland forced the Canadian government to close the fishery (see Box 20.1).

Box 20.1 *The collapse of cod stocks off the east coast of Newfoundland*

The collapse of cod stocks off the east coast of Newfoundland during 1992 forced the Canadian government to close the fishery. Job losses affected around 18,000 fishermen and some additional 30,000 employees in fish processing industries (Ruitenbeek, 1996). Following a booming industry during the 1950s, concern grew that cod catches in the Northern Atlantic, especially on the Newfoundland Grand Banks, were diminishing due to overfishing and disruption of habitats through the use of trailer nets. Although the exact reasons for the collapse of stocks in 1992 has been extensively debated among scientists (Ruitenbeek, 1996; Parsons and Lear, 2001), it is evident that high fish mortality due to excessive or unrealistic fishing quotas being set in policy was a major cause (Myers et al, 1996).

It was generally assumed that once the pressure of overfishing stopped, the system would be sufficiently resilient to self-restore and cod stocks would be replenished after a reasonable period. The 'environmental surprise' was that, contrary to expectations and although the ecosystem showed a capacity for restoration, it never returned to its earlier state. Cod stock disappeared, replaced by other species with less or no value for fishermen. A marine species, which was for generations an integral part of the region's food system, as a staple food for human consumption and a foundation for a strong cultural tradition throughout fishing communities from both sides of the Northern Atlantic, has now become a luxury commodity.

Different 'knowledge systems': Surprises for some but not for others

Events or changes might be anticipated by one set of actors but be experienced as a surprise by another. For example, while climate events may be predicted by science, this knowledge – or the ability to engage with it – may remain outside the realm of the farmers on which it has a direct impact. Kates and Clark (1996) succinctly refer to 'imaginable surprises', which they define as 'an event or process that departs from the expectations of some definable community, yet is a concept related to, but distinct from risk and uncertainty'. This extends to the more abstract notion of scientific and knowledge-generation paradigms, in which paradigmatic frames of reference have a powerful influence on the focus of events that are anticipated or taken seriously among multiple actors – including scientists from different disciplines, policy-makers with varying agendas, etc. A simple example here is the impact of political persuasion on policy choices – such as 'green' politics influencing choices that are perceived by some to be 'environmentally friendly' and by others to be economically foolish.

These surprises that are exogenous versus endogenous to the thinking paradigms (or knowledge systems) of actors within the food system may mean that decisions that build resilience or anticipatory action are taken up in highly varied ways across the food system.

It is clear that the resilience of food systems is integrally determined by the decisions and actions of people at all levels; thus the importance assigned to the 'science–decision-maker interface' in Figure 20.1 in shaping the anticipation, or the outcomes, of surprises (the science–decision-maker interface and adaptive policy-making are discussed below).

Predictability: 'Speculative possibilities'

Some changes or events may be speculated to be possible, but not certain; or as being certain, but their impacts are only considered as a possibility. These here labelled as 'speculative possibilities' and examples of these already abound in the context of food systems. One such 'speculative possibility' concerned the comprehensive assessment of the likely global environmental effects of a major nuclear conflict (>100Mt exchange) (Harwell and Hutchinson, 1985). A significant finding concerned the consequences on the global climate over the following months, triggering a predicted cooling of a few degrees Celsius over the entire northern hemisphere and the tropical zone, and a likely reduction of precipitation and solar radiation. The subsequent shortening and disruption of the growing season for the main crops (including wheat, corn and rice) would jeopardize agricultural production worldwide. Coupled with the likely collapse of all major food circulation and distribution systems, this would mean starvation for the majority of the world's population, in both combatant and non-combatant countries alike. Fortunately, this did not happen, perhaps in part because of the illustration of the consequences of uninformed action.

Many speculative possibilities have their roots in technological advances, which can be powerful drivers of change, and since change can have unforeseen consequences, technology can yield surprises. There is a long history of both positive and negative surprises and impacts as a result of technological change on commodity supply, consumption patterns, and on environmental, social and cultural systems. There are a number of aspects to technologically driven surprises and sometimes technology simply has unforeseen scope or even spinoffs. Examples are discussed below.

Other examples of speculative possibilities are the impact of biofuels on food systems (see Chapter 19; are predictive models sufficient, accurate?); and land degradation and desertification, the impacts of which generated widespread concern in the 1980s. The reality of land degradation is no surprise to many at a local level, but regional and global science and policy are still insufficiently linked to deal with the issue. Aside from immediate and direct threats to food production due to natural resource degradation, land degradation will have indirect implications for food security elsewhere through factors such as migration and conflict, and impacts on climate through, for example, changes in albedo and dust entrainment.

The multiple impacts of infectious diseases may also give rise to 'speculative possibilities'. An unfortunate example is the major social challenge of HIV and AIDS, which have the potential to drastically modify food security due to demographic changes in the workforce throughout the food system. Some of the food system vulnerabilities associated with HIV and AIDS are discussed in Chapter 7 and further highlighted in Box 20.2.

Box 20.2 *Impacts of HIV and AIDS as speculative surprises?*

We already know that there are strong linkages between ecosystems and HIV and AIDS; HIV is a profound driver of environmental changes in agriculture, conservation and land-use patterns (Drinkwater, 2005; Erskine, 2005). The environmental impacts of HIV accrue from factors such as interruption of knowledge transfer between generations about agricultural practices and conservation, and resulting declining institutional memory because it is the productive sector of the population which AIDS most effects. The impact on human capital (e.g. knowledge, labour) also has major impacts on the ability to engage in agriculture (affecting crop choices, the timing and effectiveness of agricultural practices, etc.), and in other food systems activities such as processing, distributing and retailing food. The kinds of livelihood choices that are made in the context of an HIV and/or AIDS affected household (or community) may include diversification of livelihood strategies (pursuing a range of activities to secure food and other basic needs), as well as de-diversification (frequently because assets such as livestock have to be disposed of in the face of competing needs) (Niehof, 2004). In the face of HIV and AIDS, both these strategies often represent 'coping' strategies that can have long-term implications for local agriculture but also more widely on food systems.

Some of the social and demographic impacts of HIV and AIDS are already being observed or can at least be predicted, such as rising current or future mortality among those historically or currently infected. Others are as yet unknown or very difficult to predict. For example, in South Africa what are the future social and economic ramifications for food systems of the higher mortality rates evident among women compared to men? How will falling life expectancies in sub-Saharan Africa shape food systems (and hence food security which they underpin) in terms of agricultural production, food distribution and food consumption? Will the family and community social capital resources currently central to food security of many communities (and which by extension are intrinsic to the food system) survive? What are the implications for food policy? These questions flag significant possibilities or surprises in the future of the food system that have not yet been considered or addressed.

How does 'food system resilience' alter the impacts of surprises?

In the context of social–ecological systems, resilience does not necessarily imply resistance to change or stasis, but also appropriate preparation, adaptation and the ability to capitalize on change while retaining beneficial function (e.g. Folke et al, 2002; Adger et al, 2005; Marshall and Marshall, 2007). To return to Figure 20.1: it may be possible to increase the anticipatory and response capacity of the system, so that the negative impacts of future 'nasty' surprises can be lessened. There may indeed be situations where system responses can be increased so that changes can be capitalized upon to produce positive outcomes.

Importantly, some surprises will not be predictable even with the best predictive tools, and here the role of scenario planning to build adaptive capacity becomes particularly crucial (see further discussion below and Box 15.3).

The important point in distinguishing 'resilience' from 'prediction' in terms of Figure 20.1 is that the existence or building of resilience is not *necessarily* based on prediction, nor does it depend on scientific knowledge, data analysis, or on information exchange between the scientist and the decision-maker. The focus of this discussion is not on separating 'resilience' and 'prediction', but rather to highlight what is meant by resilience, and to provide some tangible examples of 'robustness' or 'resilience' in food systems.

Resilience has been defined and discussed in Part II of this book. In this chapter, we take resilience to be a positive attribute as it enhances the ability of food systems to buffer disturbance and recover from surprises. The resilience of food systems rests not only on ecosystem resilience or social resilience, but includes interacting social–environmental factors. There is an established history of work on ecological resilience (Gunderson and Holling, 2002), but its use in human–environment relationships is more recent and more controversial (Adger et al, 2005; Galaz, 2005; Strauch et al, 2008; Thrush et al, 2009; Urich et al, 2009). Importantly, neither definition requires stasis in systems for resilience to apply. This means that food system resilience might be defined in terms of functionality and the food security outcomes of the food system, rather than the maintenance of existing food system activities (see Figure 2.1b). This has important implications for all food system actors, including researchers and decision-makers.

Food systems might be resilient to the potential negative impacts of GEC because of social actions or characteristics. At the village or community level, for example, a number of social factors have been found to increase food security and, by implication, food system resilience. These include community characteristics such as informal community safety nets, as well as strong institutional or social capital (Foster, 2007; Misselhorn, 2009). Frequently, resilience comprises interacting environmental and social factors.

Adaptation to climate change involves proactive or reactive actions aimed at reducing the negative, or capitalizing on the positive, impacts of anticipated climate changes. If adaptive action has positive and sustainable food security outcomes, such as the ability to cope with a drought, it might be considered to be enhancing the food system's resilience. In a recent illustration from the Umkhanyakude District of KwaZulu-Natal, South Africa, Oxfam Australia identified climate change and climate variability as critical issues for the development programming towards food security, particularly in the context of HIV and AIDS. Recommendations to reduce climate-related vulnerabilities covered a very broad range of project activities for adaptive or mitigative action. These included investigating the potential for development of agricultural techniques (e.g. rainwater harvesting, increasing planting distances between crops, introducing short-maturing varieties of maize and other crops) and socio-economic actions (e.g. appropriate livelihood diversification, and strengthening community level social capital resources and institutions) (Misselhorn, 2008a).

What is the role of monitoring and prediction in anticipating surprises and increasing resilience?

Anticipating surprises

Figure 20.1 suggests that strong mechanisms for monitoring and prediction in food systems would move the potential surprise events or changes towards being more predictable. Being able to anticipate what would otherwise be a surprise requires the development of knowledge about multiple food system components, and sets the stage to adapt, attenuate or even optimize unfolding change. For example, Aggarwal et al (2004) argue the case for improved early warning systems for food systems in the Indo-Gangetic Plain. Indeed, better prediction tools, such as marine ecosystem models, might have led, for example, to early anticipation of the apparently permanent impacts of overfishing on cod stocks in the Northern Atlantic (discussed in Box 20.1).

There are many examples of attempts of monitoring and modelling to avoid surprises, including monitoring seismic action for predicting volcanic eruptions, atmospheric and ocean states for predicting hurricanes and tsunamis, and predictions of forest fires by monitoring and predicting dryness of forests and weather conditions. Some of these have been more successful than others. In addition, there have been significant scientific advances in the last two decades in climate prediction from one to six months in advance to help decision-makers reduce risks associated with climate variability (Hansen et al, 2006).

From the perspective of systems science, monitoring requires the observation of the state variables of a specific system. Monitoring can be continuous, at set frequencies or irregular and opportunistic, or upon specific conditions being met. Monitoring is essential for the development of early warning systems (which may minimize surprises), as well as for the parameterization, calibration and validation of simulation models for predictions (to explore potential surprises of the systems behaviour). Monitoring of natural systems can be done by automated sensors that capture the physical or chemical characteristics such as nutrient levels in a lake or meteorological variables. However, for complex systems (such as food systems), which include both socio-economic and environmental components, efficient monitoring systems that capture the key variables of all the involved processes have to be more complex and need to include socio-economic 'sensors'. So, while a wide variety of tools and methods is available for the monitoring of biophysical variables (e.g. ground-based observation methods and remote-sensing methods for monitoring weather, land cover, land use, food production and a host of environmental indicators), the monitoring of many of the socio-economic variables that describe people's food security at any time is often more difficult and costly. This often relies on household surveys (such as those on poverty conducted by the World Bank), occasional government censuses and a wide array of rapid ground-based assessment methods used for assessing food security in particular places. Such assessment methods have included household-level indices such as the FAO's Aggregate Household Food Security Index (FAO, 1997), narrower approximations using dietary diversity scores (e.g. Drewnowski et al, 1997), and are often rooted in vulnerability analysis such as the Save the Children's UK Household Food Economy Approach (SCF-UK, 2000). In the future, technological development will offer considerable opportunities for cheaper and more effective monitoring of food systems, including the ubiquity of mobile phones and increasingly widespread access to the internet.

Increasingly powerful information technology offers the prospects of better collection, collation and assimilation of data for monitoring purposes. For instance, in relation to historical trend analysis and characterization of baselines for future monitoring, much more could be done with existing data, especially in situations where physical degradation of the resource (paper data sheets, etc.) is a substantial problem. There is a huge, essentially hidden, resource of existing data and information on indicators of vulnerability, such as clinic records that may not have been centrally collated. This is particularly the case in developing countries, and much could be done with this resource in terms of its analysis and reanalysis, but this is an activity that is not often seen (by researchers or by donors) as something that is worthwhile (Devereux, 2001).

Increased harmonization of monitoring projects that span both food system activities and outcomes is needed. A multitude of organizations is actively engaged in data collection on many aspects of the food system (although mainly related to food production), resulting in potential overlap, data redundancy and sheer inefficiency. These could be reduced via harmonization of effort, particularly related to core datasets needed for food security analysis. Currently some integrated data collection efforts are being undertaken, but much more could be done. Even some of the basic data (e.g. crop and livestock distribution data for many developing countries, and land cover data in general) are often missing or highly uncertain.

There is a substantial history of food security monitoring systems. One of the earliest was the FAO's Global Information and Early Warning System, established in 1975 to monitor global food supply and demand and which combines various remote sensing approaches (combined with modelling techniques) and socio-economic analyses. The 1980s saw the establishment of the well-known Famine Early Warning System Network (FEWSNET; see Box 10.4), funded by USAID. Similar development-orientated approaches have been adopted at more local levels for the purposes of assessing vulnerability. Oxfam Australia's work in KwaZulu-Natal, South Africa, mentioned above, employed a vulnerability analysis approach that reviewed and mapped development indicators for the district by way of reflecting relative vulnerability to food insecurity in light of the major development challenges in the province (including employment, poverty, malaria and HIV), superimposed on the future impacts of climate change (Misselhorn, 2008b). The ENSEMBLES project (Hewitt and Griggs, 2004) is another example of work aimed at providing policy-relevant information on climate change and its interactions with society. The project was launched to develop an ensemble climate forecast system for use across a range of temporal levels (seasonal, decadal and longer) and spatial levels (global, regional and local), with the aim of constructing probabilistic scenarios of future climate change and climate variability for quantitative risk assessments.

The efficacy of such risk management systems is, however, highly variable and in part depends on the developmental and political context in which information is produced and how and to whom it is disseminated. Producing information is only one part of the picture; who is engaged in deciding what kind of information is 'useful', what information is gathered and from whom, and how it is communicated, are all also critically important (e.g. Lemos, 2003, 2007; Murphy et al, 2007) (and see Chapter 10). One element of this context that has been highlighted in southern Africa is the danger of such systems being driven by a 'disaster response' cycle, rather than being part of a long-vision drive towards developing structural resilience in socio-economic conditions (Holloway, 2003). The mechanism of knowledge production – in this case the early warning system per se – is only one element of successful early warning systems; to be effective they additionally require an understanding of situational risk and uncertainty and how it is perceived; contextually relevant communication and dissemination; and the ability of actors and those affected to respond (ISDR, 2006).

A prerequisite and challenge for food system monitoring is a consistent system description in order to identify key processes and their linkages that need to be observed for the purposes of promoting sustainable food systems. It is also clear that the collection and analysis of information from a range of sources and from a range of perspectives is critical for both monitoring the state of a food system and making predictions about future developments that are up to the challenge of food system adaptation to GEC. The food system approach detailed and discussed in Chapter 2 offers an effective and robust concept to serve as a framework for monitoring all the key variables.

Increasing resilience

Monitoring and prediction play a key role in 'moving' changes from surprises to better anticipated events or processes, and models are an integral part of monitoring and prediction capacities. But people's ability to understand change, such as understanding system behaviour under various social or economic scenarios, undoubtedly also enhances food security. In other words, better monitoring and prediction capacities can also 'move' the impacts of surprises 'vertically' from negative towards positive in terms of Figure 20.1. As noted above, food security is shaped by myriad factors other than monitoring and prediction, but these can nevertheless help equip society to develop more adaptive food systems. An illustration of the importance of effective monitoring is in *adaptive* policy-making; without monitoring as a key component of an ideally iterative policy-making cycle, no learning and policy improvement can take place (Swanson et al, 2009).

Figure 20.2 provides a framework – a rationale – for the critical role of integrating disciplines to inform the kinds of learning and behavioural change (such as the learning associated with adaptive policy planning referred to above) that can lead to anticipating and preparing for surprises and building resilience and security in food systems. At the centre of Figure 20.2 is the

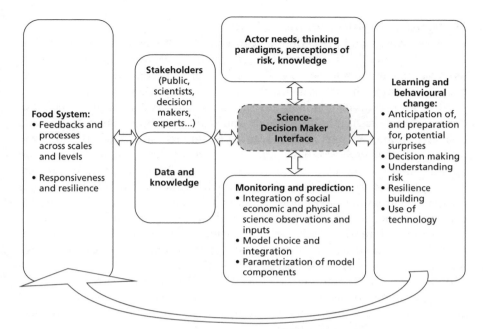

Figure 20.2 *Schematic showing the 'position' of the 'science–decision-maker interface' in better preparing for surprises in food systems through learning and behavioural change*

interface between 'science' and 'decision-maker', placed in such a way to note the two-way conduit of knowledge transfer and learning by all 'actors' in the food system (and see Chapter 10). These actors include not only scientists and policy-makers, but also wider stakeholders such as civil society and non-government organizations (see Chapter 11). Many of these are decision-makers in at least some aspects of the food system, and for whom learning and behavioural change are necessary to deal better with surprises and build resilience in food systems.

The interface between science and decision-making is discussed further below and is the subject of Chapter 10, but the 'instrument' of science, coalesced in Figure 20.2 in the box entitled 'monitoring and prediction', is a key element of an effective interface, and is used to encompass the full array of scientific tools. Scenario analysis (see Box 11.7) helps to highlight many of the key considerations leading to more effective understanding, learning and behavioural change.

In essence, through highly developed instruments and tools, research is capable of generating vast amounts of information. But for learning and behavioural change to take place such that food systems become responsive to anticipated change, as well as more resilient, this information must be developed with input from stakeholders, and matched to the needs of decision-makers. Learning is also a central element in the development of adaptive capacity and resilient food systems (e.g. Adger et al, 2005; Galaz, 2005).

Scenario planning and modelling

The 'Monitoring and prediction' box of Figure 20.2 may be unpacked a little further through a brief consideration of possible tools of prediction. A widely used tool in climate science, as well as the social and economic sciences, is scenarios planning. This is discussed further in Chapters 3, 11 and 15.

Effective scenarios planning for systems science is dependent on collaboration across scientific disciplines, and can tap both quantitative and qualitative data. A range of data sources for inputs into scenarios and storylines, and the understanding and selection of key driving forces and parameters for scenarios, is also seen to be critical. Importantly, however, scenarios planning arguably provides for an exploration of the future that falls outside the capabilities of more traditional models. Expert judgement, for example, is increasingly being used to evaluate risks and probabilities in the absence of quantitative models, or to provide an alternative understanding of the risks associated with applying probabilities to complex models with large uncertainties (Morgan and Keith, 1995; Keith, 1996; Lazo et al, 2000; Morgan et al, 2001; Van der Fels-Klerx et al, 2002; Arnell et al, 2005). The Delphi technique is one such method of synthesizing expert judgement, and can also synthesize inputs across multiple disciplines and sectors (Mitchell, 1991; Clayton, 1997; Misselhorn, 2006).

The integration of modelling with scenarios planning is a growth area of research, particularly with regard to quantifying future risks to food systems, as well as raising food system resilience. One way to pursue collaboration across disciplines is to design integrated model systems that explicitly parameterize the major components of food systems. Earth system models that include fully-coupled crop growth and development are one such emerging tool in relation to crop production (Osborne et al, 2007). These models could contribute to the analysis of system behaviour in order to assess risks to the functioning of the food system in the medium and long term. They could also provide instruments for developing and testing strategies to strengthen resilience. A significant challenge for such a modelling system is to retain sufficient complexity to both represent underlying processes and inherent uncertainties, while avoiding an excess of tuneable parameters. In practice, at least within climate and impacts modelling, limited computer power forces a tradeoff between complexity, spatial resolution and representation of uncertainty. Access to data for calibration and evaluation of models is extremely important in this context, since it can be used to constrain model parameters and thus ensure that models are getting the right answer for the right reason. Once a model is thus deemed reliable, it can be used to develop adaptation options for GEC, such as maintaining high crop productivity under climate change through a change in crop variety (Challinor, 2009). Risk-based methodologies of this type are also available, thus allowing the quantification of the risk of failed adaptation (Challinor et al, 2009a).

Increased integration of different models has already begun, with, for example, ensemble climate methods being used for a range of climate impacts such as health, hydrology and food production (Challinor et al, 2009b, 2009c). Integrated assessment models (IAMs) are more comprehensive still, since they include economic modelling (Fischer et al, 2005). However, IAMs are currently not specifically designed for the analysis of food systems, but rather for the impacts of climate change on agriculture. They thus often omit or oversimplify important drivers of surprises elsewhere in the food system, and often reduce human behaviour to economic behaviour.

The purposeful integration of natural and social sciences within a quantitative modelling framework is particularly challenging (McCown, 2002; Crane et al, 2009; Jakku and Thorburn, 2009; Meinke et al, 2009). Since the socio-economic and biophysical aspects of crop productivity, for example, have traditionally been examined and used separately and at different spatial levels, state-of-the-art methods from each of these fields are not often used in the same study, or at the same spatial or temporal level. Integrating models from different disciplines using quantitative methods may not always be possible or useful, so that a combination of quantitative and qualitative methods is needed. Further, while the need for integrated models exists, many efforts now use a suite of models to provide information that is then more broadly assimilated through discussions, scenarios, etc.

Model choice is further dependent on the target audience; sometimes the information best suited to users is based on computer-intensive systems (e.g. ensemble prediction systems, long-range climate change forecasting), while at other times less high-tech systems such as observational networks and capacity building are likely to be most beneficial. In Africa, for example, a prudent way to address the threat of climate change may be to focus on strategies for coping with climate variability (Washington et al, 2006). This may mean a greater focus on in-situ and remotely sensed observations, as well as consideration of the multiple stresses that act on food security (Haile, 2005; Verdin et al, 2005). In contrast, a greater focus on modelling may be needed at the regional level, possibly with a greater emphasis on longer-term resilience planning.

Consideration of potential approaches to prediction for some of the 'surprises' discussed in this chapter highlights this context-dependency. Clearly an integrated numerical model of potential surprises linked with HIV-AIDS and food systems (Box 20.2) is not possible. The case of overfishing of cod (Box 20.1) illustrates the ongoing dialogue between measurements, modelling and theory.

Finally, as with reflections above on risk management systems, models can only be viewed as one among many instruments that need to be drawn on in developing ways to build 'surprise-robust' food systems. Models may provide some information for part of the picture for some actors. Further, underlying structural resilience remains foundational to successful anticipation of, or responses to, surprises in the context of food systems.

What is the role of the 'science–decision-maker interface' in shaping surprises?

The many actors in food systems (individuals, community-based organizations, non-government organizations, government bodies, private sector and academia) (Chapters 11 and 18) are in a better position to mitigate negative outcomes of surprises and/or capitalize on change if their anticipatory and predictive capacities are enhanced and they are able to communicate more effectively among themselves. However, Figure 20.2 frames the importance of an interface between science and decision-maker if the necessary actors are able to take up relevant information in a timely manner; in other words, for modelling and prediction to have any 'real-world' impact. Many different activities may be needed at this interface, including information exchange, communication and learning. Figure 20.2 illustrates the role of the science–decision-maker interface in drawing on, and communicating, knowledge and information for the ultimate purpose of learning and behavioural change towards anticipation of and preparation for potential surprises, and more resilient food systems. Chapter 10 discusses the nature of this interface, while Chapter 11 discusses stakeholder engagement in practice.

What kind of information would flow at this interface?

Researchers need to translate scientific knowledge on environmental and socio-economic surprises and their effects on food security into policy-relevant information, and then into expert advice to support food management and policy decisions. This translation (and sometimes interpretation; see Chapter 10) is required to make the science credible to policy while ahead of policy demands, practical and – ultimately – operationally valuable. However, similarly there needs to be an exchange between decision-makers at multiple levels and researchers – not only to articulate needs, and to share 'world views', perspectives and paradigms, but also to open avenues of new information and data (such as traditional knowledge) for science. Unless a range of factors that influence decision-making is considered and communicated, research into risk-based thinking for food systems may fail to have a positive policy (or real world) impact. For example, researchers will often increase the predictability of the system without giving much consideration to influencing and changing human behaviour; the focus is often on research output rather than on impacts. Behavioural lessons often come later through iterative experience and lessons learned. Ultimately it is behavioural and institutional change that builds food security, or allows the opportunities of change to be capitalized upon. Not least among these processes is exchange between different knowledge systems, such as between traditional or indigenous local knowledge and scientific knowledge, in order to develop behaviour that can cope with future change (Blanco, 2006).

What is the role of technology in generating, anticipating and preparing for food system surprises?

Technology fulfils multiple roles in all food system activities, from commodity supply (e.g. high-yielding crop varieties) to consumption patterns (fast food and ready meals). Given the great range of technologies, impacts may be positive or negative, or often both. Technology is a factor that cuts across all elements of Figures 20.1 and 20.2; it is a driver of surprises as well as resilience, and it is fundamental to most modelling and predicting of the impacts of change.

There are a number of disparate technological advances that might also play a role in building the resilience of food systems. These include advances in production systems (e.g. drought tolerance in crops, disease tolerance in cattle, genetic engineering), altered distribution and marketing systems, mobile phone technology for up-to-date commodity prices, and nano- and bio-technology related to food processing and increasing shelf life. While these are all being developed with clear goals in mind, they might also bring surprises.

Many technological impacts take the form of complete surprises. Some are the surprises caused by unforeseen spinoffs from other technology development. Then there are surprises caused by the unforeseen consequences of using (or not using) technology; an example here is the impact of stopping the use of DDT, with the result of increasing the incidence of malaria in many countries in Africa (caused by surges in the development of mosquitoes in standing water, which before had been controlled). There are also technological surprises caused by path dependency. An often-cited example is the case of videocassette recorders (VCRs) for home use: Sony's Betamax format was generally regarded as being better (higher quality), but the format was defeated in market competition by VHS in the 1980s because of 'bandwagon' and 'network' effects, leading to complete vendor lock-in to VHS (Cusumano et al, 1992). There are many other examples of path dependency from economics, history, software and biology, and lock-in to specific technology can be a powerful source of inertia in systems that are changing.

Technology has a considerable role to play in the development of better prediction systems. Methods in biotechnology are leading to huge gains in understanding the location and function of specific genes, and this understanding is being incorporated into increasingly realistic and comprehensive models. Genome maps for plant and animal species open possible advances in evolutionary biology, crop and animal breeding (as in genomic selection), and animal models for human diseases (Lieschke and Currie, 2007; Lewin, 2009). At the same time, developments in computing continue unabated, and notions such as distributed computing and cloud computing, coupled with rapidly falling costs of data storage, mean that for many applications, computing and storage demands are no longer any fundamental constraint. These are all things that could lead to substantial improvements in prediction systems in the future as well as to surprises.

Technology will continue to have substantial impacts on the monitoring of food systems. Cheaper remote sensing data, coupled with improved spatial and temporal resolutions, are already affecting the availability of land-use and land-cover information, in addition to climate and weather monitoring and many other applications. Internet-based tools, such as Google maps with appropriate add-ons, are already being used for participatory, large-scale information gathering and collation, and these could be expanded in the future to help with activities such as 'ground-truthing' of land-cover maps, disease surveillance and food security earlywarning (Fritz et al, 2009). Another technology that could have enormous impacts on many aspects of food systems is radio-frequency identification (RFID). This is an electronic tag that is attached to a product or an animal and is becoming increasingly prevalent as the price of the technology decreases; it is already widely used in the USA by supermarkets, food transportation firms and food suppliers to improve the logistics of food supply and distribution, to improve food safety and to better understand and cater for consumer demand (e.g. Jedermann and Lang, 2007; Estrada-Flores, 2008; Wen et al, 2009). The tracking of food products could also form the basis for providing consumers with information concerning the origin, age, and carbon, energy and water footprints of specific products.

Technology could also have an increasing role to play in the science–policy interface. E-learning technology such as web-based distance education is already widely available, but there are several different types of e-learning being developed that could revolutionize education in the future (e.g. Chatti et al, 2007). One example is innovative re-useable learning modules developed for learning in sustainable watershed management and also agriculture (Grunwald, 2008; Chandrasekan et al, 2009). There is considerable research on communication tools (such as visualization tools and virtual reality games) that could lead to better understanding the processes of decision-making and how science can more effectively be inserted into these processes (Sheppard, 2005).

The interplay between surprises, resilience, prediction, technology and the science–decision-maker interface in the case of global phosphorous depletion is illustrated in Box 20.3.

Box 20.3 *The case of global phosphorous depletion*

The era of relatively cheap fertilizers is often associated with cheap energy, and the impact of increases in energy costs on fertilizer cost and food prices is well documented (Mitchell, 2008). Crop production is, however, dependent not only on nitrogenous fertilizers, but also on phosphorus (in addition to some other nutrients). Phosphorus for fertilizer production is derived from phosphate rock, which – unlike nitrogen (which can be extracted from the atmosphere) – is a non-renewable and limited resource. Most phosphorus fertilizer is used for food production, although increasing amounts are being used for crop-based biofuel production. There is no substitute for phosphorus in food production. The demand for phosphorus is projected to increase by 50–100 per cent by 2050, and a recent study indicates that peak phosphorus production may occur by 2030 or so, and that current global reserves may be depleted in 50–100 years (Cordell et al, 2009). The quality of the remaining phosphate rock stock is also decreasing and production costs are increasing; fertilizer prices are bound to rise, with consequent negative impacts on smallholder farmers in many developing countries (Brunori et al, 2008). Future access to phosphorus has received very little attention in recent global assessments, and Cordell et al (2009) make a compelling case for including long-term phosphorus scarcity on the priority agenda for global food security. There are obvious geopolitical issues involved because phosphate rock reserves are under the control of only a few countries such as Morocco,

China and the USA; Western Europe and India, for example, are dependent on imports.

Current food production systems are far from being resilient to phosphorus depletion; food and feed price increases are inevitable as stocks and/or crop yields decline (given that there is no known substitute), with impacts across the board on consumers, whether rich or poor. However, there are several technologies that could help to meet future phosphate fertilizer needs for global food production, related to recycling by recovery of phosphorus from municipal and other waste products. The re-use of waste water for agricultural purposes carries serious potential health risks, but urine (via urine-diverting toilets to separate it from faeces) can be safely used as it is essentially sterile, and it could provide more than half the phosphorus required to fertilize cereal crops (WHO, 2006).

There are other options that could reduce the demand for phosphorus in food production. The efficiency of phosphorus use in cropping systems could be increased in various ways, such as through precision agriculture, the addition of microbial inoculants to increase soil phosphorus availability, and more efficacious application of fertilizer that minimizes unnecessary accumulation in soils and runoff to water bodies, for instance. Some of these technologies need further development, but some can be applied now.

While options exist for recycling phosphorus and reducing demand, the problem first needs to be acknowledged and systemic implications assessed before it can be acted upon. Technology can play a role in this assessment as well as in finding solutions to the problem, and in the monitoring of the phosphorus situation, which will need to be undertaken if targets are to be defined and acted upon. There are also institutional issues that will need to be addressed, as there are no international organizations or governance structures whose focus is the long-term, equitable use and management of phosphorus resources in the global food system. Finally, enhanced communication and learning activities will be needed to ensure that the phosphorus issue is fully internalized in policy and socio-cultural debates in both developed and developing countries. Without these various considerations, the phosphorus issue could turn into a problem with extremely serious consequences for global food systems in the coming decades.

Looking ahead: Considerations for science and society

Learning and behavioural change ultimately allows for appropriate societal responses in the context of food system surprises. Learning and behavioural change is a product of effective interactions between food system stakeholders. This interaction necessarily includes the science–decision-maker interface, as well as inter-disciplinary research. The latter presents both great challenges and great rewards, with many recent advances in GEC research coming from such endeavours (Leemans et al, 2009). New thinking and new tools are required in order to ensure that we move beyond the current paradigm of multidisciplinary research (which is often based on the sum of the component disciplines) to a more interdisciplinary approach (which requires an active interaction between disciplines) (Meinke et al, 2009).

Within GEC research, a paradigm has also been called for whereby sustainability is tackled in a problem-based fashion and science is made society-oriented (Robinson, 2008). Barriers to such an approach include the influence of 'world view' and disciplinary training. This leads to varying perceptions of risks that are associated with different thinking paradigms or frames of reference, as highlighted in Figure 20.2 in the box labelled 'Actor needs, thinking paradigms, perceptions of risk, knowledge'.

Interdisciplinary work is also central to the incorporation of the multiple variables of key food system processes in early warning systems (to minimize surprises), as well as for the parameterization, calibration and validation of simulation models for predictions (to explore potential surprises of the systems behaviour). There is arguably a call for investment in monitoring systems to allow for a wider collaboration among scientists in them.

There is also a need for further investment in developing and evaluating models that can better address uncertainty in food systems, so that the probabilities of events can be assessed better, particularly extreme events in the tails of distributions. Models currently used in assessments of economic and environmental impacts are often deterministic. There are a few recent examples of models that predict outcomes as probability distributions in different fields, including hydrology and crop models (Monod et al, 2006; Challinor et al, 2009b, 2009c; Muñoz-Carpena et al, 2010). However, much more needs to be done to emphasize uncertainties in parameters and model structure and how these translate to predicted output distributions. Bayesian and Markov chain–Monte Carlo methods that are used for estimating parameters in recent modelling studies are helpful in characterizing uncertainties, but these are not widely used at this time (Wallach et al, 2006; Challinor et al, 2009a; Iizumi et al, 2009; He et al, 2010).

Another major limitation of most models for anticipating surprises is that they may address only individual components of the food system and thus are unable to analyse the interactive effects and feedbacks among components. As these are important determinants of thresholds and tipping points (Betts, 2005), there is a need to integrate environment and food system models to increase the ability of impact assessments. However, which models should be integrated, at what spatial level, and at what degree of detail? Combinations of models could lead to complexities that make their use impractical. Furthermore, the incorporation of uncertainty into integrated models in order to predict distributions of outcomes needs to be considered. There is a major research need to determine best methods for model integration in order to address food security and environmental impacts at a range of temporal and spatial levels.

Finally, there is a need to assess the role of technology in the broader context of 'surprises'. The positive outcomes of technologies are usually already intended in their development, but decision-making does not necessarily take into consideration assessments of hidden – and potentially negative – impacts. The potential for unforeseen negative aspects of given technologies need to be assessed against the broader context of food system components and feedbacks, and scenario analysis can help in this regard.

Note

1   'Knowledge system' here is used to refer to the knowledge and perceptions generated, framed and held by a particular social group; such groups might include
indigenous groups, civil society, minority social groups, politicians and scientists.

References

Adger, W. N. (2006) 'Vulnerability', *Global Environmental Change,* 16, 268–81

Adger, W. N., T. P. Hughes, C. Folke, S. R. Carpenter and J. Rockstram (2005) 'Social-
ecological resilience to coastal disasters', *Science (New York, N.Y.),* 309, 5737,
1036–39

Aggarwal, P. K., Joshi, P. K., Ingram, J. S. I. and Gupta, R. K. (2004) 'Adapting food
systems of the Indo-Gangetic plains to global environmental change: key information needs to improve policy formulation', *Environmental Science & Policy,* 7,
487–98

Arnell, N. W., E. L. Tompkins and W. N. Adger (2005) 'Eliciting information from
experts on the likelihood of rapid climate change', *Risk Analysis,* 25, 1419–31

Betts, R. (2005) 'Integrated approaches to climate-crop modelling: needs and challenges', *Philosophical Transactions of the Royal Society Biological Sciences,* 360,
2049–65

Blanco, A. V. R. (2006) 'Local initiatives and adaptation to climate change', *Disasters,*
30, 1, 140–47

Brown, L. R. (2005) *Outgrowing the Earth: The Food Security Challenge in an Age of
Falling Water Tables and Rising Temperatures,* London, Earthscan

Brunori, G., J. Jiggins, R. Gallardo and O. Schmidt (2008) New challenges for agricultural research: climate change, food security, rural development, agricultural
knowledge systems. *The Second SCAR Foresight Exercise, Synthesis Report,*
Brussels, EU Commission Standing Committee on Agricultural Research (SCAR)

Challinor, A. J. (2009) 'Developing adaptation options using climate and crop yield
forecasting at seasonal to multi-decadal timescales', *Environmental Science and
Policy,* 12, 4, 453–65

Challinor, A. J., T. R. Wheeler, D. Hemming and H. D. Upadhyaya (2009a) 'Crop
yield simulations using a perturbed crop and climate parameter ensemble: Sensitivity
to temperature and potential for genotypic adaptation to climate change', *Climate
Research,* 117–27

Challinor, A. J., F. Ewert, S. Arnold, E. Simelton and E. Fraser (2009b) 'Crops and
climate change: progress, trends, and challenges in simulating impacts and informing adaptation', *Journal of Experimental Botany,* 60, 10, 2775–89

Challinor, A. J., T. Osborne, A. Morse, L. Shaffrey, T. Wheeler and H. Weller (2009c)
'Methods and resources for climate impacts research: achieving synergy', *Bulletin of
the American Meteorological Society,* 90, 6, 825–35

Chandrasekaran, K., V. Pirabu and S. Grunwald (2009) Electronic agriculture.
EcoLearnIT Reusable Learning Object System. http://ecolearnit.ifas.ufl.edu/
viewer.asp?rlo_id=241 (accessed: undated)

Chatti, M. A., M. Jarke and D. Frosch-Wilke (2007) 'The future of e-learning: a shift
to knowledge networking and social software', *International Journal of Knowledge
and Learning,* 3, Nos 4/5

Clayton, M. J. (1997) 'A technique to harness expert opinion for critical decision making tasks in education', *Education Psychology,* 17, 4, 373–87

Cordell, D., J.-O. Drangert and S. White (2009) 'The story of phosphorus: Global food
security and food for thought', *Global Environmental Change,* 19, 2, 292–305

Crane, T., C. Roncoli, J. Paz, N. Breuer, K. Broad, K. Ingram and G. Hoogenboom (2009) 'Forecast skill and farmers' skills: Seasonal climate forecasts and agricultural risk management in the southeastern United States', *Weather, Climate, and Society*, in press

Cusumano, M. A., Y. Mylonadis and R. S. Rosenbloom (1992) 'Strategic manoeuvring and mass-market dynamics: The triumph of VHS over Beta', *The Business History Review*, 66, 1, 51–94

Devereux, S. (2001) Food security information systems. In Devereux, S. and S. Maxwell (eds) *Food Security in Sub-Saharan Africa*, London, ITDG Publishing

Drewnowski, A., S. Ahlstrom Henderson, A. Driscoll and B. Rolls (1997) 'The dietary variety score: Assessing dietary quality in healthy young and older adults', *Journal of American Dietetic Association*, 97, 266–71

Drinkwater, M. (2005) 'HIV/AIDS and agriculture in southern Africa: What difference does it make?', *IDS Bulletin New Directions for African Agriculture*, 36, 2, 36–40

Erskine, S. (2005) *HIV/AIDS and Ezemvelo KZN Wildlife*, Durban, Health Economics and HIV/AIDS Research Division, University of Kwa-Zulu Natal

Estrada-Flores, S. (2008) 'Technology for temperature monitoring during storage and transport of perishables', *Chain of Thought – The Newsletter of Food Chain Intelligence*, 2–5

FAO (1997) *Agriculture, Food and Nutrition for Africa. A Resource Book for Teachers of Agriculture*, Rome, FAO

Fischer, G., M. Shah, F. N. Tubiello and H. van Velhuizen (2005) 'Socio-economic and climate change impacts on agriculture: an integrated assessment, 1990–2080', *Philosophical Transactions of the Royal Society Biological Sciences*, 360, 2067–83

Folke, C., S. Carpenter, T. Elmqvist, L. Gunderson, C. S. Holling, B. Walker, J. Bengtsson, F. Berkes, J. Colding, K. Danell, M. Falkenmark, L. Gordon, R. Kasperson, N. Kautsky, A. Kinzing, S. Levin, K. G. Mäler, F. Moberg, L. Ohlsson, P. Olsson, E. Ostrom, W. Reid, J. Rockström, H. Savenije and U. Svedin (2002) Resilience and Sustainable Development: Building Adaptive Capacity in a World of Transformations. *Scientific Background Paper on Resilience for the process of the World Summit on Sustainable Development on behalf of The Environmental Advisory Council to the Swedish Government*

Foster, G. (2007) 'Under the radar: Community safety nets for AIDS-affected households in sub-Saharan Africa', *AIDS Care*, 19, 54–63

Fritz, S., I. McCallum, C. Schill, C. Perger, R. Grillmayer, F. Achard, F. Kraxner and M. Obersteiner (2009) 'The use of crowdsourcing to improve global land cover', *Remote Sensing*, 1, 3, 345–54

Galaz, V. (2005) 'Social-ecological resilience and social conflict: institutions and strategic adaptation in Swedish water management', *Ambio*, 34, 7, 567–72

Gregory, P. J., J. S. I. Ingram and M. Brklacich (2005) 'Climate change and food security', *Philosophical Transactions of the Royal Society B*, 360, 1463, 29 November 2005, 2139–48

Grunwald, S. (2008) Sustainable watershed management: EcoLearnIT Reusable Learning Object Learning System. http://ecolearnit.ifas.ufl.edu/viewer.asp?rlo_id =87 (accessed: 10 May 2010)

Gunderson, L. H. and C. S. Holling (eds) (2002) *Panarchy: Understanding Transformations in Human and Natural Systems*, Washington and London, Island Press

Haile, M. (2005) 'Weather patterns, food security and humanitarian response in sub-Saharan Africa', *Philosophical Transactions: Biological Sciences*, 360, 1463, 2169–82

Hansen, J., A. J. Challinor, A. Ines, T. R. Wheeler and V. Moron (2006) 'Translating climate forecasts into agricultural terms: advances and challenges', *Climate Research*, 33, 27–41

Harwell, M. A. and T. C. Hutchinson (eds) (1985) *Environmental Consequences of Nuclear War*, Chichester, Wiley

He, J., J. W. Jones, W. D. Graham and M. D. Dukes (2010) 'Influence of likelihood function choice for estimating crop model parameters using the generalized likelihood uncertainty estimation method', *Agricultural Systems*, accepted

Hewitt, C. D. and D. J. Griggs (2004) 'Ensembles-based predictions of climate changes and their impacts', *Eos*, 85, 566

Holloway, A. (2003) 'Disaster risk reduction in Southern Africa: Hot rhetoric, cold reality', *African Security Review*, 12, 1

Iizumi, T., M. Yokozawa and M. Nishimori (2009) 'Parameter estimation and uncertainty analysis of a large-scale crop model for paddy rice: Application of a Bayesian approach', *Agricultural and Forest Meteorology*, 149, 2 February 2009, 333–48

ISDR (2006) *Developing Early Warning Systems: A Checklist*, EWC III Third International Conference on Early Warning: From concept to action, 27–29 March 2006, Bonn, Germany

Jakku, E. and P. Thorburn (2009) 'A conceptual framework for guiding the participatory development of agricultural decision support systems', *Socio-Economics and the Environment in Discussion (SEED) CSIRO Working Paper*, Clayton South, CISRO

Jedermann, R. and W. Lang (2007) 'Semi-passive RFID and beyond: steps towards automated quality tracing in the food chain', *International Journal of Radio Frequency Identification Technology and Applications*, 1, 3, 247–58

Kates, R. W. and W. C. Clark (1996) 'Environmental surprise: expecting the unexpected', *Environment*, 38, 28–34

Keith, D. W. (1996) 'Assessing uncertainty in climate change and impacts – When is it appropriate to combine expert judgments? An editorial essay', *Climatic Change*, 33, 2, 139–43

Korenromp, E. L., B. G. Williams, S. J. De Vlas, E. Gouws, C. F. Gilks, P. D. Ghys and B. L. Nahlen (2005) 'Malaria attributable to the HIV-1 epidemic, sub-Saharan Africa', *Emerging Infectious Diseases*, 11, 9, 1410–19

Lazo, J. K., J. C. Kinnell and A. Fisher (2000) 'Expert and layperson perceptions of ecosystem risk', *Risk Analysis*, 20, 2, 179–93

Leemans, R., G. Asrar, A. Busalacchi, J. Canadell, J. Ingram, A. Larigauderie, H. Mooney, C. Nobre, A. Patwardhan, M. Rice, F. Schmidt, S. Seitzinger, H. Virji, C. Vörösmarty and O. Young (2009) 'Developing a common strategy for integrative global environmental change research and outreach: the Earth System Science Partnership (ESSP)', *Current Opinion in Environmental Sustainability*, 1, 1, 4–13

Lemos, M. C. (2003) 'A tale of two policies: The politics of climate forecasting and drought relief in Ceara, Brazil', *Policy Sciences*, 36, 101–23

Lemos, M. C. (2007) Drought, governance and adaptive capacity in North East Brazil: A case study of Ceará. *Human Development Report 2007/2008. Fighting climate change: Human solidarity in a divided world*, New York, UNDP

Lewin, H. A. (2009) 'It's a bull's market', *Science*, 323, 478–79

Lieschke, G. J. and P. D. Currie (2007) 'Animal models of human disease: zebrafish swim into view', *Nature Reviews Genetic*, 8, 353–67

Marshall, N. A. and P. A. Marshall (2007) 'Conceptualizing and operationalizing social resilience within commercial fisheries in northern Australia', *Ecology and Society*, 12, 1

McCown, R. L. (2002) 'Changing systems for supporting farmers' decisions: problems, paradigms, and prospects', *Agricultural Systems*, 74, 179–220

Meinke, H., S. M. Howden, P. C. Struik, R. Nelson, D. Rodriguez and S. C. Chapman (2009) 'Adaptation science for agriculture and natural resource management – urgency and theoretical basis', *Current Opinion in Environmental Sustainability*, 1, 69–76

Misselhorn, A. (2006) Food Insecurity in Southern Africa: Causes and emerging response options from evidence at regional, provincial and local scales. Unpublished Doctoral Thesis, University of the Witwatersrand, Johannesburg

Misselhorn, A. (2008a) *Adapting to Climate Change in Umkhanyakude District, Kwa-Zulu Natal, South Africa*, Carlton, Victoria, Oxfam Australia

Misselhorn, A. (2008b) *Vulnerability to Climate Change in Umkhanyakude District, KwaZulu-Natal, South Africa*, Carlton, Victoria, Oxfam Australia

Misselhorn, A. (2009) 'Is a focus on social capital useful in considering food security interventions? Insights from KwaZulu-Natal', *Development Southern Africa*, 26, 2, 189–208

Mitchell, D. (2008) A note on rising food prices, *Policy Research Working paper 4682*, Washington, DC, World Bank

Mitchell, V. W. (1991) 'The Delphi Technique: An exposition and application', *Technology Analysis and Strategic Management*, 3, 4, 333–58

Monod, H., C. Naud and D. Makowski (2006) Uncertainty and sensitivity analysis for crop models. In Wallach, D., D. Makowski and J. W. Jones (eds) *Working with Dynamic Crop Models*, Amsterdam, Elsevier

Morgan, M. G. and D. W. Keith (1995) 'Climate change – Subjective judgments by climate experts', *Environmental Science and Technology*, 29 A468–76

Morgan, M. G., L. F. Pitelka and E. Shevliakova (2001) 'Elicitation of expert judgments of climate change impacts on forest ecosystems', *Climatic Change*, 49, 279–307

Muñoz-Carpena, R., G. A. Fox and G. J. Sabbagh (2010) 'Parameter importance and uncertainty in predicting runoff pesticide reduction with filter strips', *Journal of Environmental Quality*, 39, 1, 1–12

Murphy, K., C. Packer, A. Stevens and S. Simpson (2007) 'Effective early warning systems for new and emerging health technologies: Developing an evaluation framework and an assessment of current systems', *International Journal of Technology Assessment in Health Care*, 23:3, 3, 324–30

Myers, R. A., J. A. Hutchings and N. J. Barrowman (1996) 'Hypotheses for the decline of cod in the North Atlantic', *Marine Ecology Progress Series*, 138, 293–308

Niehof, A. (2004) The significance of diversification for rural livelihood systems, *Food Policy*, 29, 4, 321–38

O'Brien, K. L., R. M. Leichenko, U. Kelkar, H. Venema, G. Aandahl, H. Tompkins, A. Javed, S. Bhadwal, S. Barg, L. Nygaard and J. West (2004) 'Mapping vulnerability to multiple stressors: climate change and globalization in India', *Global Environmental Change*, 14, 303–13

Osborne, T. M., D. M. Lawrence, A. J. Challinor, J. M. Slingo and T. R. Wheeler (2007) 'Development and assessment of a coupled crop-climate model', *Global Change Biology*, 13, 169–83

Parsons, L. S. and W. H. Lear (2001) 'Climate variability and marine ecosystem impacts: a North Atlantic perspective', *Progress In Oceanography*, 49, 1–4, 167–88

Robinson, J. (2008) 'Being undisciplined: Transgressions and intersections in academia and beyond', *Futures*, 40, 1, 70–86

Ruitenbeek, H. J. (1996) 'The great Canadian fishery collapse: some policy lessons', *Ecological Economics*, 19, 2, 103–6

SCF-UK (2000) *Riskmap Report, Swaziland*, London, Save the Children Fund UK

Sheppard, S. R. J. (2005) 'Landscape visualisation and climate change: the potential for influencing perceptions and behaviour', *Environmental Science and Policy*, 8, 637–54

Simon, F. T. (2009) 'Forecasting the limits of resilience: integrating empirical research with theory', *Proceedings B of the Royal Society*, 276, 3209–17

Strauch, A. M., J. M. Muller and A. M. Almedom (2008) 'Exploring the dynamics of social-ecological resilience in East and West Africa: Preliminary evidence from Tanzania and Niger', *African Health Sciences*, 8, Special Edition, S28–35

Swanson, D., S. Barg, S. Tyler, H. D. Venema, S. Tomar, S. Bhadwal, S. Nair, D. Roy and J. Drexhage (2009) Chapter 2: Seven guidelines for policy-making in an uncertain world. In Swanson, D. and S. Bhadwal (eds) *Creating Adaptive Policies: A Guide for Policy-making in an Uncertain World*, London, Sage/IDRC

Thrush, S. F., J. E. Hewitt, P. K. Dayton, G. Coco, A. M. Lohrer, A. Norkko, J. Norkko and M. Chiantore (2009) 'Forecasting the limits of resilience: integrating empirical research with theory', *Proceedings B of the Royal Society*, 276, 1671, 3209–17

Urich, P. B., L. Quirog and W. G. Granert (2009) 'El Nino: an adaptive response to build social and ecological resilience', *Development in Practice*, 19, 6, 766–76

Van der Fels-Klerx, I. H. J., L. H. J. Goossens, H. W. Saatkamp and S. H. S. Horst (2002) 'Elicitation of quantitative data from a heterogeneous expert panel: Formal process and application in animal health', *Risk Analysis*, 22, 1, 67–81

Verdin, J., C. Funk, G. Senay and R. Choularton (2005) 'Climate science and famine early warning', *Philosophical Transactions of the Royal Society B: Biological Sciences*, 360, 1463, 2155–68

Wallach, D., D. Makowski and J. W. Jones (2006) *Working with Dynamic Crop Models: Evaluation, Analysis, Parameterization, and Applications*, Amsterdam, Elsevier

Washington, R., M. Harrison, D. Conway, E. Black, A. Challinor, D. Grimes, R. Jones, A. Morse, G. Kay and M. Todd (2006) 'African climate change: taking the shorter route', *Bulletin of the American Meteorological Society*, October 2006, 1355–66

Wen, L., S. Zailani and Y. Fernando (2009) Determinants of RFID 'Adoption in supply chain among manufacturing companies in China: A discriminant analysis', *Journal of Technology Management & Innovation*, 4, 1, 22–32

WHO (2006) Guidelines for the safe use of wastewater, excreta and greywater. *Excreta and Greywater Use in Agriculture Volume 4*, Geneva, World Health Organisation

21
Part V: Main Messages

1 The close interactions among increasingly globalized food commodity markets have accelerated the transfer of risk between multiple regions and communities; global environmental change will increase the potential risk of associated food-related violence. Current institutional frameworks are either inadequate or inflexible to alleviate this risk. Three contrasting dominant discourses about food need to be resolved: food as a global commodity; food as a product of environmental services; and food security as a basic human right. The tensions between these discourses are embodied in the current institutional arrangements for food systems, setting the scene for potential conflict. In regions of the world that are already food insecure these tensions can – and do – spill over into violence. New forms of governance are urgently needed to better manage these tensions as both globalization and global environmental change continue apace.

2 The emergence of many private sector and non-state actors in food systems suggests that new hybrid forms of food system governance are emerging. Thus the roles of key food system actors are being redefined, particularly in the context of global environmental change; some of these roles are potentially contradictory. In general, there is increasing interdependence between public and private forms of authority, due to a flattening and fragmentation of regulation and the subsequent rise of multinational corporations.

3 Hybrid forms of governance including networks of supply chains and NGOs are becoming the norm. Unresolved questions are 'how well will these new governance forms accommodate the enhanced risk of violence created by global environmental change?' and 'to what extent can these hybrid forms of governance accommodate increased risk, maintain ecosystem services and enhance food security at multiple levels of governance and over time?'.

4 At present, the twin goals of feeding 9 billion people and lowering the environmental footprint of food systems are only weakly linked and at times conflicting in practice. The multiple pathways to achieve greater synergy

between enhanced food security and environmental outcomes require more coordination than exists at present. Three key challenges include:

(i)   improving the understanding of interactions of food systems operating at local through regional levels with global food requirements;
(ii) addressing mismatches between the capacity of current institutions to manage for both food security and environmental goals; and
(iii) creating an approach to respond to these issues that is sufficiently sophisticated and nuanced, but not so complex as to be unachievable.

5 Good judgement and choices about tradeoffs and synergies between and within food security and environmental outcomes will be needed. These relate to food choices and diets; food institutions and governance decisions; and food system management and technologies.

6 Surprises usually impact food systems deleteriously. They fall into a number of types: climate and ecosystem surprises; those caused by differences in knowledge systems (and priorities) across scales and levels; and speculative possibilities that cannot be predicted. Building greater resilience in food systems is the best way to minimize the negative impacts of such surprises. A key approach to building such resilience is bringing together a range of science disciplines and decision-makers, so as to co-create the potential for learning and behavioural change. Modelling is a useful tool to help understand the complexity of systems and identify potential options for enhancing their resilience, including the use of new monitoring and predicting technologies.

22
Reflections on the Book

Thomas Rosswall

In 2001, the 'Challenges of a Changing Earth: Global Change' Open Science Conference resulted in the Amsterdam Declaration, which stated that the participants 'recognise that, in addition to the threat of significant climate change, there is growing concern over the ever-increasing human modification of other aspects of the global environment and the consequent implications for human well-being. Basic goods and services supplied by the planetary life support system, such as food, water, clean air and an environment conducive to human health, are being affected increasingly by global change', and 'a new system of global environmental science is required'. As a consequence, the four signatories to the Declaration, the global change programmes sponsored by the International Council for Science (ICSU), DIVERSITAS, IGBP, IHDP and WCRP, agreed to form the Earth System Science Partnership (ESSP) to, through scientific research, elucidate how significant components of the life support system of Planet Earth will be affected by global environmental change (GEC) processes. The ESSP established a few key projects integrating the best natural and social science competencies of its parent organizations to address issues of GEC and human well-being. Thus, projects were developed to consider water, human health, carbon and food. Results from the food project, Global Environmental Change and Food Systems (GECAFS), complemented by results from many other food security-related projects, are reported in this book.

The need for a truly interdisciplinary view of the links between global change processes and human well-being was also stressed by ICSU in its report to the World Summit on Sustainable Development (WSSD, Johannesburg, 2002) (ICSU, 2002), on behalf of the international science and technology community. In its key statement, ICSU committed the international science community to addressing the challenge of integrating the environmental, social and economic dimensions in its programmes. In preparing for the conference, Kofi Annan challenged the government delegations to address five key issues: water, energy, health, agriculture and biodiversity (WEHAB). The ESSP Joint Projects thus address the key issues of sustainable development as identified by the UN Secretary-General, with biodiversity being addressed by DIVERSITAS.

However, addressing GEC issues *and* development challenges in a research context is not a simple task. Development needs to be addressed in a local context and global issues must be understood at the level of Planet Earth. In addition to bringing together the natural and social sciences communities, it was also necessary to provide a platform where GEC and development researchers could meet. GECAFS provided the first, and very successful, example of how such bridges can be built. By deciding to focus on the regional level, it provided the necessary context for addressing GEC and development issues within a synthetic framework.

At WSSD, ICSU also argued that the scientists needed to climb down from their ivory towers and engage in societal issues engaging relevant stakeholders in setting the scientific agenda with the aim of strengthening international science for the benefit of society. A transdisciplinary approach is necessary to address urgent societal issues. ESSP, and its Joint Projects (including GECAFS), has tried to break down the disciplinary silos that haunt many research environments, particularly universities. Through its focus on food security (vis-à-vis just food production), and the development of the integrated food system concept, GECAFS has succinctly outlined an approach to truly transdisciplinary research involving prominent scientists from all the necessary scientific disciplines. It also utilized the necessary participatory approaches through its scenario exercises involving the users of scientific information, especially in a policy context. In this, GECAFS has been truly innovative and an excellent example for the international science community, not only for those engaged in GEC issues.

However, it is not only scientific research that is mostly developed in disciplinary silos. Also, the policy environment is 'siloed' and there has been, for example, very little contact between the negotiators dealing with the UN Framework Convention on Climate Change, the UN Convention to Combat Desertification (CCD) and the Convention on Biodiversity (CBD). At the national level, issues are handled by various ministries, often with insufficient collaboration. This fragmentation of the current knowledge and policy systems must change if policies are to be developed that will lead to sustainable development.

The approach of GECAFS is also highly relevant for the follow-up to the Millennium Ecosystem Assessment (MA, 2005), which addressed the needs of four international conventions, including CCD and CBD. The MA conceptual framework outlined an approach for addressing linked ecological–social systems in the context of ecosystem services for human well-being. Agricultural production systems are vulnerable not only to climate change but also to many other external factors, which makes it necessary for farmers to manage a variety of risks. They have done so ever since agriculture was 'invented', but the challenge now is that the climate will change outside of the historical variability, where farmers could rely on traditional knowledge. Now, more than ever before, and as food systems also become more complex, farmers – and all other actors in the food system – are increasingly dependent on scientific knowledge. Similarly, researchers need to understand how farmers and others have coped with risks in the past. The approach of GECAFS will again help to build the necessary bridges. 'When the winds of change blow, some people build windmills, others build walls' (old Chinese proverb). The challenge is now to build windmills based on research presented in this book in order to prepare for current and future challenges. As was pointed out during the Agriculture and Rural Development Day during the climate negotiations in Copenhagen in December 2009, farmers are part of the problem but must also be part of the solution; agriculture and forestry worldwide contribute about one-third of the annual global greenhouse gas emission (CGIAR, 2009). At the same time, the poorest smallholder farmers in developing countries will be the hardest hit.

One of the targets of Millennium Development Goal 1 is to halve, between 1990 and 2015, the proportion of people who suffer from hunger. However, over the past few years the number of undernourished people has increased and more than 1 billion currently go to bed hungry. It is necessary to increase global food production by at least 70 per cent until 2050 to feed the growing world population. When the world leaders meet in September 2010 to review progress, they will note that progress in reducing hunger is being eroded by the worldwide increase in food prices, although there has been a drop since the most recent peak in 2007–8. Escalating prices were being driven partly by supply disruptions, but mostly by rising demand due to changing diets, economic growth, an expanding world population, urbanization, use of food crops for biofuel and inappropriate agricultural policies, including subsidies in developed countries.

In addition to this comes the threat of climate change, plus other major changes in the global environment, which will seriously affect many aspects of our food systems and especially the agricultural sector, and make food security for all a difficult goal to achieve. It is clear that more scientific research is essential to guide the development of society towards a world which is food secure and the results of GECAFS point the way forward for the scientific community to address this challenge.

The book provides a challenging synthesis of various aspects of the interactions between GEC and food systems. In Parts I and II, it is argued that many, if not most, previous attempts to address food security in the context of GEC have been too restricted. The analysis on why one should adopt a food systems approach is very convincing. This makes the scientific challenge of developing true transdisciplinarity more daunting and complex but the expected results will be more policy relevant at various spatial levels than past efforts. The typology of the food system space is illuminating and a conclusion is certainly that achieving food security without such a holistic approach will not be possible. The chapters also stress the need for considering the tradeoffs among improved food security and livelihoods, while ensuring protection of the natural environment through sustainable use of the natural resources base.

Part I also analysed the assessment landscape. To develop international research, the results of which become part of the global public good, it needs to be complemented by appropriate assessment processes. Scenario exercises that have been a core part of GECAFS are necessary components of assessments, but also provide an important participatory approach to scientific agenda setting. As is stated in Part II, there is a need for broader sectoral engagement in assessing the state of the food system and, as demonstrated in this book, scenarios can provide a vehicle for achieving buy-in from relevant stakeholders.

Engaging stakeholders is analysed in more depth in Part III and GECAFS has taken on board the ICSU recommendations from the Johannesburg Summit in truly engaging stakeholders in addressing the issues around food systems and their importance for food security. To do this at the global level is often not very helpful and the regional approach that GECAFS has adopted, and which is further reported on in Part IV, provides the essential platform for considering global processes with major local effects. Most previous GEC research has addressed the science–policy interface at the global level only. As Part III discusses, policy is developed at many different spatial levels and different approaches to strengthen the dialogue between scientists and policy-makers must take this into account. There is a need to inform scientists about the policy-making process and not only transmit best available scientific understanding to the policy-makers. Boundary organizations and knowledge brokers are becoming increasingly important. Both sides must understand the realities of how both science and policy-making progress and this is an area of research in its own right, as explained in Part III.

The final part of the book outlines the challenges of feeding an increasing global population with, in many regions, rapidly changing consumption patterns, increasing pressures on the system from climate change and other global changes, as well as competition for land and rural to urban migration. It will be a truly daunting challenge to increase food production by 70 per cent by 2050. Food riots in many countries initiated by soaring food prices in 2007–8 caused societal unrest, and national conflicts can also develop, triggered by lack of access to food. It is only through relevant research, technological innovations and an informed policy environment that this challenge will be able to be properly addressed. As pointed out in this final part, trade is also an important driver of change and local food production can become decoupled from food security in a global context. The final part also addresses the tradeoffs at various spatial levels and the importance of adaptive approaches that can deal with surprises. The future is not, and will never be, what it has been.

GECAFS will leave behind an important scientific legacy, but it has also been an inspiration for the development of a new major initiative developed in partnership between ESSP and the Consultative Group on International Agricultural Research (CGIAR). They have jointly developed the science plan for the Climate Change, Agriculture and Food Security (CCAFS) Challenge Programme. This is a unique initiative in that it builds on the global change science community in partnership with CGIAR, a key player in addressing science as an important basis for feeding the global population. Climate change and agriculture will also be a key component in a current restructuring and refocus of the CGIAR and its 15 international centres (Lele, 2010). The new initiative will focus on smallholder farmers and how they can manage the risks posed by climate change. It will focus on both adaptation to, and mitigation of, climate change.

The first Global Conference on Agricultural Research for Development (GCARD, March 2010) included discussions on climate change and agriculture. However, unlike GECAFS it did not focus much on the importance of food systems to achieve food security. The strength of the GECAFS holistic approach in addressing how to feed a global population of ca. 9 billion in 2050, while lowering the environmental footprint of the food system, is well documented in the this book and I hope it will be read by many engaged in the current food security debate and that it will continue to inspire a young generation of scientists that are ready to address one of the key challenges for humanity: how to achieve food security in the face of global changes with an increasing population and changing consumption patterns.

References

CGIAR (2009) *Climate change and Food Security: A Strategy for Change*, Washington, DC, CGIAR

ICSU (2002) Science and Technology at the World Summit on Sustainable Development. *Series on Science for Sustainable Development No. 11,* Paris, International Council for Science

Lele, U. (2010) 'Food security for a billion poor', *Science,* 326, 1554

MA (2005) *Our Human Planet: Summary for Decision-Makers*, Washington, DC, Island Press

Index